Operations Research and Manageme

Operations Research and Management

Edited by **Courtney Hoover**

NY RESEARCH
P R E S S

New York

Published by NY Research Press,
23 West, 55th Street, Suite 816,
New York, NY 10019, USA
www.nyresearchpress.com

Operations Research and Management
Edited by Courtney Hoover

International Standard Book Number: 978-1-63238-499-7 (Hardback)

The publisher's policy is to use permanent paper from mills that operate a sustainable forestry policy. Furthermore, the publisher ensures that the text paper and cover boards used have met acceptable environmental accreditation standards.

Trademark Notice: Registered trademark of products or corporate names are used only for explanation and identification without intent to infringe.

Printed in the United States of America.

Contents

Permissions

List of Contributors

Preface

Operations research mainly focuses on providing professionals with the tools and techniques that facilitate better decision making. It uses mathematical analysis, statistics and mathematical modeling for this purpose. Organizations often face the dilemma of selecting an optimum solution among several lucrative choices; operations research provides them with the tools to compare these options. Operations management on the other hand deals with redesigning business processes, scheming, controlling and managing production. While operations research deals with the quantitative analysis, operations management is the combination of both qualitative and quantitative aspects. This book studies, analyses and uphold the pillars of these fields and their utmost significance in modern times. For all readers who are interested in these disciplines, the case studies included in this book will serve as an excellent guide to develop a comprehensive understanding. This text is a compilation of chapters that discuss the most vital concepts and emerging trends in the field of operations research and management.

This book is a comprehensive compilation of works of different researchers from varied parts of the world. It includes valuable experiences of the researchers with the sole objective of providing the readers (learners) with a proper knowledge of the concerned field. This book will be beneficial in evoking inspiration and enhancing the knowledge of the interested readers.

In the end, I would like to extend my heartiest thanks to the authors who worked with great determination on their chapters. I also appreciate the publisher's support in the course of the book. I would also like to deeply acknowledge my family who stood by me as a source of inspiration during the project.

Editor

Ordering and Pricing Strategies for Fresh Products with Multiple Quality Levels Considering Consumer Utility

Peiqi Ma

School of Management, Jinan University, Guangzhou, China
Email: Page-Mar@foxmail.com

Abstract

In this paper, considering a scenario in which there are two quality levels of fresh products and introduction of consumer utility function, we studied the optimal ordering and pricing strategies under certain quantity. Our results showed that, facing the two quality levels of fresh products, retailers would not benefit from sales of lower quality of fresh products with the deterministic demand. In the pursuit of profit maximization, the initial order quantity is smaller than the potential demand for market.

Keywords

Fresh Products, Consumer Utility, Multiple Quality Level, Pricing and Ordering

1. Introduction

Given the technological advancements and logistical capabilities over the last seven decades or so, perishables have become a large part of supermarket retailing sales. For instance, a report by FMI (2009) indicates that of the $430 billion plus in US supermarket sales registered for the year 2008, 81% or over $348 billion is due to sales of groceries directly related to foods and beverages (F & B). And there is no doubt that fresh foods have played a dramatically important role in china.

Motivated by the common practice, retailers usually divide fresh products, which may decay or deteriorate during the sale process, into different quality levels, and sell them in a separate way. In this paper, considering a scenario in which there are two quality levels of fresh products and introduction of consumer utility function, we studied the optimal ordering and pricing strategies under certain quantity. Then, we discussed how environmental factors, such as demand volatility and ordering costs, affected retailers' decision. By considering consumer utility to study the optimal ordering and pricing decisions of multi-quality fresh products, we found corresponding answers of the above questions, and gave the management suggestions.

2. Literature Review

Ordering and pricing of fresh products are within the scope of supply chain management, which has become a hot issue for discussion and research. Goyal *et al.* (2001), Li *et al.* (2010), and Bakker *et al.* (2012) provided a comprehensive introduction about deteriorating items inventory management research from different perspectives [1]-[3]. Under certain quantity, Xiao *et al.* (2010) assumed that there are two quality levels of fresh products; retailer balances supply and demand of fresh products with price discount to study the optimal selling strategies, but they do not take characteristics of fresh product into account [4]. Bai and Kendall (2008) proposed a single-period inventory and shelf-space allocation model for fresh produce. The demand rate is assumed to be deterministic and dependent on both the displayed inventory and the items' freshness condition [5]. Avinadav Tal *et al.* (2009) presents an extension of the classical EOQ model for items with a fixed shelf life and a declining demand rate due to a reduction in the quality of the item in the course of its shelf-life [6]. However, Perishable products are subject to both obsolescence and deterioration, but above researches that study both types of loss are limited. Dan and Chen (2008) created an exponential function with downward slope, trying to denote valuable loss with greenness, and studies the coordination in two-level fresh agricultural supply chain [7]. Cai *et al.* (2010) considered a supply chain in which a distributor procures from a producer a quantity of a fresh product [8]. During the transportation process, the distributor has to make an appropriate effort to preserve the freshness of the product, and his success in this respect impacts on both the quality and quantity of the product delivered to the market. In reality, each consumer has different preferences to fresh products, which may influence retailer's ordering and pricing strategies. Above two references do not model consumer behavior. Some scholars analyzed pricing decision that involves modeling customer behavior. Ferguson and Koenigsberg (2007) have presented a two-period model where the quality of the leftover inventory is often perceived to be lower by customers, and the firm can decide to carry all, some, or none of the leftover inventory to the next period [9]. This is also a model involving quality drop and quantity change. Akcay *et al.* (2010) considered a dynamic pricing problem facing a firm that sells given initial inventories of multiple substitutable and perishable products over a finite selling horizon. They modeled this multiproduct dynamic pricing problem as a stochastic dynamic program and analyzed its optimal prices [10]. Li *et al.* (2012) studied the joint pricing and inventory control problem for perishables when a retailer does not sell new and old inventory at the same time [11]. Sainathan (2013) considered pricing and ordering decisions faced by a retailer selling a perishable product with a two-period shelf life over an infinite horizon [12]. Those scholars considered multiple quality levels of deteriorating or decaying products, however, they are not on the background of fresh products.

The remainder of this article is organized as follows. In Section 2, I review the related literature. In Section 3, I describe demand model and the retailer's profit model, and then I find the solution of base model. In Section 4, I examine how demand affects the retailer's problem. In Section 5, I give a sensitivity analysis for the results.

3. Base Model

Considering a scenario in which a single retailer sells one category fresh product throughout the whole sale period with two stages, there may be residual at the end of one stage, and the retailer can sell the low quality product at the next stage. We assumed that there exist two different quality levels (high and low) to discuss effectiveness of retailers selling decision, ordering and pricing strategies. At the start of period, the retailer procure Q unit fresh product with per unit procurement cost of ω from producer, and the procurement lead time is assumed to be zero. In sales process, retailer can do preservation effort at φ rate, $\varphi \in [0,1]$, then its cost is assumed to be $C = \frac{1}{2}Q\varphi$. Then I describe the choice process among customers. Any customer who visits the retailer has three choices: buy one unit of the low quality product, buy one unit of the high quality product, or do not buy anything. The high and low quality products compete among customers in their attributes and prices. Each customer selects his preferred choice based on his utility from purchasing a unit of product type i, which is given by

$$U_H = u\alpha_H - P_H, \quad U_L = u\alpha_L - P_L \tag{1}$$

where u is a customer's quality sensitivity, $u \sim U[0,1]$, α_i is product type$i$'s quality factor that is a measure of its desirability (therefore $\alpha_H > \alpha_L$), and P_i is its price. Then, I define R_i to be the probability that a customer buying one unit of the high (low) quality product. Next, I discuss how demand certainty impacts the retailer's optimal strategy

Based on the utility model, I first derive expression for R_H and R_L as functions of the prices and quality factors. The percentage of a customer buying the high quality level R_H, and that of buying the old product, R_L, are as follows:

$$R_H = P_r\{U_H \geq U_L, U_H \geq 0\} = P\left\{u \geq \max\left(\frac{P_H}{\alpha_H}, \frac{P_H - P_L}{\alpha_H - \alpha_L}\right)\right\} \tag{2}$$

$$R_L = P_r\{U_L \geq U_H, U_L \geq 0\} = P\left\{\frac{P_L}{\alpha_L} \leq u \leq \frac{P_H - P_L}{\alpha_H - \alpha_L}\right\} \tag{3}$$

In the expression 2, there is need to compare $\frac{P_H}{\alpha_H}$ with $\frac{P_H - P_L}{\alpha_H - \alpha_L}$, so we have two following cases:

Case 1 When $\frac{P_H}{\alpha_H} > \frac{P_H - P_L}{\alpha_H - \alpha_L}$, i.e. $\frac{P_H}{\alpha_H} < \frac{P_L}{\alpha_L}$, which means all customer only buy high quality products. And after it clears out, customer will buy low quality products. Accordingly, we can draw an expression:

$R_H = P\left\{\frac{P_H}{\alpha_H} \leq u \leq 1\right\} = 1 - \frac{P_H}{\alpha_H}, R_L = 0$.

Case 2 When $\frac{P_H}{\alpha_H} < \frac{P_H - P_L}{\alpha_H - \alpha_L}$, i.e. $\frac{P_H}{\alpha_H} > \frac{P_L}{\alpha_L}$, which means customer will buy fresh products between high and low quality. The expression for R_H and R_L are then obtained by finding the corresponding probabilities by using the fact that $u \sim U[0,1]$:

$$R_H = \max\left(1 - \frac{P_H - P_L}{\alpha_H - \alpha_L}, 0\right), \quad R_L = \min\left(\frac{P_H - P_L}{\alpha_H - \alpha_L}, 1\right) - \frac{P_L}{\alpha_L} \tag{4}$$

Suppose the total number of customers is m. The demand for the high and low quality products is deterministic and is given by mR_H and mR_L, respectively. Therefore, the retailer's profit then is given by:

$$\pi_{P_H,P_L}\left(P_H, P_L | M = m, Q\right)$$
$$= P_H \min\left(m \cdot R_H, Q\right) + P_L \min\left(m \cdot R_L, (Q - m \cdot R_H)^+\right) - \omega Q - \frac{1}{2}Q\varphi \tag{5}$$

The symbol "+" means the expression, $Q - m \cdot R_H$ is positive.

4. Optimal Ordering and Pricing Decision

In this section, we solve the optimal ordering and pricing decision according to the retailer's demand and profit model. As the same, there is different relationship between $\frac{P_H}{\alpha_H}$ and $\frac{P_H - P_L}{\alpha_H - \alpha_L}$ when maximizing retailer's profit. So here are two scenarios as following:

Scenario 1: Suppose that $\frac{P_H}{\alpha_H} > \frac{P_H - P_L}{\alpha_H - \alpha_L}$, i.e. $\frac{P_H}{\alpha_H} < \frac{P_L}{\alpha_L}$. Then $R_H = 1 - \frac{P_H}{\alpha_H}, R_L = 0$, and the profit becomes $\pi_{P_H,P_L}\left(P_H, P_L | M = m, Q\right) = P_H \cdot \min\left(m \cdot R_H, Q\right) - \omega Q - \frac{1}{2}Q\varphi$. We assume that Q is piecewise linear, the retailer has two optimal choice for Q.

$$Q = 0 \quad \text{and} \quad \pi^1_{P_H,P_L}\left(P_H, P_L | M = m, Q\right) = 0;$$

$Q = m \cdot R_H$ and $\pi^1_{P_H,P_L}\left(P_H,P_L \mid M=m,Q\right) = \left(P_H - \omega - \frac{1}{2}\varphi\right) \cdot m \cdot R_H = m\left(P_H - \omega - \frac{1}{2}\varphi\right)\left(1 - \frac{P_H}{\alpha_H}\right)$, then I solve

the profit function derivative on the high quality product price and I obtain $P_H = \dfrac{\left(\alpha_H + \omega + \frac{1}{2}\varphi\right)}{2}$, which results

in a positive profit and is hence better than producing 0 units.

Senario 2: Suppose that $\dfrac{P_H}{\alpha_H} \le \dfrac{P_H - P_L}{\alpha_H - \alpha_L}$, i.e. $\dfrac{P_H}{\alpha_H} \ge \dfrac{P_L}{\alpha_L}$, the retailer's profit is still piecewise linear in Q,

therefore, she has three potentially optimal choices for Q:

$$Q = 0, \text{ and } \pi^2_{P_H,P_L}\left(P_H,P_L \mid M=m,Q\right) = 0 \; ;$$

$$Q = m \cdot R_H \text{ and } \pi^2_{P_H,P_L}\left(P_H,P_L \mid M=m,Q\right) = \left(P_H - \omega - \frac{1}{2}\varphi\right) \cdot m \cdot \max\left(1 - \frac{P_H - P_L}{\alpha_H - \alpha_L}, 0\right), \text{ suppose that}$$

$\dfrac{P_H - P_L}{\alpha_H - \alpha_L} > 1$, we can get the same result as scenario 1; Suppose that $\dfrac{P_H - P_L}{\alpha_H - \alpha_L} < 1$, so

$\pi^2_{P_H,P_L}\left(P_H,P_L \mid M=m,Q\right) = \left(P_H - \omega - \frac{1}{2}\varphi\right) \cdot m\left(1 - \frac{P_H - P_L}{\alpha_H - \alpha_L}\right)$. According to the presumption $\dfrac{P_H}{\alpha_H} < \dfrac{P_H - P_L}{\alpha_H - \alpha_L}$,

then $\pi^2_{P_H,P_L} < \pi^1_{P_H,P_L}$, we do not take this case into consideration.

$$Q = m \cdot (R_H + R_L) = m \cdot \left(1 - \frac{P_L}{\alpha_L}\right) \text{ and}$$

$$\pi^2_{P_H,P_L}\left(P_H,P_L \mid M=m,Q\right) = m \cdot P_H \cdot \left(1 - \frac{P_H - P_L}{\alpha_H - \alpha_L}\right) + m \cdot P_L \cdot \left(\frac{P_H - P_L}{\alpha_H - \alpha_L} - \frac{P_L}{\alpha_L}\right) - m \cdot \left(\omega + \frac{1}{2}\varphi\right)\left(1 - \frac{P_L}{\alpha_L}\right), \text{ so we only}$$

consider case (iii). The retailer's optimization problem is now given by

$$\max \pi^2_{P_H,P_L}\left(P_H,P_L \mid M=m,Q\right)$$
$$= m \cdot P_H \cdot \left(1 - \frac{P_H - P_L}{\alpha_H - \alpha_L}\right) + m \cdot P_L \cdot \left(\frac{P_H - P_L}{\alpha_H - \alpha_L} - \frac{P_L}{\alpha_L}\right) - m \cdot \left(\omega + \frac{1}{2}\varphi\right)\left(1 - \frac{P_L}{\alpha_L}\right) \tag{6}$$
$$\text{s.t. } \frac{P_H}{\alpha_H} \ge \frac{P_L}{\alpha_L} \quad P_H, P_L > 0$$

The Hessian for the objective function is given by:

$$H = \begin{pmatrix} H_1 & H_2 \\ H_3 & H_4 \end{pmatrix} = \begin{pmatrix} \dfrac{-2m}{\alpha_H - \alpha_L} & \dfrac{2m}{\alpha_H - \alpha_L} \\ \dfrac{-2m}{\alpha_H - \alpha_L} - \dfrac{2}{\alpha_L} & \dfrac{2m}{\alpha_H - \alpha_L} \end{pmatrix}$$

Which is negative definite ($H_1 < 0$,

$H_2 \cdot H_3 - H_1 \cdot H_4 = \dfrac{2m}{\alpha_H - \alpha_L} \cdot \left(\dfrac{-2m}{\alpha_H - \alpha_L} - \dfrac{2}{\alpha_L}\right) - \dfrac{4m^2}{\left(\alpha_H - \alpha_L\right)^2} = \dfrac{-4m}{\alpha_L \cdot \left(\alpha_H - \alpha_L\right)} < 0$). Because of the constraint

condition, we need construct Lagrange function

$L(P_H, P_L, \sigma) = m \cdot P_H \cdot \left(1 - \frac{P_H - P_L}{\alpha_H - \alpha_L}\right) + m \cdot P_L \cdot \left(\frac{P_H - P_L}{\alpha_H - \alpha_L} - \frac{P_L}{\alpha_L}\right) - m \cdot \left(\omega + \frac{1}{2}\varphi\right)\left(1 - \frac{P_L}{\alpha_L}\right) + \sigma\left(\frac{P_H}{\alpha_H} - \frac{P_L}{\alpha_L}\right)$. There-

fore, KKT conditions are necessary and sufficient and are given by:

$$
\begin{cases}
\dfrac{\partial \pi}{\partial P_H} = m\left(1 - 2\left(\dfrac{P_H - P_L}{\alpha_H - \alpha_L}\right)\right) + \dfrac{\sigma}{\alpha_H} \le 0, P_H > 0, P_H \dfrac{\partial \pi}{\partial P_H} = 0 \\[3mm]
\dfrac{\partial \pi}{\partial P_L} = m\left(2(\dfrac{P_H - P_L}{\alpha_H - \alpha_L}) + \dfrac{\omega + \frac{1}{2}\varphi - 2P_L}{\alpha_L}\right) - \dfrac{\sigma}{\alpha_L} \le 0, P_L > 0, \ P_L \dfrac{\partial \pi}{\partial P_L} = 0 \\[3mm]
\dfrac{\partial \pi}{\partial \sigma} = \dfrac{P_H}{\alpha_H} - \dfrac{P_L}{\alpha_L} \ge 0, \sigma \ge 0, \sigma \dfrac{\partial \pi}{\partial \sigma} = 0
\end{cases}
\tag{7}
$$

Suppose that $\sigma = 0$, we obtain $P_L = \dfrac{\omega + \frac{1}{2}\varphi + \alpha_L}{2}$ and $P_H = \dfrac{\omega + \frac{1}{2}\varphi + \alpha_H}{2}$. However, we find these values

are not feasible because $\dfrac{P_H}{\alpha_H} > \dfrac{P_L}{\alpha_L}$. Therefore, suppose that $\sigma > 0$, and hence $\dfrac{\partial \pi}{\partial \sigma} = \dfrac{P_H}{\alpha_H} - \dfrac{P_L}{\alpha_L} = 0$, *i.e.*

$\dfrac{P_H}{\alpha_H} = \dfrac{P_L}{\alpha_L}$. And the retailer's profit becomes $\pi_{P_H, P_L}\left(P_H, P_L \mid M = m, Q\right) = m\left(P_H - \omega - \dfrac{1}{2}\varphi\right)\left(1 - \dfrac{P_H}{\alpha_H}\right)$. We have

the optimal pricing and ordering quantity of high and low quality fresh product:

$$
\begin{cases}
P_H = \dfrac{\omega + \frac{1}{2}\varphi + \alpha_H}{2} \\[3mm]
P_L = \dfrac{\omega + \frac{1}{2}\varphi + \alpha_L}{2} \\[3mm]
Q = \dfrac{\left(\alpha_H - \left(\omega + \frac{1}{2}\varphi\right)\right)}{2\alpha_H} m
\end{cases}
\tag{8}
$$

$$
\pi_{P_H, P_L}\left(P_H, P_L \mid M = m, Q\right) = \dfrac{\left(\alpha_H - \left(\omega + \frac{1}{2}\varphi\right)\right)^2}{4\alpha_H} m
\tag{9}
$$

5. Numerical Analysis

Giving a fresh product as an example, we assume that u is a customer's quality sensitivity, $u \sim U[0,1]$. There are discussions about the optimal decision of retailer and fluctuation of the potential demand for market. We have assumptions about some values of parameters as following tables (**Table 1** and **Table 2**).

Proposition 1: Facing the two quality levels of fresh products, retailers will not benefit from sales of lower quality of fresh products with the deterministic demand.

Proof 1: From the optimal retailer's profit $\dfrac{\left(\alpha_H - \left(\omega + \frac{1}{2}\varphi\right)\right)^2}{4\alpha_H} m$, we can infer that the retailer's profit is only

related to high quality factor, ordering and preserving costs. This is because retailer sells low quality products at a discount price, which still needs to undertake ordering and preserving costs. Therefore, the retailer sells low quality fresh products only to minimum the loss and does not make profit from it.

Proposition 2: To realize the optimal profit, the retailer's initial ordering quantity is smaller than the potential demand for market.

Table 1. Value of parameters.

α_L	α_H	ω	φ
	1.5		
0.3	2.5	1	0.2
	3.5		

Table 2. Sensitivity analysis.

m	Q	P_H	P_L	π_{P_H,P_L}
	13.3	1.3		2.7
100	28	1.8	0.7	19.6
	34.3	2.3		41
	26.6	1.3		5.4
200	56	1.8	0.7	39.2
	68.2	2.3		82
	39.9	1.3		7.1
300	74	1.8	0.7	58.8
	102.9	2.3		123

Proof 2: From $Q = \dfrac{\left(\alpha_H - \left(\omega + \frac{1}{2}\varphi\right)\right)}{2\alpha_H}m$, we can obtain $0 < \dfrac{\left(\alpha_H - \left(\omega + \frac{1}{2}\varphi\right)\right)}{2\alpha_H} < 1$, *i.e.* $Q < m$. This is mainly because the retailer sells high quality fresh product at a higher price to make more profit. As long as the retailer has low ordering and preservation costs, it will have the motive to repeat order, which is consistent with zero inventory and rapid inventory turnover in operation management.

6. Conclusion and Future Research

This paper focuses on multiple quality fresh products and considers consumer utility to analyze the retailer's optimal ordering and pricing strategies. Our results showed that, facing the two quality levels of fresh products, retailers would not benefit from sales of lower quality of fresh products with the deterministic demand. In the pursuit of profit maximization, the initial order quantity is smaller than the potential demand for market. Possible extensions of this paper involve relaxing some of assumptions, for example, considering random customer arrival process and demand substitution of high and low quality products.

References

[1] Goyal, S.K. and Giri, B.C. (2001) Recent Trends in Modeling of Deteriorating Inventory. *European Journal of Operational Research*, **134**, 1-16. http://dx.doi.org/10.1016/S0377-2217(00)00248-4

[2] Li, R., Lan, H. and Mawhinney, J.R. (2010) A Review on Deteriorating Inventory Study. *Journal of Service Science and Management*, **3**, 117-129. http://dx.doi.org/10.4236/jssm.2010.31015

[3] Bakker, M., Riezebos, J. and Teunter, R.H. (2012) Review of Inventory Systems with Deterioration since 2001. *European Journal of Operational Research*, **221**, 275-284. http://dx.doi.org/10.1016/j.ejor.2012.03.004

[4] Xiao, Y., Wu, P. and Wang, Y. (2010) Pricing Strategies for Fresh Products with Multiple Quality Levels Based on Customer Choice Behavior. *Chinese Journal of Management Science*, **18**, 58-65.

[5] Bai, R. and Kendall, G. (2008) A Model for Fresh Produce Shelf-Space Allocation and Inventory Management with Freshness-Condition-Dependent Demand. *INFORMS Journal on Computing*, **20**, 78-85. http://dx.doi.org/10.1287/ijoc.1070.0219

[6]   Tal, A. and Teijo, A. (2009) An EOQ Model for Items with a Fixed Shelf-Life and a Declining Demand Rate Based on Time-to-Expiry Technical Note. *Asia-Pacific Journal of Operational Research*, **26**, 759-767. http://dx.doi.org/10.1142/S0217595909002456

[7]   Dan, B. and Chen, J. (2008) Coordinating Fresh Agricultural Supply Chain under the Valuable Loss. *Chinese Journal of Management Science*, **5**, 6.

[8]   Cai, X., Chen, J., Xiao, Y., *et al.* (2010) Optimization and Coordination of Fresh Product Supply Chains with Freshness-Keeping Effort. *Production and Operations Management*, **19**, 261-278. http://dx.doi.org/10.1111/j.1937-5956.2009.01096.x

[9]   Ferguson, M.E. and Koenigsberg, O. (2007) How Should a Firm Manage Deteriorating Inventory? *Production and Operations Management*, **16**, 306-321. http://dx.doi.org/10.1111/j.1937-5956.2007.tb00261.x

[10]   Akcay, Y., Natarajan, H.P. and Xu, S.H. (2010) Joint Dynamic Pricing of Multiple Perishable Products under Consumer Choice. *Management Science*, **56**, 1345-1361. http://dx.doi.org/10.1287/mnsc.1100.1178

[11]   Li, Y., Cheang, B. and Lim A. (2012) Grocery Perishables Management. *Production and Operations Management*, **21**, 504-517. http://dx.doi.org/10.1111/j.1937-5956.2011.01288.x

[12]   Sainathan, A. (2013) Pricing and Replenishment of Competing Perishable Product Variants under Dynamic Demand Substitution. *Production and Operations Management*, **22**, 1157-1181.

2

Investigating Supply Chain Integration Effects on Environmental Performance in the Jordanian Food Industry

Zu'bi M. F. Al-Zu'bi, Ekhleif Tarawneh, Ayman Bahjat Abdallah, Mahmoud A. Fidawi

School of Business, The University of Jordan, Amman, Jordan
Email: z.alzubi@ju.edu.jo

Abstract

The purpose of this study was to examine the effects of supply chain integration on environmental performance of food manufacturing companies in Jordan. Data for this study were collected from one hundred and nineteen food companies. To answer the study questions, and to verify its hypotheses, descriptive statistical tools and linear regression tests were used. The study results indicated that supply chain integration positively affected environmental performance. Additionally, the results showed that supply chain integration positively affected environmental control and pollution management.

Keywords

Supply Chain Integration, Environmental Performance, Environmental Control, Pollution Management, Empirical Research

1. Introduction

Over the past 15 years, awareness concerning environmental changes was apparently increasing among population and industries. International agencies and national governments have increased their efforts concerning natural resources depletion, ozone depletion, gas emissions and waste reduction. Such efforts entail significant adjustments and changes in production processes and supply chain planning. Therefore, in order to maintain the same level of production, in an environmentally and sustainable way, the traditional way of managing manufacturing operations has to be modified in order to respond to environmental concerns [1].

Supply chain describes the various processes and procedures starting from acquiring the needed raw materials from suppliers and ending with the delivery of finished goods to final consumers. Stock and Boyer [2] defined

SCM as "The management of a network of relationships within a firm and between interdependent organizations and business units consisting of material suppliers, purchasing, production facilities, logistics, marketing, and related systems that facilitate the forward and reverse flow of materials, services, finances and information from the original producer to final customer with the benefits of adding value, maximizing profitability through efficiencies, and achieving customer satisfaction". Integration was defined as "the quality of the state of collaboration that exists among departments that are required to achieve unity of effort by the demands of the environment" [3]. The definition of supply chain integration is extended in our study to include upstream suppliers and downstream customers. Supply chain integration has been a challenging task for many companies, despite the strong consensus over the strategic importance of its implementation (e.g. [4]-[6]).

While reviewing the supply chain related literature, we found that most of the published research focused on investing the relationship between supply chain integration and operational or business performance (e.g. [7]-[10]) as well as the linkages between SCI and other operational or business practices (e.g. [11]-[14]). Although sufficient literature existed concerning green supply chain management, a limited number of research papers attempted to discuss the linkages between supply chain integration and environmental performance. Additionally, we failed to find published research concerning this relationship in developing countries. Therefore, this study attempts to address this research gap by investigating the impact of supply chain integration on environmental performance in a developing country, Jordan.

2. Literature Review

2.1. Environmental Situation in Jordan

Jordan produces significant amount of waste each year. Per capita daily waste in Jordan is around 1kg, and more than 2 million tons of municipal solid waste is produced per year [15]. Several landfills exist in Jordan. Main ones are Rusaifeh and Al Ghabawi; they receive daily waste of 60 and 3000 tons respectively. Al Ghabawi is a sanitary one where gas produced by waste is used for electrical production. The annual medical and hazardous industrial waste is estimated to be 4000 and 15,000 tons respectively. The recycling rate in Jordan is 5% of the waste, while in Europe it is around 60%. Jordan imports 98% of its energy needs, at a cost of 25% of its GDP. Zarqa Governorate industrial plants produce 75% of the total pollution in Jordan [15].

With a 500 million meters deficiency, Jordan is counted the fourth most water poor country. Jordan laws lack waste management texts, but efforts seem underway. For instance, the Gulf of Aqaba Environmental Action Plan (GAEAP) calls to audit of power plant, update plans for oil spills, improve air and marine water quality and manage protected marine areas. The Jordanian wildlife is threatened by industrial pollution. This water shortage led to the damage of Azraq Oasis. The annual estimated water required for Jordan is approximately one billion cubic meters. This number is expected to reach 1.5 billion by year 2020. Cities of Amman, Zarqa, Balqa and Madaba generate 68% of solid chemical waste. Most of the plants in the central region are located in the country's capital city of Amman and produce chemical products such as detergent and paint. Southern Jordan is home to only eight chemical plants, these produce 31.5% of the country's chemical waste. The plants based in southern cities such as Karak, Maan, Tafileh, and Aqaba produce mainly paint and fertilizers [15] [16]. Jordan is known for its mineral resources, such as phosphate, potash, uranium, copper, and others. Mining and processing these minerals, is an internationally significant industry in Jordan. Products obtained from the extraction of these minerals are increasingly transformed to chemicals then exported. Possessing these resources can be a basis for downstream cluster of chemical industry [17].

2.2. Supply Chain Integration

SCI has been defined as "the degree to which a manufacturer strategically collaborates with its supply chain partners and collaboratively manages intra- and inter-organization processes" [5].

Integration between a buying firm and its suppliers and customers occurs in order to improve operations in the buying firm as well as in the supply network. There is a wide agreement among researchers that internal integration is as important as external integration (e.g. [5] [7] [18] [19]).

One way to view SC integration is based on two perspectives that can be referred to as logistical integration and technological integration [20]. Both perspectives can include suppliers and customers.

Logistical integration has been investigated under different terms such as supply management [21], supplier cooperation [7], and supplier involvement [22]. While several factors can be considered in viewing this perspective of integration, including both informational and delivery aspects [23], many researchers has asserted that the idea of information flow between supply chain partners is the main predictor of delivery integration and performance [24]-[26]. Thus, logistical integration is an effective and reliable information exchange mechanism in the supply chain that facilitates delivery activities. Dyer and Nobeoka [27] referred to this type of information as explicit since it involves easily transferable knowledge. Therefore, high logistical integration is characterized by timely and effectively information sharing. Moreover, high logistical integration enhances the flexibility and responsiveness to unexpected situations and facilitates the adaptability of supply chain partners to external requirements and pressures. Information integration is expected to bring different benefits to supply chain partners. Such benefits include, but not limited to, improved communication, commonly used performance measures and improvement goals, ability to adapt changes to product specifications, timely received updates concerning delivery status, improved coordination, supported task completion mechanism, and improved technical infrastructure [7] [19] [28] [29]. Examples of shared information among supply chain partners may include production planning, inventory levels, demand status, lead times, forecasting issues, operating procedures, resources, and logistics [7] [30]-[32].

Technological integration can be defined as an implicit knowledge sharing between two or more partners in the supply chain in strategic aspects [20]. Such aspects may include mass customization [33]-[35] and new product development and design [4] [36]. Information sharing in this type of integration is usually related to specific projects rather than targeting regular operational activities. This shared information facilitates modifications related to process and product aspects as well as to managerial and administrative aspects. As such, technological integration can be regarded as an effective tool to promote innovative activities [37]. Therefore, the degree of such integration depends heavily on the quality and amount of shared information as well as on the degree of involving supply chain partners in process design and new product development activities. Trust and long-term strategic partnership are major prerequisites for such integration [11] [38]. Technological integration offers mutual benefits for supply chain partners. For example, supplier having a sound expertise may assist in process re-design and new product development resulting in increased process effectiveness and shortened new product launch time [39]. On the other hand, the buying firm may assist its suppliers in achieving a high level of quality [22] [34], set up time reduction [40], and in development efforts directed towards strategic manufacturing goals [41] [42]. Process integration is often viewed as a vital aspect of SCI and is expected to contribute to internal operational performance and to external performance to customers and society. It helps in analyzing and developing operational activities among the supply chain partners to meet certain requirements such as customer needs, regulations, and mass customization requirements [11] [41] [43] [44].

2.3. Environmental Performance

There is an agreement among researchers that environmental performance lacks a common measure and structure despite the fact that the different measures of environmental performance applied by researchers reflect a common understanding of the concept to mean the environmental effect of a firm [45]. Environmental performance can be viewed as an outcome of Environmental Management System (EMS) [46]. EMS is defined as "part of the overall management system that includes organizational structure, planning activities, responsibilities, practices, procedures, processes and resources for developing, implementing, achieving, reviewing and maintaining the environmental policy" [47]. Sroufe [48] asserted that EMS attempts to enhance compliance and decrease waste. He pointed to compliance as having effective environmental control that allows the firm to comply with legal standards. He further indicated that waste reduction is concerned with pollution prevention starting from the design phase. Whitelaw [49] argued that environmental control includes the identification of all the environmental sides of the firm's activities, prioritizing them using logic, and focusing the efforts to find ways to improve the environmental situation by reducing the environmental effects of significant activities. According to Washington State Department of Ecology [50], pollution prevention was defined as "the use of processes or practices that reduce or eliminate the use of hazardous substances and the generation of pollutants or wastes at the source". They further indicated that successful pollution prevention strategy requires effective environmental control system that includes prevention policy, execution, and measurement and monitoring.

Papadopoulos and Giama [51] indicated that effective environmental performance should include operative measures related to waste and pollution as well as management measures related to control procedures by the management. Similarly, Chien and Shih [52] measured environmental performance in terms of management performance concerned with control procedures and operational performance related to pollution and waste disposal.

Based on the above review, we measure environmental performance in terms of the two following dimensions:

Environmental control: Systematic actions undertaken by the firm to promote environmental awareness in order to recognize, control, and reduce acts causing environmental harm so that to ensure environmental compliance to regulations.

Pollution management: Systematic actions undertaken by the firm to reduce and prevent all sources of waste that may potentially generate pollution so that the levels of pollution are minimized.

2.4. Supply Chain Integration and Environmental Performance

Dubey *et al.* [53] using a sample of 174 firms in rubber goods manufacturing companies in India found that supplier integration positively affects environmental performance. Green *et al.* [54] found that green supply chain practices implemented by manufacturing companies positively affect environmental performance as well as organizational performance. Hsu and Hu [55] investigated the impact of GSCM on environmental performance in Taiwanese electronic manufacturing companies. The results revealed positive relationship between GSCM and environmental performance. Their findings pointed to the crucial role of environmental control procedures such as product testing reports, management commitment, and the existence of environmental database in enhancing environmental performance.

Another study conducted in Taiwan by Chien and Shih [52] found that GSCM practices such as green procurement, environmental management of suppliers, and green manufacturing positively affect environmental performance of electrical and electronic companies. Muma *et al.* [56] investigated the effect of GSCM on environmental performance in tea processing companies in Kenya. They found that green procurement, green distribution, and green manufacturing were insignificantly related to environmental performance.

3. Theoretical Framework

3.1. Research Framework

The suggested framework for this study is shown in **Figure 1**. The framework illustrates the effect of SCI on environmental performance in terms of environmental control and pollution management.

3.2. Research Hypotheses

Green and sustainable supply chain management have emerged due to environmental pressures on manufacturing companies to become environmentally friendly. SCI represents the key factor for such survival strategy. Internal environmental operations, green design, and integration with supply chain partners are regarded as main green supply chain practices [57]. Green purchasing has been regarded as a vital enabler to facilitate green production [58]. Govindasamy [59] found that environmental purchasing and sustainable packaging were the main sustainable supply chain management practices to prove their effects on the outcome of sustainable supply chain performance. Although green supply chain has to be initiated internally, it cannot be successfully implemented without integrating key suppliers. Supplier integration assists in protecting resources and controlling negative

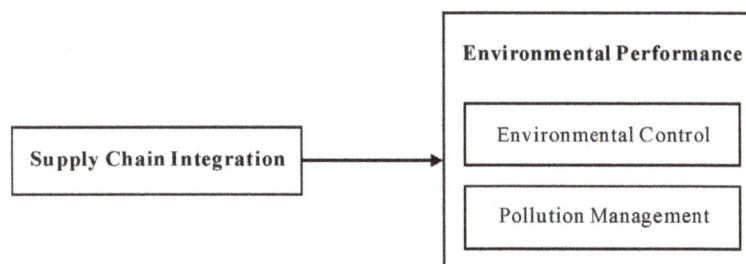

Figure 1. Research framework.

effects of production operations. The adoption of green or clean technology should be expanded to include suppliers. Such green technology along the supply chain is expected to reduce pollution levels and to encourage the use of pollution control facilities. Hall [60] asserted that supplier integration represents key success factor to achieve environmental change. Such integration will require collaboration in order to assist suppliers in developing environmental capabilities. However, it will require monitoring system in order to inspect suppliers' operational practices and delivered materials [20]. Awareness of environmental issues should be shared with suppliers in order to assure their understanding of environmental regulations and their willingness to comply with green supply chain requirements. Such integration is expected to improve environmental control and facilitate the pollution management resulting in increased environmental performance. Additionally, Internal and cross-functional integration of information and resources are regarded among main drivers to enhance environmental practices and, thus, environmental performance [48].

Supply chain integration increases company's ability to maximize the usage of green resources while minimizing production waste that leads to pollution [61]. Additionally, green and environmentally-friendly supply chain practices reduce the levels of energy usage, decrease raw materials consumption, reduce inventory levels, shorten production steps, and result in higher accuracy and lower costs [62] [63]. Moreover, supplier integration facilitates the reuse and recycling of raw materials, minimizes the risk of violating regulation with regard to environmental issues, and eliminates the production of hazardous components and waste [52].

H1: Supply chain integration is positively related to the environmental performance.

H1a: Supply chain integration is positively related to the Environmental control.

H1b: Supply chain integration is positively related to the Pollution Management.

4. Research Method

4.1. Sample

The population of the study represented all food manufacturing companies in Jordan. The survey questionnaire method was used to obtain data for this study. The measurement scales were adopted from previous related studies after conducting a thorough review of the published literature. A total of one hundred and fifty companies were visited personally by the authors in order to improve the representation level. Companies in Jordan usually do not pay attention to survey questionnaires sent by mail or e-mail, therefore we selected the personal visit approach. One manager was selected from each company to fill out the questionnaire. Those managers included operations managers, purchasing managers, supply chain managers, and store managers. Some managers promised to fill out the questionnaires later due to their busy schedules; however, not all of them returned the questionnaires. Other managers declined to participate. Ultimately, one hundred and twenty eight questionnaires were collected. Nine questionnaires were excluded due to different reasons. The number of final usable questionnaires used in the subsequent analysis was one hundred and nineteen representing a response rate of 79.3%. This response rate is regarded high and very satisfactory compared with other studies in Jordan that selected the personal visit approach to companies. For instance, Abdallah [64] and Obeidat *et al.* [65] received response rates of 59.5% and 52% respectively.

4.2. Instruments

The question items used in our survey were adapted from some previous studies [5] [7] [24] [45] [52]. Our focus was on ensuring high reliability and validity of our measurement instrument. We used eight question items to measure supply chain integration. Four question items were used to measure environmental control and five question items to measure pollution management. We required respondents to answer the question items using five-point Likert scale where 1 indicated strong disagreement and 5 indicated strong agreement. Next, we used factor analysis with principal component's analysis to evaluate the validity of our constructs. Our criterion was to keep question items that loaded onto one factor with factor loading greater than 0.40 and with eigenvalue for the factor greater than 1. Two question items did not meet these requirements and were deleted from subsequent analysis. We used Cronbach's α-coefficient to assess the reliability of our constructs. All our scales met the recommended standard of $\alpha \geq 0.60$ suggesting that they were internally consistent [66].

Table 1 shows the mean values, standard deviations, and Cronbach's α-coefficients for our measurement scales.

5. Results

We started our analysis by testing hypothesis H1 which indicated that supply chain integration is positively related to environmental performance. Simple regression analysis was used to test the hypothesis. The results of the regression analysis show that supply chain integration, the independent variable, is highly correlated with environmental performance, the dependent variable (R = 0.726) as shown in **Table 2**. The table shows that the value of R-square, which is the explained variance, is 0.527. This means that 52.7% of the variance (R-square) in the environmental performance has been explained by the supply chain integration. The t-value of 16.576 is significant at P < 0.01 level as it is greater than the tabulated value of 1.96. Thus, hypothesis H1 is accepted.

Next, we tested hypothesis H1a which indicated that supply chain integration is positively related to environmental control. Simple regression analysis was used to test the hypothesis. The results of the regression analysis show that supply chain integration, the independent variable, is highly correlated with environmental control, the dependent variable (R = 0.646) as shown in **Table 3**. The table shows that the value of R-square is 0.418. This means that 41.8% of the variance in the environmental control has been explained by the supply chain integration. The t-value of 13.309 is significant at P < 0.01 indicating that the hypothesis is accepted.

Finally, we tested hypothesis H1b concerning the impact of supply chain integration on pollution management. In a similar manner to previous hypotheses, simple regression analysis was used. The results of the regression analysis show that supply chain integration, the independent variable, is highly correlated with pollution management, the dependent variable (R = 0.686) as shown in **Table 4**. The table shows that the value of R-square is

Table 1. Cronbach's α-values, means, and standard deviations table type styles.

Measurement scale	Cronbach's alpha value	Mean	Standard deviation
Supply chain integration	0.683	4.35	0.554
Environmental performance	0.806	3.89	0.842
Environmental control	0.745	3.70	0.961
Pollution management	0.615	4.07	0.724

Table 2. Regression analysis of environmental performance.

Variable entered	Model coefficients	R	R²	Adj. R²	Std. error of the estimate	t	Sig.
(Constant)	−0.406*					−1.544	0.124
SC integration	0.726**	0.726	0.527	0.525	0.52246	16.576	0.000

*Unstandardized coefficient; **Standardized coefficient.

Table 3. Regression analysis of environmental control.

Variable entered	Model coefficients	R	R²	Adj. R²	Std. error of the estimate	t	Sig.
(Constant)	−1.176*					−3.180	0.002
SC integration	0.646**	0.646	0.418	0.415	0.73505	13.309	0.000

*Unstandardized coefficient; **Standardized coefficient.

Table 4. Regression analysis of pollution management.

Variable entered	Model coefficients	R	R²	Adj. R²	Std. error of the estimate	t	Sig.
(Constant)	0.172*					0.646	0.519
SC integration	0.686**	0.686	0.470	0.468	0.52805	14.806	0.000

*Unstandardized coefficient; **Standardized coefficient.

0.470. This means that 47% of the variance in the pollution management has been explained by the supply chain integration. The t-value of 14.806 is significant at $P < 0.01$ indicating that hypothesis H1b is accepted.

6. Discussion and Conclusion

This study attempted to investigate the effect of supply chain integration on environmental performance in food companies in Jordan. While previous research indicated a positive and significant influence of GSCM on environmental performance in developed countries, lack of studies in developing countries was evident in the published literature. Our results are consistent with the results of other studies conducted in the developed countries (e.g. [67]-[69]). The results imply that awareness concerning environmental issues is taking place in developing countries. As environmental pressures on manufacturing companies increase, the challenge will be to select suppliers who are devoted and keen to improve their environmental performance and adapt their operations to green requirements. Such suppliers should bear in mind that environmentally-friendly manufacturing and supply chain integration should not imply that quality and cost can be sacrificed [70]. Food companies willing to achieve high environmental performance have to assist their suppliers in improving their environmental performance. Such capable and aware suppliers can potentially be integrated into design projects to develop green products [71].

Our findings were inconsistent with the findings of Muma *et al.* [56] who found insignificant impact of GSCM on environmental performance in Kenya. Such inconsistency could be attributed to the lack of environmental control in Kenyan companies as opposite to Jordanian food companies. Environmental control allows companies to enhance environmental performance and set control systems and procedures that assign environmental-related responsibilities to different departments and managers. Prior training to managers and workers is required in order to establish an effective accountability policy [72].

The results also indicate that pollution management has been significantly explained by the supply chain integration. A major problem in Jordan is that the recycling rate is 5% of the waste compared with a rate of 60% in Europe [15]. While related governmental agencies face real challenges to increase this low rate, SC integration represents an essential approach for food companies to reduce their waste and overall pollution. SC integration promotes environmental practices such as solid return policy, usage of less hazardous materials in the products, and improvements of the usage of recycled materials. The involvement and education of employee and supply chain partners stimulate the overall environmental management system and contribute to higher levels of environmental performance.

The limitation of our study is that only food companies are included in our sample. Generally, the environmental situation and awareness in these companies are regarded favorable compared with other manufacturing sectors. Future studies are needed with a sample representing different manufacturing sectors so that to improve the generalizability of the results. Another limitation is that contextual factors such as company size, age, and type are not considered in our study. Such factors may influence the proposed relationship; therefore further similar studies with contextual factors are needed. Additional limitation exists in this study concerning the validity of the findings. Data were collected from single respondent in each food company. Although this method of data collection is widely used in empirical studies, multiple respondents each of whom addresses her particular field of expertise are expected to enhance the validity of the results.

References

[1]   Vachon, S. and Klassen, R.D. (2008) Environmental Management and Manufacturing Performance: The Role of Collaboration in the Supply Chain. *International Journal of Production Economics*, **111**, 299-315.
       http://dx.doi.org/10.1016/j.ijpe.2006.11.030

[2]   Stock, J. and Boyer, S. (2009) Developing a Consensus Definition of Supply Chain Management: A Qualitative Study. *International Journal of Physical Distribution & Logistics Management*, **39**, 690-711.
       http://dx.doi.org/10.1108/09600030910996323

[3]   Lee, H. and Whang, S. (2001) E-Business and Supply Chain Integration, Stanford University. *Global Supply Chain Management Forum SGSCMF-W*220, November 2001, 17-26.

[4]   Al-Abdallah, G.M., Abdallah, A.B. and Bany Hamdan, K. (2014) The Impact of Supplier Relationship Management on Competitive Performance of Manufacturing Firms. *International Journal of Business and Management*, **9**, 192-202.
       http://dx.doi.org/10.5539/ijbm.v9n2p192

[5]   Flynn, B.B., Huo, B. and Zhao, X. (2010) The Impact of Supply Chain Integration on Performance: A Contingency and

Configuration Approach. *Journal of Operations Management*, **28**, 58-71. http://dx.doi.org/10.1016/j.jom.2009.06.001

[6]   Tsinopoulos, C. and Al-Zu'bi, Z.M.F. (2014) Lead Users, Suppliers, and Experts: The Exploration and Exploitation Trade-Off in Product Development. *International Journal of Technology Marketing*, **9**, 6-20. http://dx.doi.org/10.1504/IJTMKT.2014.058080

[7]   Abdallah, A.B., Obeidat, B.Y. and Aqqad, N.O. (2014) The Impact of Supply Chain Management Practices on Supply Chain Performance in Jordan: The Moderating Effect of Competitive Intensity. *International Business Research*, **7**, 13-27. http://dx.doi.org/10.5539/ibr.v7n3p13

[8]   Al-Ettayyem, R. and Al-Zu'bi, Z.M.F. (2015) Investigating the Effect of Total Quality Management Practices on Organizational Performance in the Jordanian Banking Sector. *International Business Research*, **8**, 1-12. http://dx.doi.org/10.5539/ibr.v8n3p79

[9]   Lee, C.W., Kwon, I.G. and Severance, D. (2007) Relationship between Supply Chain Performance and Degree of Linkage among Supplier, Internal Integration, and Customer. *Supply Chain Management: An International Journal*, **12**, 444-452. http://dx.doi.org/10.1108/13598540710826371

[10] Sezen, B. (2008) Relative Effects of Design, Integration and Information Sharing on Supply Chain Performance. *Supply Chain Management: An International Journal*, **13**, 233-240. http://dx.doi.org/10.1108/13598540810871271

[11] Abdallah, A. and Matsui, Y. (2009) The Impact of Lean Practices on Mass Customization and Competitive Performance of Mass-Customizing Plants. *Proceedings of the* 20*th Annual Production and Operations Management Society* (*POMS*) *Conference*, Orlando, 1-4 May 2009.

[12] Abdallah, A. and Phan, C.A. (2007) The Relationship between Just-in-Time Production and Human Resource Management, and Their Impact on Competitive Performance. *Yokohama Business Review*, **28**, 27-57.

[13] Al-Zu'bi, Z.M.F. and Tsinopoulos, C. (2012) Suppliers versus Lead Users: Examining Their Relative Impact on Product Variety. *Journal of Product Innovation Management*, **29**, 667-680. http://dx.doi.org/10.1111/j.1540-5885.2012.00932.x

[14] Vanichchinchai, A. and Igel, B. (2009) Total Quality Management and Supply Chain Management: Similarities and Differences. *The TQM Journal*, **21**, 249-260. http://dx.doi.org/10.1108/17542730910953022

[15] Zafar, S. (2014) Biomass Energy, Middle East, Renewable Energy. http://www.ecomena.org/tag/bioenergy/

[16] Hadadin, N. and Tarawneh, Z. (2007) Environmental Issues in Jordan, Solutions and Recommendations. *American Journal of Environmental Sciences* **3**, 30-36. http://dx.doi.org/10.3844/ajessp.2007.30.36

[17] Ramadna, A. (2012) Mining Sector in Jordan: Current Solution and Investment Opportunities. Ministry of Industry and Trade Jordan. http://ec.europa.eu/enterprise/policies/raw-materials/files/docs/euromed_presentations/ramadna_mining_sector_jourdan_en.pdf

[18] Al-Zu'bi, Z.M.F., Heizer, J. and Render, B. (2013) Operations Management. 1st Arab World Edition, Pearson, London.

[19] Li, S., Ragu-Nathan, B., Ragu-Nathan, T.S. and Subba Rao, S. (2004) The Impact of Supply Chain Management Practices on Competitive Advantage and Organizational Performance. *Omega*, **34**, 107-124. http://dx.doi.org/10.1016/j.omega.2004.08.002

[20] Vachon, S. and Klassen, R.D. (2006) Extending Green Practices across the Supply Chain. *International Journal of Operations & Production Management*, **26**, 795-821. http://dx.doi.org/10.1108/01443570610672248

[21] Shin, H., Collier, D.A. and Wilson, D.D. (2000) Supply Management Orientation and Supplier/Buyer Performance. *Journal of Operations Management*, **18**, 317-333. http://dx.doi.org/10.1016/S0272-6963(99)00031-5

[22] Phan, A.C., Abdallah, A.B. and Matsui, Y. (2011) Quality Management Practices and Competitive Performance: Empirical Evidence from Japanese Manufacturing Companies. *International Journal of Production Economics*, **133**, 518-529. http://dx.doi.org/10.1016/j.ijpe.2011.01.024

[23] Frohlich, M.T. and Westbrook, R. (2001) Arcs of Integration: An International Study of Supply Chain Strategies. *Journal of Operations Management*, **19**, 185-200. http://dx.doi.org/10.1016/S0272-6963(00)00055-3

[24] Chen, I.J. and Paulraj, A. (2004) Towards a Theory of Supply Chain Management: The Constructs and Measurements. *Journal of Operations Management*, **22**, 119-150. http://dx.doi.org/10.1016/j.jom.2003.12.007

[25] Stock, G.N., Greis, N.P. and Kasarda, J.D. (2000) Enterprise Logistics and Supply Chain Structure: The Role of Fit. *Journal of Operations Management*, **18**, 531-547. http://dx.doi.org/10.1016/S0272-6963(00)00035-8

[26] Tsinopoulos, C. and Al-Zu'bi, Z.M.F. (2012) Clockspeed Effectiveness of Lead Users and Product Experts. *International Journal of Operations & Production Management*, **32**, 1097-1118. http://dx.doi.org/10.1108/01443571211265710

[27] Dyer, J. and Nobeoka, K. (2000) Creating and Managing a High-Performance Knowledge-Sharing Network: The

Toyota Case. *Strategic Management Journal*, **21**, 345-367.
http://dx.doi.org/10.1002/(SICI)1097-0266(200003)21:3<345::AID-SMJ96>3.0.CO;2-N

[28] Al-Zu'bi, Z.M.F., Alshurideh, M., Abuhamed, A. and Ghannajeh, A.M. (2015) A Qualitative Analysis of Product Innovation in Jordan's Pharmaceutical Sector. *European Science Journal*, **11**, 474-503.

[29] Gimenez, C. and Sierra, V. (2013) Sustainable Supply Chains: Governance Mechanisms to Greening Suppliers. *Journal of Business Ethics*, **116**, 189-203. http://dx.doi.org/10.1007/s10551-012-1458-4

[30] Al-Zu'bi, Z.M.F. and Tsinopoulos, C. (2012) An Outsourcing Model for Lead Users: An Empirical Investigation. *Production Planning and Control*, **24**, 1-10.

[31] Sakka, O. and Botta-Genoulaz, V. (2009) A Model of Factors Influencing the Supply Chain Performance. *Proceedings of CIE*, Troyes, 6-9 July 2009, 913-918. http://dx.doi.org/10.1109/iccie.2009.5223796

[32] Singh, R. K. (2013) Prioritizing the Factors for Coordinated Supply Chain Using Analytic Hierarchy Process (AHP). *Measuring Business Excellence*, **17**, 80-98. http://dx.doi.org/10.1108/13683041311311383

[33] Abdallah, A. and Matsui, Y. (2008) Customer Involvement, Modularization of Products, and Mass Customization: Their Relationship and Impact on Value to Customer and Competitiveness. *Proceedings of the 3rd World Conference on Production and Operations Management*, Tokyo, 5-8 August 2008.

[34] Al-Zu'bi, Z.M.F., Al-Lozi, M., Dahiyat, S., Alshurideh, M. and Al Majali, A. (2012) Examining the Effect of Quality Management Practices on Product Variety. *European Journal of Economics, Finance and Administrative Sciences*, **1**, 1-19.

[35] Al-Zu'bi, Z.M.F., Dahiyat, S.E., Warrad, T., Shannak, R.O. and Masa'deh, R.M. (2012) Investigating the Effect of Foreign Direct Investment Technology Transfer on Mass Customization Capability in Jordan's Manufacturing Sector. *International Research Journal of Finance and Economics*, **94**, 79-90.

[36] Al-Zu'bi, Z.M.F. and Khamees, B.A. (2014) Activity-Based Costing vs. Theory of Constraints: An Empirical Study into Their Effect on the Cost Performances of NPD Initiatives. *International Journal of Economics and Finance*, **6**, 157-165. http://dx.doi.org/10.5539/ijef.v6n12p157

[37] Abdallah, A.B., Phan, C.A. and Matsui, Y. (2015) Investigating the Effects of Managerial and Technological Innovations on Operational performance and Customer Satisfaction of Manufacturing Companies. *International Journal of Business Innovation and Research*, in Press. http://www.inderscience.com/info/ingeneral/forthcoming.php?jcode=ijbir

[38] Alkalha, Z., Al-Zu'bi, Z.M.F., Al-Dmour, H., Alshurideh, M. and Masa'deh, R. (2012) Investigating the Effect of Human Resource Policies on Organizational Performance: An Empirical Study on Commercial Banks Operating in Jordan. *European Journal of Economics, Finance and Administrative Sciences*, **1**, 94-102.

[39] Kaufman, A., Wood, C. and Theyel, G. (2000) Collaboration and Technology Linkages: A Strategic Supplier Typology. *Strategic Management Journal*, **21**, 649-663.
http://dx.doi.org/10.1002/(SICI)1097-0266(200006)21:6<649::AID-SMJ108>3.0.CO;2-U

[40] Abdallah, A. and Matsui, Y. (2007) Just in Time Production and Total Productive Maintenance: Their Relationship and Impact on JIT and Competitive Performances. *Proceedings of the 9th International Decision Sciences Institute Conference*, Bangkok, 11-15 July 2007.

[41] Abdallah, A.B., Phan, A.C. and Matsui, Y. (2009) Investigating the Relationship between Strategic Manufacturing Goals and Mass Customization. *Proceedings of the 16th International Annual European Operations Management Association (EurOMA)*, Goteborg, 14-17 June 2009.

[42] Al Hasan, R. and Al-Zu'bi, Z.M.F. (2014) Evaluating the Relationships between Lean Manufacturing Dimensions and Radical Product Innovation in the Jordanian Pharmaceutical Sector. *European Scientific Journal*, **10**, 230-258.

[43] Al-Zu'bi, Z.M.F. (2010) Collaboration in Mass Customisation: Exploring the Impacts of Suppliers and Lead Users. 1st Edition, VDM Verlag, Saarbrücken.

[44] Zairi, M. (1997) Business Process Management: A Boundary Less Approach to Modern Competitiveness. *Business Process Re-Engineering & Management Journal*, **3**, 64-80. http://dx.doi.org/10.1108/14637159710161585

[45] Afagachie, C. (2013) The Relationship between Corporate Environmental Performance and Corporate Financial Performance—Using the Framework of Corporate Environmental Policy Implementation. Unpublished Dissertation, University of Bristol, Bristol.

[46] Halkos, G. and Sepetis, A. (2007) Can Capital Markets Respond to Environmental Policy of Firms? Evidence from Greece. *Ecological Economics*, **63**, 578-587. http://dx.doi.org/10.1016/j.ecolecon.2006.12.015

[47] Gelber, M. (2004) BS 8555 Case Study. *BEST Conference*, Brussels, 10 February 2004.

[48] Sroufe, R. (2003) Effects of Environmental Management Systems on Environmental Management Practices and Operations. *Production and Operations Management*, **12**, 416-431. http://dx.doi.org/10.1111/j.1937-5956.2003.tb00212.x

[49] Whitelaw, K. (2004) ISO 14001 Environmental Systems Handbook. 2nd Edition, Elsevier Butterworth-Heinemann, Oxford.

[50] Washington State Department of Ecology (1997) Hazardous Waste and Toxics Reduction Program. Publication #97-401. https://fortress.wa.gov/ecy/publications/publications/97401.pdf

[51] Papadopoulos, A.M. and Giama, E. (2007) Environmental Performance Evaluation of Thermal Insulation Materials and Its Impact on the Building. *Building and Environment*, **42**, 2178-2187. http://dx.doi.org/10.1016/j.buildenv.2006.04.012

[52] Chien, M.K. and Shih, L.H. (2007) An Empirical Study of the Implementation of Green Supply Chain Management Practices in the Electrical and Electronic Industry and Their Relation to Organizational Performances. *International Journal of Environmental Science and Technology*, **4**, 383-394.

[53] Dubey, R., Gunasekaran, A. and Ali, S. (2015) Exploring the Relationship between Leadership, Operational Practices, Institutional Pressures and Environmental Performance: A Framework for Green Supply Chain. *International Journal of Production Economics*, **160**, 120-132. http://dx.doi.org/10.1016/j.ijpe.2014.10.001

[54] Green, K.W., Pamela, Z., Meacham J. and Bhadauria, S.V. (2012) Green Supply Chain Management Practices: Impact on Performance. *Supply Chain Management: An International Journal*, **17**, 290-305. http://dx.doi.org/10.1108/13598541211227126

[55] Hsu, C.W. and Hu, A.H. (2008) Green Supply Chain Management in the Electronic Industry. *International Journal of Environmental Science and Technology*, **5**, 205-216. http://dx.doi.org/10.1007/BF03326014

[56] Muma, B., Nyaoga, R., Matwere, R. and Nyambega, E. (2014) Green Supply Chain Management and Environmental Performance Among Tea Processing Firms in Kericho County-Kenya. *International Journal of Economics, Finance and Management Sciences*, **2**, 270-276. http://dx.doi.org/10.11648/j.ijefm.20140205.11

[57] Zhu, Q. and Sarkis, J. (2006) An Inter-Sectoral Comparison of Green Supply Chain Management in China: Drivers and Practices. *Journal of Cleaner Production*, **14**, 472-486. http://dx.doi.org/10.1016/j.jclepro.2005.01.003

[58] Min, H. and Galle, W.P. (2001) Green Purchasing Practices of U.S. Firms. *International Journal of Operations & Production Management*, **21**, 1222-1238. http://dx.doi.org/10.1108/EUM0000000005923

[59] Govindasamy, V. (2010) Sustainable Supply Chain Management (SSCM) Practices: Antecedents and Outcomes on Sustainable Supply Chain Performance. Unpublished Master's Thesis, Universiti Sains Malaysia, Penang.

[60] Hall, J. (2000) Environmental Supply Chain Dynamics. *Journal of Cleaner Production*, **8**, 455-471. http://dx.doi.org/10.1016/S0959-6526(00)00013-5

[61] Gopalakrishnan, K., Yusuf, Y.Y., Musa, A., Abubakar, T. and Ambursa, H.M. (2012) Sustainable Supply Chain Management: A Case Study of British Aerospace (BAe) Systems. *International Journal of Production Economics*, **140**, 193-203. http://dx.doi.org/10.1016/j.ijpe.2012.01.003

[62] Luthra, S., Kumar, V., Kumar, S. and Haleem, A. (2010) Green Supply Chain Management Issues: A Literature Review Approach. *Journal of Information, Knowledge and Research in Mechanical Engineering*, **1**, 12-20.

[63] Porter, M.E. and Van der Linde, C. (1995) Green and Competitive: Ending the Stalemate. *Harvard Business Review*, **73**, 120-134.

[64] Abdallah, A.B. (2013) The Influence of "Soft" and "Hard" Total Quality Management (TQM) Practices on Total Productive Maintenance (TPM) in Jordanian Manufacturing Companies. *International Journal of Business and Management*, **8**, 1-13. http://dx.doi.org/10.5539/ijbm.v8n21p1

[65] Obeidat, B.Y., Masa'deh, R.M. and Abdallah, A.B. (2014) The Relationships among Human Resource Management Practices, Organizational Commitment, and Knowledge Management Processes: A Structural Equation Modeling Approach. *International Journal of Business and Management*, **9**, 9-26. http://dx.doi.org/10.5539/ijbm.v9n3p9

[66] Nunnally, J. (1978). Psychometric Theory. 2nd Edition, McGraw-Hill, New York.

[67] Hsu, C.W. and Hu, A.H. (2009) Applying Hazardous Substance Management to Supplier Selection Using Analytic Network Process. *Journal of Cleaner Production*, **17**, 255-264. http://dx.doi.org/10.1016/j.jclepro.2008.05.004

[68] Ku, C.Y., Chang, C.T. and Ho, H.P. (2010) Global Supplier Selection Using Fuzzy Analytic Hierarchy Process and Fuzzy Goal Programming. *Quality & Quantity*, **44**, 623-640. http://dx.doi.org/10.1007/s11135-009-9223-1

[69] Van Hoof, B. and Lyon, T.P. (2013) Cleaner Production in Small Firms Taking Part in Mexico's Sustainable Supplier Program. *Journal of Cleaner Production*, **41**, 270-282. http://dx.doi.org/10.1016/j.jclepro.2012.09.023

[70] Luthra, S., Garg, D. and Haleem, A. (2013) Identifying and Ranking of Strategies to Implement Green Supply Chain Management in Indian Manufacturing Industry Using Analytical Hierarchy Process. *Journal of Industrial Engineering and Management*, **6**, 930-962. http://dx.doi.org/10.3926/jiem.693

[71] Rao, P. (2002) Greening the Supply Chain: A New Initiative in South East Asia. *International Journal of Operations*

and Production Management, **22**, 632-655. http://dx.doi.org/10.1108/01443570210427668

[72] Dasgupta, S., Hettige, H. and Wheeler, D. (1998) What Improves Environmental Performance? Evidence from Mexican Industry. Development Research Group, World Bank.
http://web.worldbank.org/archive/website01004/WEB/IMAGES/106506-3.PDF

Impacts of Exchange Rate Volatility and FDI on Technical Efficiency—A Case Study of Vietnamese Agricultural Sector

Nguyen Khac Minh[1], Pham Van Khanh[2], Nguyen Viet Hung[3]

[1]Economics and Management Faculty, Thang Long University, Hanoi, Vietnam
[2]Institute of Economics and Corporate Group, Hanoi, Vietnam
[3]Faculty of Economics, National Economics University, Hanoi, Vietnam
Email: van_khanh1178@yahoo.com

Abstract

The objective of this research is to examine impacts of exchange rate volatility and FDI on efficiency of the Vietnamese agricultural sector at the provincial level for the period 1998-2011. Due to the characteristic of high uncertainty in agricultural production, the chance-constrained programming model would be used to estimate efficiency of the agricultural production sector. In order to study impacts of exchange rate volatility and FDI, we employ the two-stage model. In the first stage, we use the chance-constrained programming model to measure technical efficiency and ARIMA model to quantify exchange rate volatility. In the second stage, we use the fixed effect model to evaluate impacts of exchange rate volatility and FDI on efficiency of agricultural production in poor and rich provinces. The estimated results show that fluctuation in exchange rate volatility would reduce efficiency in agricultural production but FDI has an insignificant impact on the efficient production in Vietnam agricultural sector.

Keywords

Chance-Constrained Programming, DEA, Exchange Rate Volatility, FDI, Technical Efficiency

1. Introduction

After economic transformation, Vietnam, which was the rice importer before 1980, has become the third largest rice exporter in the world in 1989. The volume of rice exports increased from 2 million tons in 1990 to over 3.6 million tons in 2000. The major rice importers of Vietnam consist of China, Malaysia, Singapore, Indonesia,

Hong Kong (China), and East Timor. In 2010, India and Thailand were the two largest rice exporters in the world while Vietnam was at the third position. In 2011, Vietnam exceeded Thailand in terms of volume of rice exports to become the second largest after India. However, in terms of value of rice exports, Vietnam is still behind Thailand because the quality of exported rice from Vietnam is lower than one of Thailand, and thereby the price is also lower.

Red river and Mekong river deltas are two major areas of rice exportation in Vietnam. Even though these two deltas account for only 15% of total area square of the whole country, the volume of rice production is two third of the one of the whole country. The Red river delta contributes to 16.7% of total volume of rice exportation of Vietnam.

The volume of rice exportation increases over time; however, the value of exportation is not quite high. One of the main reasons is the low quality of Vietnamese rice which could not satisfy the requirement of customers in the world. There have been several researches of efficiency in food production in Vietnam. For instance, Minh and Long (2009) in [1] have employed stochastic frontier approach and non-parametric method to estimate efficiency in agricultural production in Vietnam, using the data at the provincial level in the period 1990-2005. They came to the conclusion that the average technical efficiency, allocation efficiency, and economic efficiency in agricultural production were not high. Minh and Khanh (2011) in [2] have employed the chance-constrained programming model to study productivity growth and efficiency at the provincial level in Vietnam in the period 1995-2007. These researches, however, still separated the efficiency in agricultural production from the overall economic environment. The authors have not taken the characteristic of being rice exporter of Vietnam into consideration. Therefore, they ignored the impacts of exchange rate volatility on rice exporters (or provinces exporting rice) as well as the fact that Vietnam is one of countries attracting a large amount of FDI. This research, thereby, would be different from earlier researches by examining the environment of agricultural production in three aspects:

1) The first issue is that the output of agricultural production is uncertain. The agricultural production can be influenced by several stochastic factors such as weather condition and they can cause a serious damage to agricultural production. This research would not bring the factors of climate change into the model but it still considers the output of agricultural production being uncertain and models them under the chance-constraints.

2) The second issue is that the efficiency of agricultural production could be significantly affected by exchange rate volatility. This argument implies that surprising changes in exchange rate would have impact on decisions of rice exportation because of direct impacts of exchange rate on inputs such as fertilizer, pesticide, seed... and on supply of rice for exportation; hence they would affect production efficiency. This argument is quite close to ones of Ethier (1973) in [3], Clark (1973) in [4], Demers (1991) [5] while they study impacts of exchange rate volatility on trade flows. However, our goal here is to study in more depth impacts of exchange rate volatility on agricultural production, and more particularly rice production. So, our analysis and model derive from the fact that Vietnam is both the importer of input factors for agricultural production and rice exporter. Thereby, exchange rate volatility would have impact on agricultural production through two channels: indirect and direct impacts. The indirect impact on output of agricultural production is through the channel of imported inputs for agricultural production while the direct impact is the change in price of exported rice, which in turn influences the supply capacity of rice of farmers. So, changes in exchange rate can have negative effects on efficiency of agricultural production. The rice production for exportation is mostly concentrated in two areas: Red river delta and Mekong river delta. Therefore, in the model, we must take this characteristic into consideration to study impacts of exchange rate on efficiency of agricultural production among provinces.

3) The third issue is to examine differences in terms of effects of FDI flows on efficiency in agricultural production between rich and poor provinces.

As known so far, many countries try to attract FDI by undertaking different packages of policies because they expect that FDI inflow could bring new technology, know-how and contribute to improvement in productivity and competitiveness of domestic industries. They grant foreign enterprises favorable treatments such as subsidy or favorable tax treatment because policymakers believe that FDI can bring positive externalities to the economy (Caves (1974, 1996) in [6] and [7], Findlay (1978) in [8]) through spread of production know-how and technology (Borensztein *et al.* (1998) in [9], Rappaport (2000) in [10]), thereby, promoting the development of the host country. FDI helps increase amount of capital and employment in the economy (Noorzoy, 1979 in [11]); encouraging technology advance by absorbing foreign technology and the spread of production know-how and technology can occur through granted agreement, imitation, labor training, and application of new processes and products. It is believed that FDI would enhance the existing knowledge in the economy through labor training, absorbing and spreading skills (Van Loo 1997 in [12]; Borenstein, De Gregorio and Lee 1998 in [9]; de Mello 1999i [13]...). FDI contributes to bring new more efficient method of management and organization of production process. FDI also helps increase exports of the host country. Therefore, FDI can play an important role in modernizing the economy and promoting the economic development. However, there also has been point of views that FDI can have negative impacts on economic growth due to weak absorbability, the crowding-out effect on domestic investment, more dependence on foreign investors, severe competition between multinational and domestic enterprises, or break of balance of payments when profit could be returned back to the home country. Because of low FDI into agriculture, spill-over effects in agricultural sector, if have, would be small. This feature can be checked with data of Vietnam in the period 1998-2012.

All issues above would be brought in different ways into the model.

2. Methodology

All issues mentioned above would be presented in this section. Firstly, we would model the uncertainty of output in agricultural sector by chance-constrained data envelopment model (CCDEA) to estimate efficiency, productivity growth in agriculture of sixty provinces in Vietnam. Then, models to evaluate impacts of exchange rate volatility and foreign direct investment on efficiency of agricultural production sector in Vietnam would be given.

2.1. A Programming Model with Chance-Constraints to Estimate Efficiency

We consider a set of N provinces, each consuming amounts of P inputs to produce M outputs. Assuming that each provinces has at least one positive input and one positive output and we can construct the production frontier

$$H = \left\{ (X,Y) : X \geq \sum_{r=1}^{N} X_{ir} z_r, Y \leq \sum_r^N Y_{ir} z_r, \sum_{r1}^N z_r = 1, z \geq 0, r = 1,2,\cdots,60 \right\}$$

H satisfies the free disposability of inputs and outputs and includes all provinces.

2.1.1. Data Envelopment Analysis

The CRS input-oriented measured of technical efficiency for the r_0 th province is calculated as solution to the following mathematical programming problem:

$$\lambda_c^{r_0} = \text{Min}_{\lambda,z} \lambda$$

$$s.t. \ y_{i,r_0} \leq \sum_{r=1}^N y_{i,r} z_r; \ i = 1,2,\cdots,M$$

$$\sum_{r=1}^N x_{j,r} z_r \leq \lambda x_{j,r_0}, \ j = 1,2,\cdots,P \tag{1}$$

$$z \in R_+^N$$

The scalar value λ represents a proportional reduction in all inputs such that $0 \leq \lambda \leq 1$, and $\lambda_c^{r_0}$ is the minimum value of λ so that $\lambda_c^{r_0} x_r$ represents the vector of technically efficient inputs for the r_0th provinces.

Maximum technical efficiency is achieved when $\lambda_c^{r_0}$ is equal to utility. In other words, according to the DEA result when $\lambda_c^{r_0}$ is equal to utility, a province is operating at best-practice and cannot improve on this performance under the condition of the existing set of observations.

2.1.2. Data Envelopment Analysis with Stochastic Constraints (Chance-Constrained Programming Model-CCDEA)

There are several papers about efficiency which employ chance-constrained programming model. Land et al. (1993) in [14] examines the basic chance-constrained programming model to estimate production efficiency in the case that inputs and outputs are stochastic. Copper et al. (2004) in [15] employs the programming approach with stochastic constraints to study congestion in stochastic DEA. Chen (2004) in [16] uses the DEA approach in deterministic and stochastic forms to measure technical efficiency of commercial banks in Taiwan in the period of financial crisis. The author finds that the score of technical efficiency from the deterministic form of DEA approach is higher than one from the stochastic form. Chen (2005) also uses the stochastic DEA approach and stochastic frontier analysis to measure technical efficiency of 39 commercial banks in Taiwan. The author shows that there is a significant difference between efficiency estimated from DEA model with stochastic constraints and one from stochastic frontier production function. Minh *et al.* (2011) in [2] uses the programming approach with stochastic constraints to decompose total factor productivity change (tfpch) into efficiency change (effch) and technical change (techch) in the Vietnamese agriculture sector for the period 1995-2007. The DEA model with stochastic constraints would have three following assumptions:

Assumption 1: Stochastic output.

Based on DEA model mentioned above, we assume that the probability that the best practical agricultural output exceeds the observed output must not be less than α. In other words, the constraint

$$y_{i,r_0} \leq \sum_{h=1}^{N} y_{i,r} z_r, \ i = 1, 2, \cdots, M \tag{2}$$

can be converted to the following constraint:

$$p\left[y_{i,r_0} \leq \sum_{h=1}^{N} y_{i,r} z_r \right] \geq \alpha, \ i = 1, 2, \cdots, M. \tag{3}$$

Thus, corresponding chance-constrained efficiency measure is calculated as:

$$\lambda_c^{r_0} = \mathrm{Min}_{\lambda, z} \lambda$$

$$s.t. \ prob\left[\sum_{h=1}^{N} y_{i,r} z_r \geq y_{i,r_0} \right] \geq \alpha; \ i = 1, 2, \cdots, M \tag{4}$$

$$\sum_{r=1}^{N} x_{j,r} z_r \leq \lambda x_{j,r_0}, \ j = 1, 2, \cdots, P$$

$$z \in R_+^N$$

Assumption 2: $y_{m,k}$ has normal distribution with mean being $E(y_{m,k})$ and variance being $\mathrm{var}(y_{m,k})$.

Assumption 3: All outputs of provinces are randomly independent and $\mathrm{var}(y_{m,k}) = 1$ (for all $k = 1, 2, \cdots$, M) and $\mathrm{cov}(y_{m,h}, y_{m,k}) = 0$ for $h \neq k$. We apply the method introduced by Minh *et al.* (2011) to solve problems (4) and to measure TE.

2.1.3. Test Difference between the Average Efficiency Scores of CCDEA and DEA

We use two Banker's asymptotic DEA efficiency tests to test for inefficiency differences between two different efficiency scores:

1) The two inefficiency scores $(1 - \theta_{\mathrm{CCDEA}})$ and $(1 - \theta_{\mathrm{DEA}})$ follow the exponential distribution.

The test statistic is: $\dfrac{\dfrac{\sum_{i=1}^{N_{\mathrm{CCDEA}}} (1 - \theta_{i,\mathrm{CCDEA}})}{N_{\mathrm{CCDEA}}}}{\dfrac{\sum_{i=1}^{N_{\mathrm{DEA}}} (1 - \theta_{i,\mathrm{DEA}})}{N_{\mathrm{DEA}}}}$ evaluated relative to the *F*-distribution with $(2N_{\mathrm{CCDEA}}, 2N_{\mathrm{DEA}})$ de-

grees of freedom.

2) The two inefficiency scores $(1 - \theta_{\mathrm{CCDEA}})$ and $(1 - \theta_{\mathrm{DEA}})$ follow the half-normal distribution, the test sta-

tistic is $\dfrac{\dfrac{\sum_{i=1}^{N_{\mathrm{CCDEA}}} (1 - \theta_{i,\mathrm{CCDEA}})^2}{N_{\mathrm{CCDEA}}}}{\dfrac{\sum_{i=1}^{N_{\mathrm{DEA}}} (1 - \theta_{i,\mathrm{DEA}})^2}{N_{\mathrm{DEA}}}}$ evaluated relative to the *F*-distribution with $(N_{\mathrm{CCDEA}}, N_{\mathrm{DEA}})$ degrees of freedom.

2.2. Impacts of FDI Flow on Efficiency of Agricultural Sector

To analyze impacts of FDI on technical efficiency in agricultural production, two variables are included: FDI represents the FDI inflow to each province in the period 1998-2011. This variable can reflect impacts of vertical spillover of FDI to agricultural production of provinces. The second variable FDI*G is the product of dummy variable G (taking value of 1 if that is a rich province and 0 if not). The reason for constructing this variable includes: 1) firstly, it can be shown that rich provinces have more FDI enterprises than poor ones have, 2) theoretically, richer provinces have more advantages in capacity of absorbing spillover effects of FDI.

2.3. Impacts of Exchange Rate Volatility on Efficiency of the Agricultural Sector

There have been a lot of researches concerned with impacts of exchange rate volatility on trade flows. Many measurements of exchange rate volatility have been used. For instance, Thursby and Thursby in [17] use the absolute percentage change of exchange rate, i.e. $\{(V_t = |(e_t - e_{t-1})|/e_{t-1})$, in which e_t is the spot rate and t represents the period} as the measurement of exchange rate volatility. McIvoi [18] use residuals from ARIMA model as the measurement of exchange rate volatility. McKenzie [19] use ARCH model to quantify the volatility of exchange rate.

In this research, exchange rate volatility is measured by residuals from ARIMA model. The exchange rate volatility is estimated from quarterly data, however data used to estimate the model of evaluating impacts of exchange rate on efficiency are only available at the annual frequency. Thereby, we would use the two-step procedure to select proxy variable for exchange rate volatility.

1) Estimation of ARIMA (p, q, d)

$$\Delta^d ex_t = \alpha_1 \Delta^d ex_{t-1} + \cdots + \alpha_p \Delta^d ex_{t-p} + u_t + m_1 u_{t-1} + \cdots + m_q u_{t-q} + CD_t \tag{5}$$

in which p is the autoregressive order, q is the moving average order, and d is the degree of difference.

ex_t is the real exchange rate and Δ is the difference operator defined as follow:

$$\Delta^d ex_t = \Delta\left(\Delta^{d-1} ex_t\right), d \geq 2; \quad \Delta ex_t = ex_t - ex_{t-1}$$

The vector D_t consists of deterministic terms such as constant, linear trend, dummy variables… and u_t is the white noise process with expected value 0 and variance σ_u^2.

2) From residuals e_t of the estimated model ARIMA (p, q, d), we pick up three values each year

$$e_{t,max} = \text{Max}\left\{e_{tI}, e_{tII}, e_{tIII}, e_{tIV}\right\}, \text{ denoting } \sigma_{e1}; \tag{6a}$$

$$e_{t,min} = \text{Min}\left\{e_{tI}, e_{tII}, e_{tIII}, e_{tIV}\right\}, \text{ denoting } \sigma_{e2}; \tag{6b}$$

$$e_{t,average} = \text{Average}\left\{e_{tI}, e_{tII}, e_{tIII}, e_{tIV}\right\}, \tag{6c}$$

denoting σ_{e3}, in which e_{tI}, e_{tII}, e_{tIII}, e_{tIV} are residuals e_t of the model ARIMA (p, q, d) in year t, quarter I, quarter II, quarter III and quarter IV respectively.

Then, we can establish the series $\{\sigma_{e1}\}$, $\{\sigma_{e2}\}$ and $\{\sigma_{e3}\}$. These series correspond to three models, in which the models differ from each other at the way to measure exchange rate volatility. The exchange rate volatility in each model can be estimated from a pair of variables σ_e and σ_{eN}. The variable σ_e $\{(\sigma_{e1}), (\sigma_{e2}), \text{ and } (\sigma_{e3})\}$ shows exchange rate volatility while σ_{eN} is the product of two variables e and n, in which n is the dummy variable taking value of 1 in the case of poor province, i.e. having income lower than the average level of the whole country. This variable ($\{\sigma_{e1n}\}$, $\{\sigma_{e2n}\}$ and $\{\sigma_{e3n}\}$) would help us examine impacts of exchange rate on poor provinces.

2.4. Effects of Exchange Rate Volatility, FDI and Other Factors on Efficiency in Agricultural Production

To study impacts of exchange rate volatility and FDI on efficiency in agricultural production, we employ fixed effect model to evaluate impacts of exchange rate volatility (σ_e), FDI, FDIG and σ_N on variables TECCDEA. The model can be specified as follow:

$$\text{TECCDEA}_{i,t} = \alpha_i + \alpha_1 \text{FDI}_{i,t} + \alpha_2 \left(\frac{K}{L}\right)_{i,t} + \alpha_3 \sigma_{et} + \alpha_4 \text{FDIG}_{i,t} + \alpha_5 \sigma_{eN_{i,t}} + \varepsilon_{i,t} \tag{7}$$

with i denoting provinces, and t denoting time periods. The dependent variable $\text{TECCDEA}_{i,t}$ is the annual technical efficiency of ith province at time t, estimated from CCDEA model. $\left(\dfrac{K}{L}\right)_{i,t}$ is the capital-labor ratio of ith province at time t. $\text{FDI}_{i,t}$ is foreign direct investment in ith province at time t. σ_{et} (residuals) represents exchange rate volatility in year t, estimated from ARIMA(p, q, d) model. $\text{FDIG}_{i,t}$ is foreign direct investment in the rich province at year t. $\sigma_{eN_{i,t}}$ expresses the impacts of exchange rate volatility on poor provinces at time t.

There are three forms of model (7). Model (7a) ((7b) and (7c)) is model (7) with σ_e in (6a) ((6b) and (6c)).

The parameter α_i may have two different interpretations and two different models may be distinguished according to this interpretations. If the α_i is assumed to be fixed parameters, Equation (6) is a fixed effect panel data model. Conversely, if the α_i are assumed to be random, Equation (6) is a random effect panel data model. In general, fixed effect model is indicated when the regression analysis is limited to a precise set of individuals. For this reason, since our data set consists of the observations over the 60 provinces, we use a fixed effect panel data model to analyze the impacts of exchange rate volatility and FDI on technical efficiency from CCDEA model in agricultural production.

3. Data Source

Data set used in this research consists of inputs and outputs of agricultural production in sixty provinces in Vietnam in the period 1998-2011. Data of some provinces must be added up due to separation or combination between provinces in this period. For instance, Ha Tay and Hanoi are two separated provinces before 2008, but then they are combined with each other to become new Hanoi. Therefore, statistical data for HaTay would not be available since then. To ensure the consistency of data, we would add up each indicator of HaTay to the corresponding indicator of Hanoi in the period 1998-2008, and now we have new data set for only one province, namely new Hanoi. We do the same transformation for two other pairs of provinces including Dac Lac and Dac Nong, Dien Bien and Lai Chau. The data set used in this research is collected from General Statistical Office and Ministry of Labor-Invalids and Social Affairs.

Most of researches about agricultural productivity in Vietnam use gross value of agricultural value as gross value of agricultural production. The gross value of agricultural value is defined as the gross value of agricultural production in following fields: cultivation, forestry, animal husbandry, fishery, and secondary works. The value of all inputs in agricultural production would also be included in the gross value of Vietnamese agricultural value. Thereby, in this research, the net value or added value of agricultural output (VAO) would be used to measure gross value of Vietnamese agricultural output. The VAO data of provinces in the period 19988-2011 would be adjusted with respect to GDP deflator, in which base year is 1994. The quarterly exchange rate (VND/$) data in the period 1995-2012 and the data of implemented FDI in the period 1998-2011 are collected from General Statistical Office.

4. Estimation Results

4.1. The Estimated Results of Efficiency in Agricultural Production of Provinces from CDEA

We employ the indicator, namely inverse of (2), to measure technology efficiency in agricultural sector for each province in Vietnam in the period 1998-2011. These values of technical efficiency are measured under the assumption of constant returns to scale within sixty provinces in the period 1998-2011.

To have a preliminary picture of technical efficiency in agricultural sector, we would firstly review estimated results of efficiency of provinces in this period. The results show that agricultural production in Ho Chi Minh city has efficiency score lying in the group of ten best provinces within 12 years, and lying on the frontier curve in 9 years. Agricultural production in Tra Vinh also has efficiency score lying in the group of ten best provinces within 12 years, but only lying on the frontier curve in 5 year. Agricultural production in Ba Ria-Vung Tau has efficiency score lying in the group of ten best provinces within 10 years, and lying on the frontier curve in 4 years. Efficiency score of agricultural production of Ben Tre belongs to the group of ten best in 9 years and lies on the frontier curve in 4 years... In contrast, the group of ten least efficient provinces in agricultural production includes provinces such as Quang Ninh and Dien Bien (in 13 years), Thai Nguyen and Quang Binh (in 12 years), Ha Giang and Phu Yen (in 11 years). Especially, Phu Tho has lowest efficiency score of agricultural production within 13 years but in 2011, it steps up to become one of the ten most efficiency in terms of agricultural production. Quang Ngai also lies in the group of ten least efficiency in 1998 but jumps up to the group of ten most efficiency in 2011.

4.2. Estimated Results of Efficiency Tests between Two Different Scores from CCDEA and DEA Model

To compare between technical efficiency estimated from CCDEA model and technical efficiency estimated from DEA model, we use two Banker's asymptotic DEA efficiency tests.

The estimated results of two Banker's asymptotic DEA efficiency tests can be given in **Table 1**.

The test statistics estimated under the assumption of the two inefficiency scores following the exponential distribution (the half-normal distribution) are reported in the second column (the fourth column) in **Table 1** and the critical value in the third column (the fifth column). The both types of test results show that there is no significant difference between the average efficiency scores of DEA and chance-constrained DEA methods. We can conclude that the chance-constrained DEA model measurement does not have a different between the DEA and chance-constrained DEA model measurement. The difference between two models can be explained as: the CCDEA model involves only the probabilistic structure of agricultural sector output, and flexible production frontier may occur.

4.3. Evaluating Impacts of Exchange Rate and FDI Flow on Efficiency

We conduct a regression analysis to determine the impacts of exchange rate volatility and FDI flow on technical efficiency in Vietnam agricultural sector. **Table 2** presents estimation results of fixed-effect model using panel data.

The first column in this table shows the name of variables in the models. The third, fourth, and fifth column give estimation results from three models in which efficiency scores estimated from CCDEA model. All three models have the same independent variables except exchange rate. In the first model, the maximum exchange rate (σ_{e1}) is used, and in the second model the minimum exchange rate (σ_{e2}) is used and in the third one, the average exchange rate (σ_{e3}) is used.

Table 1. Summary of efficiency difference test results (CCDEA vs. DEA).

Year	Exponential type	Critical value	Half-normal type	Critical value
1998	0.9602	1.5330	1.2778	1.8360
1999	0.8646	1.5330	1.2250	1.8360
2000	1.0395	1.5330	1.3134	1.8360
2001	1.063	1.5330	1.3644	1.8360
2002	1.1131	1.5330	1.3784	1.8360
2003	1.0593	1.5330	1.3501	1.8360
2004	1.0288	1.5330	1.3294	1.8360
2005	1.0179	1.5330	1.3001	1.8360
2006	1.0311	1.5330	1.3080	1.8360
2007	0.999	1.5330	1.3005	1.8360
2008	0.9565	1.5330	1.2659	1.8360
2009	0.8596	1.5330	1.2187	1.8360
2010	0.8219	1.5330	1.2043	1.8360
2011	1.0016	1.5330	1.3144	1.8360

Source: The authors estimate from the data of GSO.

Table 2. The estimation results of fixed-effect model for.

	Chance-constrained model		
	Model (7a)	Model (7b)	Model (7c)
FDI	0.00026 (0.00002)	0.000267 (0.00002)	0.000018 (0.00002)
K/L	1.3882*** (0.3118)	1.4055*** (0.3139)	1.4688*** (0.3155)
σ_e	−2.0277*** (0.3128)	−1.7632*** (0.3222)	−4.4618*** (0.7945)
FDIG	−0.00004 (0.00003)	−0.00004 (0.00003)	−0.00004 (0.00003)
σ_{en}	1.1293* (0.6271)	0.7733 (0.7869)	5.7488*** (1.6723)
_cons	0.5077*** (0.0055)	0.5003*** (0.0050)	0.4882*** (0.0044)
/sigma_u	0.2242	0.2251	0.2250
/sigma_e	0.1139	0.1148	0.1147
Rho	0.7946	0.7936	0.7936
R^2: within	0.0775	0.0641	0.0653
Between	0.0446	0.0095	0.0133
overall	0.0267	0.0169	0.0181
Number of obs	840	840	840

Source: The authors estimate from the data of GSO.

The estimation results show that in all three models, the coefficient of exchange rate variable (σ_e) takes negative value and highly statistical significance. This implies that fluctuation in exchange rate would reduce efficiency in agricultural production. This can be explained by the fact that a large proportion of agricultural production in Vietnam is rice, whose export value is very large. Exchange rate volatility can directly affect exports. This result is compatible with theoretical view that exchange rate volatility would reduce trade flow. This view claims that a sudden change in exchange rate would have effect on decision of risk-adverse traders, therefore volume of trade would decline. This view is also given in Ethier [3] published soon after the breakdown of Bretton-Woods system. Clark [4] develops a similar model for a risk-adverse firm and comes to the same conclusion. Meanwhile, the sign of variable σ_{en} is positive and statistically significant (except in the model 2), which means that exchange rate volatility in the case of provinces having no rice exports has positive impacts on efficiency. The possible reason is that exchange rate volatility often makes imported agricultural materials cheaper, so provinces without rice export would benefit due to lower input costs, so the product of these two variables would take positive value and be highly statistically significant.

The estimation results of FDI variable in all three models take positive value but insignificant at the any level. The sign of the variable GFDI is negative and insignificant. It means that FDI flow during the period of 1988-2011 have not been significant impact on the efficient production in Vietnam agricultural sector.

5. Conclusions

This research employs chance-constrained DEA approach to estimate technical efficiency in Vietnamese agriculture sector in the period 1998-2011.

This research examines effects of exchange rate volatility and FDI flows on efficiency of production in agriculture sector by using residuals from ARIMA model as proxy variable for exchange rate volatility. By dividing provinces in the country into two groups, namely rich and poor one, we can derive several interesting findings. The results show that in all three models, the coefficient of exchange rate variable is negative and highly statistically significant, that is, fluctuation of exchange rate would reduce efficiency of production in agriculture production. However, exchange rate volatility in provinces without rice exportations has a positive impact on efficiency. The coefficient of FDI variable is positive and insignificant at any level. Meanwhile, the coefficient of FDIG variable is negative and insignificant.

References

[1] Minh, N.K. and Long, G.T. (2009) Efficiency Estimates for the Agricultural Production in Vietnam: A Comparison of Parametric and Non-Parametric Approaches. *Agricultural Economics Review*, **10**, 62-78.

[2] Minh, N.K. and Khanh, P.V. (2011) A Chance-Constrained Data Envelopment Analysis Approach to Problem Provincial Productivity Growth in Vietnamese Agriculture from 1995 to 2007. *Open Journal of Statistics*, **1**, 217-235. http://dx.doi.org/10.4236/ojs.2011.13026

[3] Ethier, W. (1973) International Trade and the Forward Exchange Market. *American Economic Review*, **63**, 494-503.

[4] Clark, P.B. (1973) Uncertainty, Exchange Risk, and the Level of International Trade. *Western Economic Journal*, **11**, 302-313. http://dx.doi.org/10.1111/j.1465-7295.1973.tb01063.x

[5] Demers, M. (1991) Investment under Uncertainty, Irreversibility and the Arrival of Information over Time. *Review of Economic Studies*, **58**, 333-350. http://dx.doi.org/10.2307/2297971

[6] Caves, R. (1974) Multinational Firms, Competition and Productivity in the Host Country. *Economics*, **41**, 176-193. http://dx.doi.org/10.2307/2553765

[7] Caves, R. (1996) Multinational Enterprise and Economic Analysis. Cambridge University Press, Cambridge.

[8] Findlay, R. (1978) Relative Backwardness, Direct Foreign Investment and the Transfer of Technology: A Simple Dynamic Model. *Quarterly Journal of Economics*, **92**, 1-16. http://dx.doi.org/10.2307/1885996

[9] Borensztein, E.J., DeGregorio, J. and Lee, J.W. (1998) How Does Foreign Direct Investment Affect Economic Growth? *Journal of International Economics*, **45**, 115-135. http://dx.doi.org/10.1016/S0022-1996(97)00033-0

[10] Rappaport, J. (2000) How Does Openness to Capital Flows Affect Growth? Research Working Paper, RWP00-11, Federal Reserve Bank of Kansas City.

[11] Noorzoy, M.S. (1979) Flows of Direct Investment and Their Effects on Investment in Canada. *Economic Letters*, **2**, 257-261. http://dx.doi.org/10.1016/0165-1765(79)90032-6

[12] Van Loo, F. (1977) The Effect of Foreign Direct Investment on Investment in Canada. *Review of Economics and Statistics*, **59**, 474-481. http://dx.doi.org/10.2307/1928712

[13] De Mello Jr., L. (1999) Foreign Direct Investment-Led Growth: Evidence from Time Series and Panel Data. *Oxford Economic Papers*, **51**, 133-151. http://dx.doi.org/10.1093/oep/51.1.133

[14] Land, K.C., Knox Lovell, C.A. and Thore, S. (1993) Chance-Constrained Data Envelopment Analysis. *Managerial and Decision Economics*, **14**, 541-554. http://dx.doi.org/10.1002/mde.4090140607

[15] Copper, W.W., Deng, H.Z., Huang, Z. and Li, S.X. (2004) Chance Constrained Programming Approaches to Congestion in Stochastic Data Envelopment Analysis. *European Journal of Operation Research*, **155**, 487-501. http://dx.doi.org/10.1016/S0377-2217(02)00901-3

[16] Chen, T. (2002) A Comparison of Chance-Constrained DEA and Stochastic Frontier Analysis: Bank Efficiency in Taiwan. *The Journal of the Operational Research Society*, **53**, 492-500. http://dx.doi.org/10.1057/palgrave.jors.2601318

[17] Thursby, M.C. and Thursby, J.G. (1985) The Uncertainty Effects of Floating Exchange Rates: Empirical Evidence on International Trade Flows. Ballinger Publishing Co., Cambridge, 153-166.

[18] McIvor, R. (1995) Exchange Rate Variability and Australia's Export Performance.

[19] McKenzie, M. and Brooks, R. (1997) The Impact of Exchange Rate Volatility on German-US Trade Flows. *Journal of International Financial Markets, Institutions and Money*, **7**, 73-87. http://dx.doi.org/10.1016/S1042-4431(97)00012-7

4

Combining Lean Concepts & Tools with the DMAIC Framework to Improve Processes and Reduce Waste

Mazen Arafeh

The Department of Industrial Engineering, Faculty of Engineering & Technology, The University of Jordan, Amman, Jordan
Email: hfmazen@gmail.com

Abstract

Six Sigma DMAIC methodology has been applied to systematically apply lean manufacturing concepts and tools in order to improve productivity in a local company specialized in the manufacturing of safety and fire resistance metal doors, windows, and frames. In-depth analysis of the plant processes unfolded different critical processes, specifically foam injection process and sheet metal cutting process. Throughout the different project phases, various improvements had been implemented to reduce production cycle time from 216 min to 161 min; non-value added activities in the different processes were identified and eliminated. Plant layout and machine reconfiguration reduced backtracking and unutilized space. Percentage of defective doors (needing rework) dropped from 100% to only 15%. The successful implementation of this project is largely due to top management active involvement and participation of workers and operators in all stages of the project. Finally, new policies and mentoring programs are introduced to maintain improvements.

Keywords

Six Sigma, DMAIC, Lean Manufacturing, Value-Added Activities, Production Cycle Time, Safety and Fire Resistant Doors

1. Introduction

Heightened challenges from competitors have prompted many manufacturing firms to adopt new manufacturing approaches. Particularly salient among these is the concept of lean production [1] [2].

The concept of Lean Thinking (LT) developed from Toyota Production System (TPS) involves determining the value of any process by distinguishing valued-added activities or steps from non-value added activities or steps and eliminating waste, so that every step adds value to the process [3]. Lean is about controlling the resources in accordance with the customers' needs and about reducing unnecessary waste, including the waste of time and material [4].

Six Sigma (SS) was developed at Motorola by an engineer Bill Smith in the mid-1980s. SS is a business improvement approach that seeks to find and eliminate causes of defects or mistakes in business processes by focusing on process outputs which are critical in the eyes of customers [3]. SS started on the shop floor and then moved into the front offices [5]. Nowadays, SS is being used in many organizations and environments, for example, manufacturing, services, and healthcare [6].

The Define, Measure, Analyze, Improve and Control phases (DMAIC) present a clear strategy for the deployment of SS projects [7]. SS projects are usually initiated by the SS champion and the SS Black Belt (SSBB). Champions are generally upper managers. They serve as mentors and leaders; they support project teams, allocate resources, and remove barriers. SSBB belts lead improvement projects full time, supported by project team members [8].

This work presents an application of the Six Sigma DMAIC methodology to improve processes in a local company specialized in the manufacturing of safety and fire resistance metal doors, windows, and frames. The company suffered from inefficient processes and operations resulting in large amounts of dollar losses.

Fire resistant doors are comprised of the frames, leafs, covers, U-shape, insulation material (polyurethane-foam), and door accessories (hinges, hands, and panic device), as shown in **Figure 1**.

Specifically, this work aimed at firstly reducing manufacturing cycle time and improving production flow by smoothing the flow of operations and reducing the amount of rework, and secondly reducing material waste.

2. Methodology

The Six Sigma DMAIC methodology is adopted to execute the improvement project since it provides a proven framework for problem solving. The DMAIC methodology builds on three fundamental principles: firstly, results-focused; driven by data, facts, and metrics. Secondly, it is project-based and project-structured. Thirdly, it links a combination of tools, tasks, and deliverables that varies by step in the method [9].

In the define phase, we start by identifying improvement opportunities, clarifying scope and defining goals. Identifying improvement opportunities is accomplished by capturing the voice of the customer (VOC), the voice

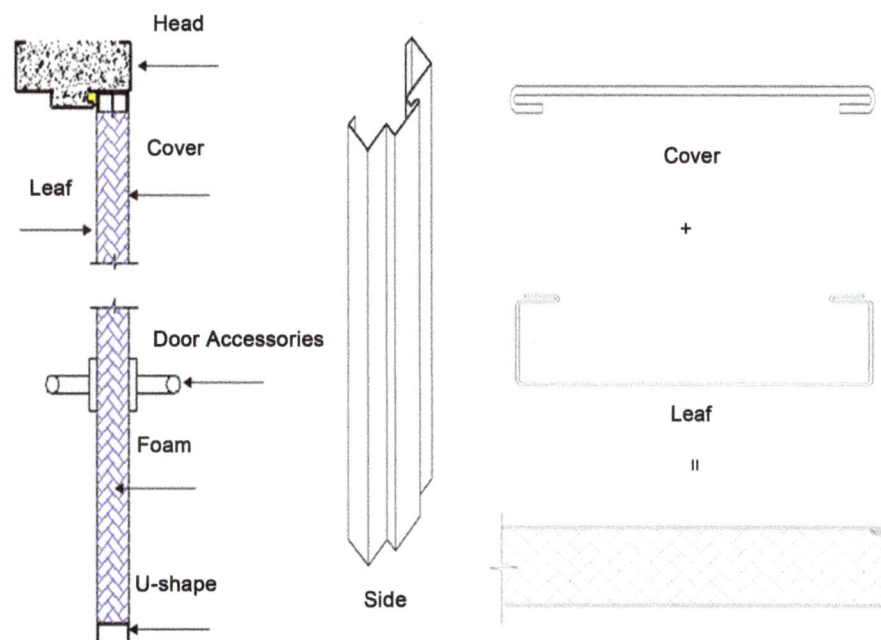

Figure 1. Door parts.

of the process (VOP), and the voice of business (VOB). Customers may be classified as internal or external. In this project we focus our attention at the internal customers, more often known as the voice of the employee (VOE) represented by those who work in the company coming from different departments. The VOB is derived from financial information and data. The VOP evaluates how well or poorly the process is performing as it relates to the VOB and VOC. By paying attention to the VOP, analysts can identify poorly performing processes and use the information to identify and prioritize potential projects.

In the measure phase, data collection is carried out and the initial performance is measured against customer requirements. The analyze phase consists of analysis of the root cause(s), accounting for the errors or defects that are quantified by the data collected. In the improve phase, alternative solutions to eliminate the root cause(s) or errors or defects are examined and the optimal one(s) are selected, the system performance is then evaluated after implementing the process improvements. The control phase entails development and implementation of a monitoring system to reduce future errors, and documentation of results and recommendations for additional action.

3. Application of the DMAIC Methodology and Discussion

3.1. Define Stage

The improvement team was led by a Six Sigma Black Belt (SSBB). The SSBB and the champion described and scoped the project. They also met with the selected team members, explained to them the project objectives and importance, discussed with them their roles and listened to their feedback.

To collect the VOC, VOP and VOB, several brainstorming sessions were carried out. Procurement department brought the attention to the waste of material from a financial point of view. Manager and workers (Punchers, welders, etc…) discussed the several problems they face while they perform their daily work. The brainstorming sessions mainly focused on three major opportunities for improvements: unsmooth flow of production processes, material waste, and time waste.

The team prepared a preliminary process map as shown in **Figure 2**. The figure presents the different manu-

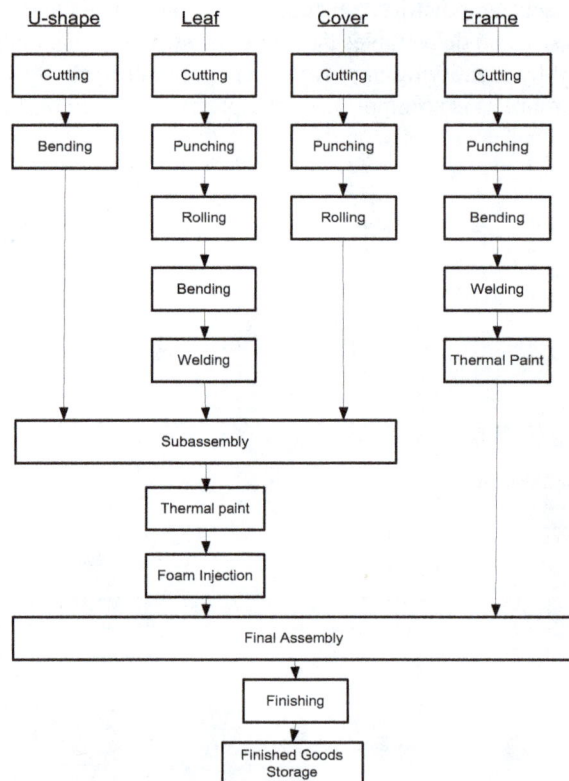

Figure 2. Process flow chart.

facturing processes. One of the main reasons for of drawing a process map is to clarify processes, roles and set a clear scope for the project. The process map enlightened the plant staff members because they have never seen their processes modeled end-to-end. They only knew fragments of the process and were unaware of how what they did fits into the overall outcome.

The foam injection step is where the insulation material (Polyurethane) is prepared and sprayed (injected) into the assembled door. The Polyurethane used in this process is a two-component reactive system mixed at a spray gun. Hence, it is known as spray polyurethane foam (SPF) materials.

When small items are being cut out from large objects, two problems arise. The first one is the assortment problem addressing the issue of choosing proper dimensions for the large objects. The second one is the trim loss problem addressing the issue of how to cut out the small items from the given large objects in such a way that wastage material will be minimized. In practice, the small items are known as order list and the large objects are known as stock material. In the cutting process the stock material can seldom be used as a whole but some residual pieces or trim loss will be produced. Since the primary objective of the cutting process is to minimize the wastage the problem is known as trim loss problem. The combination of the assortment problem and the trim loss problem is known as the cutting stock problem (CSP) [10].

3.2. Measure Phase

In order to assess the initial state of the plant, the measure phase starts with data collection, followed by analyzing the initial state. Data collection points were selected and identified after studying the plant layout and the observing the plant in operation such that minimum interference with work flow could occur. The data collected includes plant layout dimensions, machine dimensions, transportation distances traveled, backtracking, in addition to the raw material waste. Preliminary analysis of the information gathered from the brainstorming session and examining the data gathered, we identified three major problematic areas. Firstly, a poor plant layout, secondly, the foam injection process, and thirdly, the cutting process.

Optimizing any process means maximizing the proportion of value adding activities, while eliminating non-value-adding activities as far as possible or at least reducing them to a minimum [11]. The term value added refers to any activity that transforms the product or deliverable, in the view of the customer, to a more complete state. The product has been physically changed, and its value to the customer has increased. Conversely, the term non-value-added refers to activity that consumes time (people expense), material, and/or space (facilities expense), yet does not physically advance the product or increase its value [12]. Value-added flow charts are created to identify and separate value adding activating from non-value adding ones. **Figure 3** resents a value-added flow chart for cutting a frame. We note here that the only value adding activity is cutting, all other activities are non-value-adding activities.

Similar flow charts were created for the remaining parts in the making a door as shown in **Figure 4**.

Analyzing the process of making a door, we found that approximately only 45% of the processing time is value-added. The plant operated at a cycle time of 216 minutes.

Figure 5 shows a Pareto chart of the non-value added times for the different steps. Foam injection process predominantly contributes to non-value activities followed by cutting processes.

3.2.1. Foam Injection Process

We created a process flowchart to better understand the foam injection process as shown in **Figure 6**.

Orders totaling twenty doors were sampled and examined to identify their defects. The processing times, the defects with their occurrence frequency, and quantities of foam waste were recorded. There are mainly 5 defects. **Figure 7** presents a Pareto chart for these defects ordered according to their frequency of occurrence. Fifty percent of the doors were unleveled surface doors. Unleveled doors needed rework adding an average of 40 minutes to the cycle time.

Inhomogeneous foam distribution means that the foam did not spread evenly inside the door. Unmixed foam means that the mixing of two liquids material was not complete. In this case, the door is irreparable and hence, it is scraped. We can see the 15% of the production is scraped.

The amount of foam waste was also recorded and another Pareto chart was created as shown in **Figure 8**. Unleveled doors, foam leakage, and inhomogeneous foam produce about 95% of foam waste.

Value-added   Non-value-added

Transfer	0.34 min
Calculate	4.06 min
Setup	0.07 min
Cut	3.6 min
Transfer	0.36 min
Pick	0.83 min
Setup	0.17 min
Cut	2.31 min
Transfer	0.38 min
Pick	0.76 min
Setup	0.16min
Cut	2.3 min
Transfer	0.17 min
Setup	0.16 min
Cut	1.0 min

Total time of value-added activities 9.21 min. (47%)   **Total time of non-value-added activities 10.37 min. (53%)**

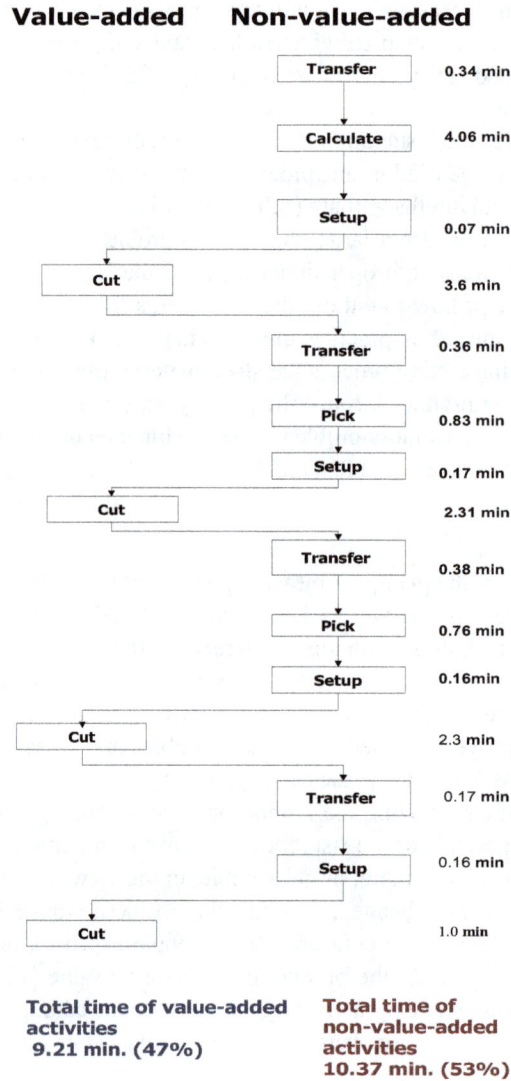

Figure 3. A value-added flow chart for preparing a frame.

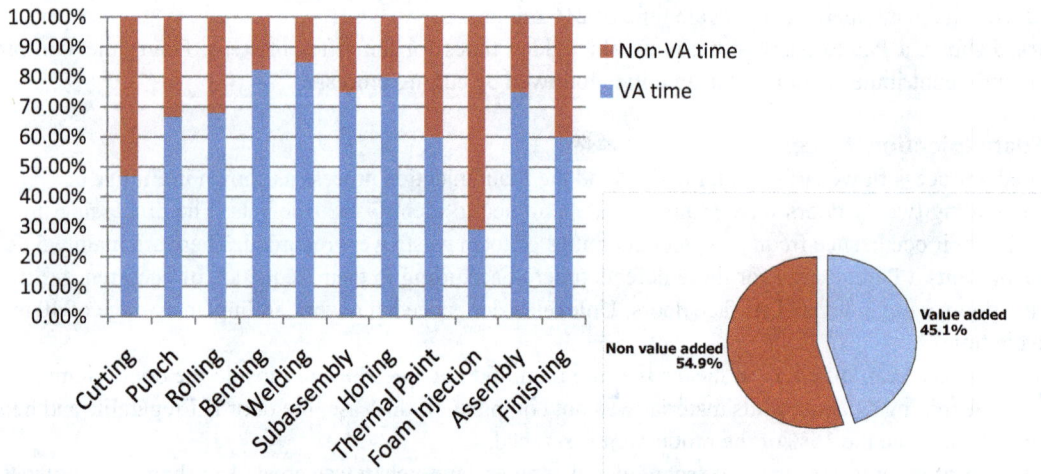

Figure 4. Value-added vs. non-value-added for steps in preparing a door.

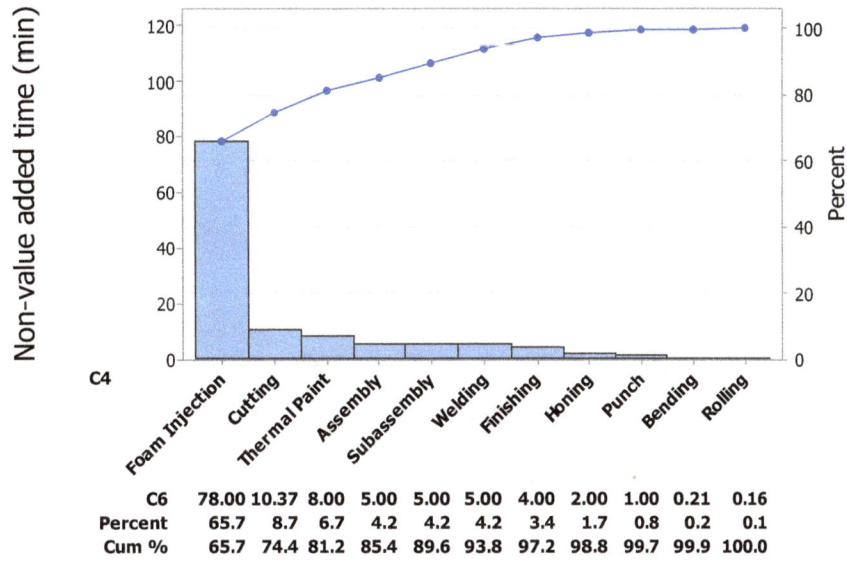

C6	78.00	10.37	8.00	5.00	5.00	5.00	4.00	2.00	1.00	0.21	0.16
Percent	65.7	8.7	6.7	4.2	4.2	4.2	3.4	1.7	0.8	0.2	0.1
Cum %	65.7	74.4	81.2	85.4	89.6	93.8	97.2	98.8	99.7	99.9	100.0

Figure 5. Pareto chart of non-value-added time.

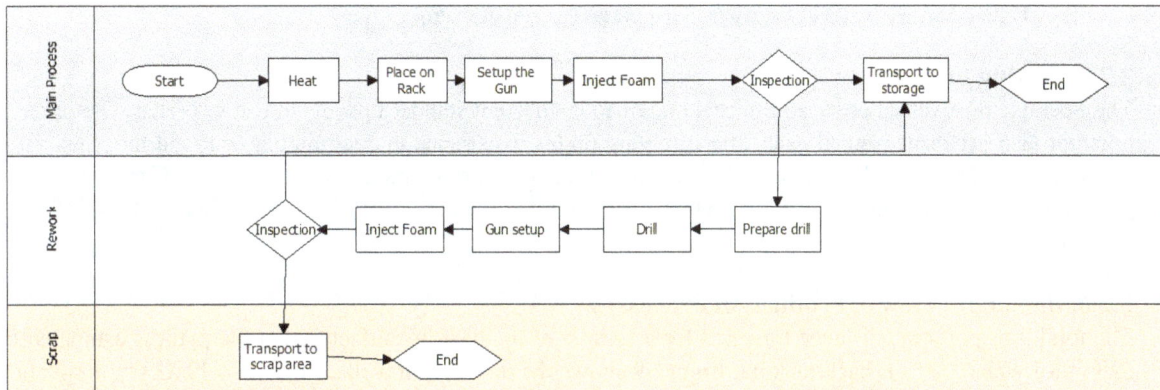

Figure 6. Foam injection process flow chart.

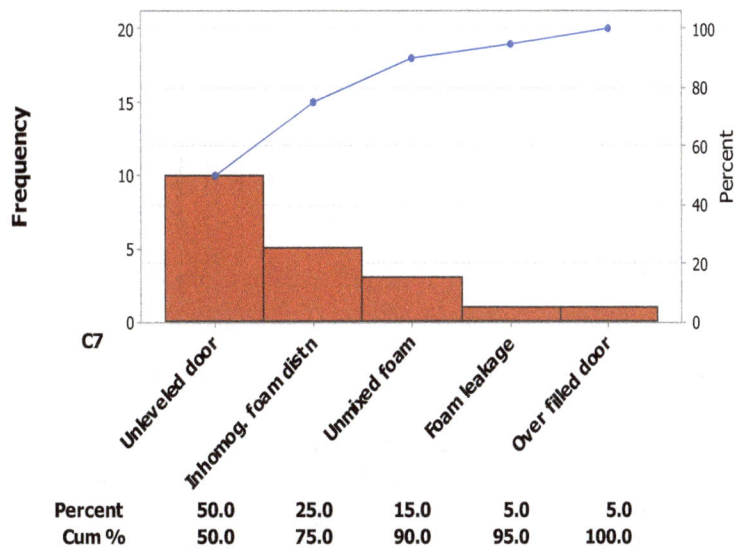

	Unleveled door	Inhomog. foam distn	Unmixed foam	Foam leakage	Over filled door
Percent	50.0	25.0	15.0	5.0	5.0
Cum %	50.0	75.0	90.0	95.0	100.0

Figure 7. Pareto chart of foam injection process (frequency of defects).

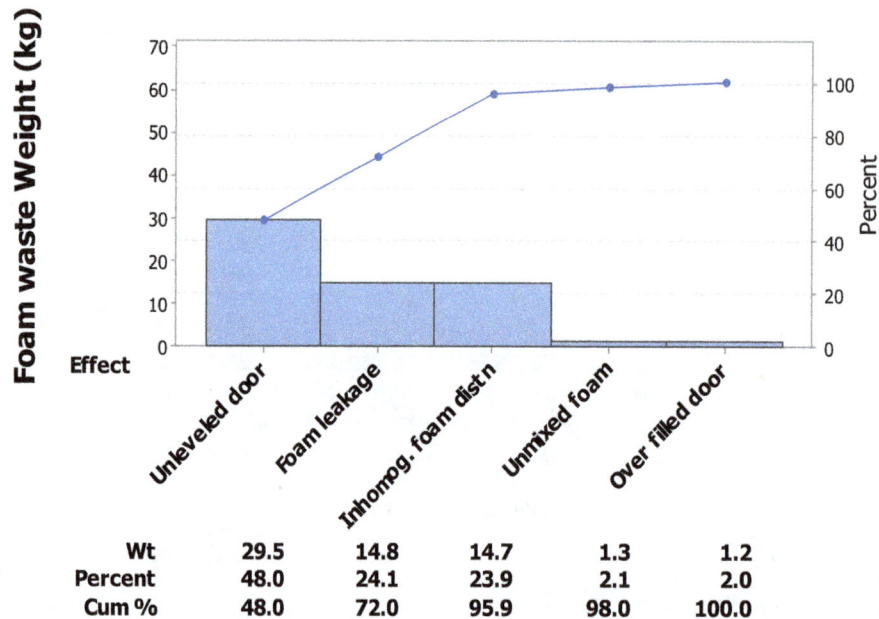

Wt	29.5	14.8	14.7	1.3	1.2
Percent	48.0	24.1	23.9	2.1	2.0
Cum %	48.0	72.0	95.9	98.0	100.0

Figure 8. Pareto chart of foam injection process (weight of foam).

3.2.2. Cutting Process

The operator receives an order and selects one of four different standard panels to use to prepare the order. The operator also performs manual calculations relying on his experience to determine how to cut the panel into the required parts. Manual calculations and experience does not yield the minimum quantity of off cuts. Twenty five percent of the cutting time is spent on this step and the total quantity of off cuts was estimated to be 24% of the original panels.

3.2.3. Unsmooth Flow of Production Processes

The total transportation distance travelled for a sample of the parts manufactured at the plant is estimated to be 323 m. of which, 62% is backtracking. **Figure 9** shows the distances travelled (drawn in black) in production of a frame and backtracking drawn in red.

Total unutilized area is estimated to be approximately 17% of the total plant area where piles of scrap, debris, and unnecessary items were accumulated.

3.3. Analyze

After observing the processes at the plant and collecting data, efforts in the analyze phase are focused on investigating the root causes of the problems in the processes. Brainstorming sessions were held to examine process flow charts, and cause and effect diagrams are created to identify the root causes of each problem.

3.3.1. Foam Injection Process

Based on the observations and the data collected in the measure phase, the defects shown in **Figure 7** are analyzed and traced back to their root causes. A cause and effect diagram was created to identify these causes as shown in **Figure 10**.

A Pareto chart was created for the causes as they relate to the quantity of foam waste resulted as shown in **Figure 11**.

The divider rack shown in **Figure 12** was installed by the operator in an effort aimed at increasing the productivity of the foam injection process; however, it turned out to cause fifty percent of the foam waste, followed by the wooden stoppers used to close the holes used for injection, and the low foam flow rate resulting from the low pressure in the spraying gun. The foam spayed into the doors would leak because the wooden stoppers were not airtight.

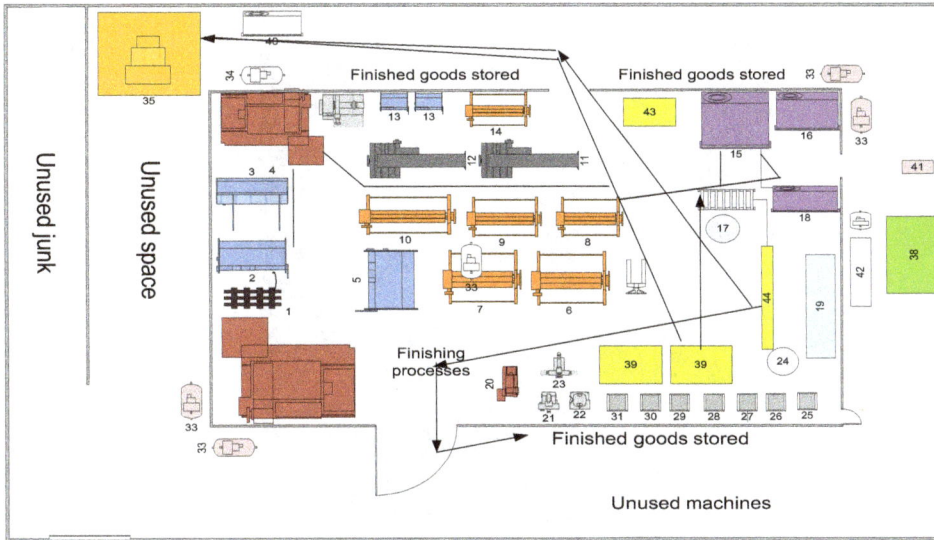

Figure 9. Frame production path.

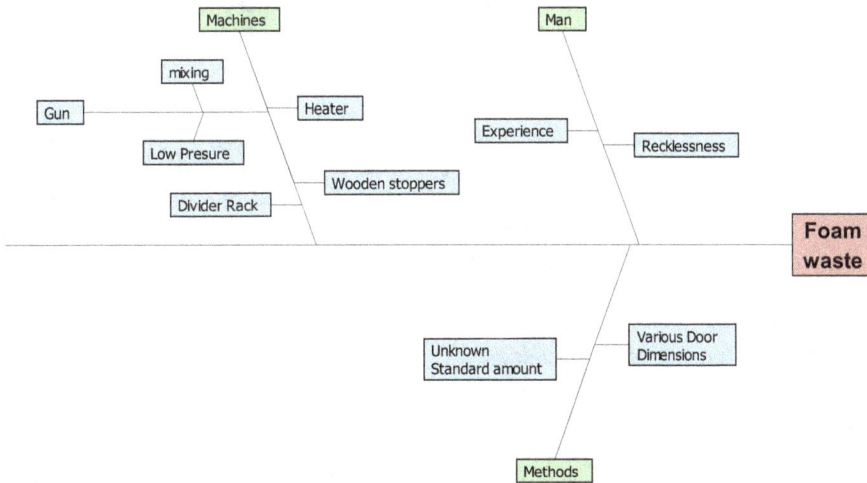

Figure 10. Foam waste cause and effect diagram.

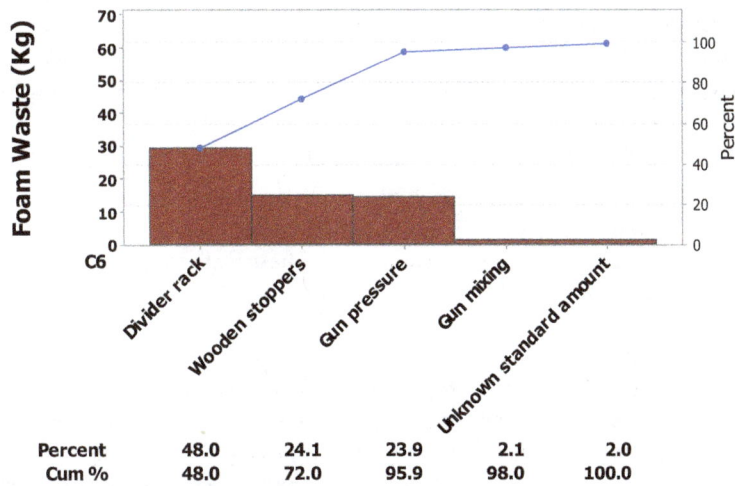

	Divider rack	Wooden stoppers	Gun pressure	Gun mixing	Unknown standard amount
Percent	48.0	24.1	23.9	2.1	2.0
Cum %	48.0	72.0	95.9	98.0	100.0

Figure 11. Pareto chart for the root cause.

Figure 12. The divider rack.

3.3.2. Cutting Process

From the analysis of the data gathered in the measure phase a cause and effect diagram for high offcuts percentage and long cutting time were created as shown in **Figure 13** and **Figure 14**, respectively.

Manual calculations appeared to be causing both long cutting time and high trim loss. The lack of standard operating procedures (SOPs) allowing each operator to rely on his own experience in performing the manual calculations. In addition, the absence of a preventive maintenance program caused many of the saw blades to wear prematurely.

3.3.3. Unsmooth Flow of Production Processes

Figure 15 presents a cause and effect diagram for unsmooth flow of operations; we can clearly see that many of the root causes presented in this C & E diagram are were also presented in the previous C & E diagrams emphasizing the connection between the causes and their effects. Specifically mentioned here, is backtracking and excess distance. Backtracking and excess transportation distance is caused by improper location of machines. Unorganized and unclean workplace is another root cause.

While analyzing root causes for the problems identified, the figures also presented clues for possible solutions. For instance, it was clear that the lack of order and tidiness regarding tools, machines, material, scrap and storage, in addition to the lack of cleanliness throughout the plant is a major cause of delay and unsmooth operations. Machine locations and arrangement and plant layout should also be considered in the improve stage.

3.4. Improve & Control Phases

The goal of the improve phase is to implement the changes to the system that are needed to improve it [13].

Several brainstorming sessions were held to identify high potential improvement opportunities. The value of the improvement methods applied is that they encompass the best techniques for driving out defects and improving process efficiency and capacity [14].

3.4.1. Foam Injection Process

The improvement efforts started by removing the divider rack. A new spraying gun was purchased and used. The new gun provided the required foam spray flow rate. Using a tape instead of the wooden stoppers to seal the injection holes stopped the leakage of foam. After carefully studying the spraying operation, markings on the valves helped stabilizing the gun mixing of the two foam materials. After these solutions were implemented, only 15% of doors needed rework and the amount of foam waste dropped by 81%.

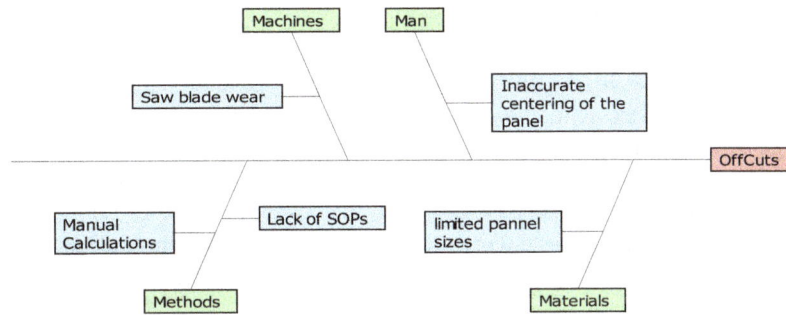

Figure 13. Cutoffs cause and effect diagram.

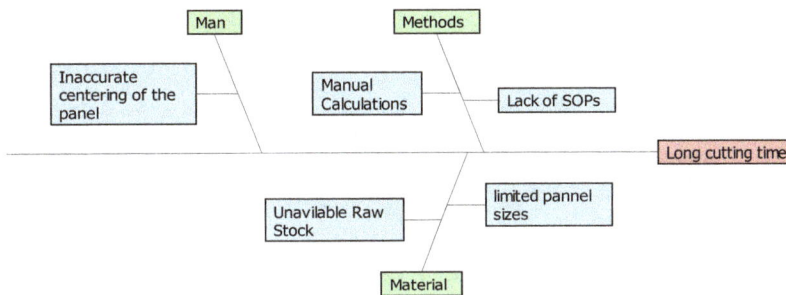

Figure 14. Long cutting time cause and effect diagram.

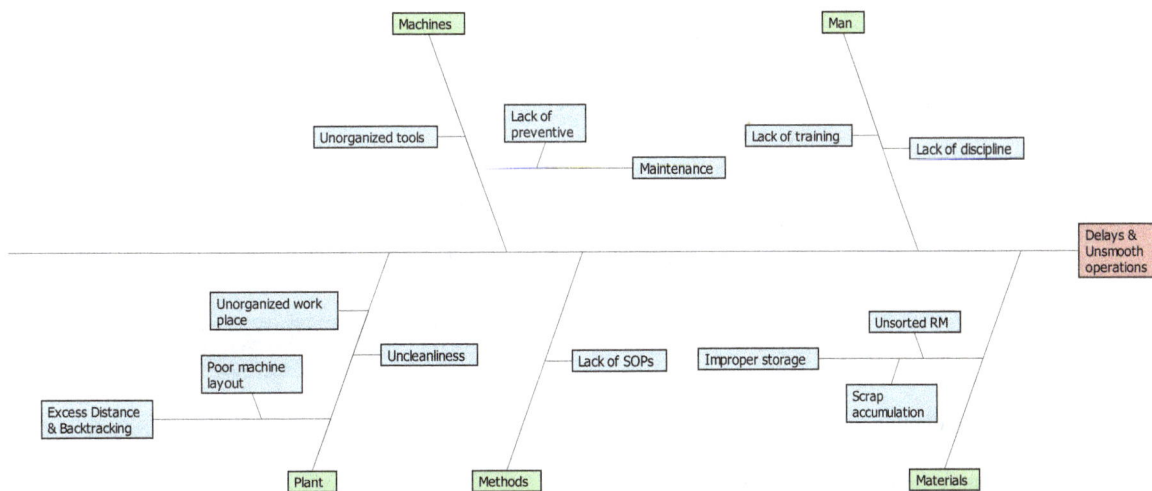

Figure 15. Unsmooth operations cause and effect diagram.

3.4.2. Cutting Process

It only seemed logical that the attempted solution should solve both problems; a simple and inexpensive, yet very efficient solution was implemented by using Smart-2D-Cutting® software. The software is used to determine the optimal standard sheet and the optimal way to cut the sheet to produce the needed sheets with minimum amount of leftovers. The software was used in the factory and data was collected for two weeks. The trim loss was reduced by 39% and the cutting time by 26%.

Periodical checks as a simple preventive maintenance program were performed, and recommendations to the upper management to have a complete preventive maintenance program as a future improvement project.

3.4.3. Unsmooth Flow of Production Processes

A Predecessor-successor processes table for the different parts manufactured similar to the one shown in **Table 1** were prepared and analyzed resulting in a new configuration of the equipment producing an efficient layout of the plant to facilitate smooth flow of processes.

The improved plant layout with the new equipment locations is shown in **Figure 16**.

A process is impacted by its environment, improvements in the general state of the work area, including access to hand tools, and so on, are essential to smooth flow of production processes, especially critical here are the cleanliness and general housekeeping status of any area. Once a process has been physically reconfigured, 5S methods are applied to its work operations. 5S is a workplace organization method that can help improve the efficiency and management of operations [15].

The term "5Ss" derives from the Japanese words for five practices leading to a clean and manageable work area: *seiri* (organization), *seiton* (tidiness), *seiso* (purity), *seiketsu* (cleanliness), and *shitsuke* (discipline) [2]. These are defined in **Table 2**.

The results of applying the 5S practices and the new equipment configuration are presented in **Table 3**.

Combined effect of all the improvement steps resulted in an improved cycle time of 161 minutes.

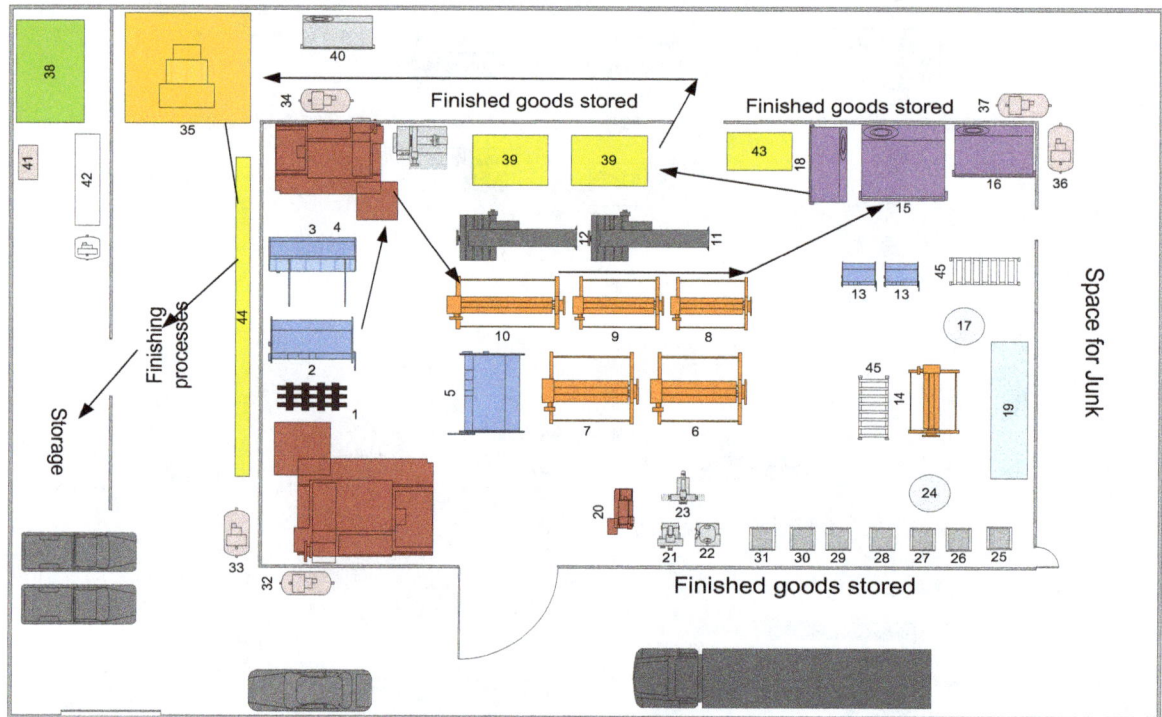

Figure 16. Improved frame path.

Table 1. Frame processing.

Process	Predecessor process	Successor process
Cutting	–	Punching
Punching	Cutting	Bending
Bending	Punching	Welding
Welding	Bending	Thermal paint
Thermal paint	Welding	Final assembly
Final assembly	Thermal paint	Finishing
Finishing	Final assembly	Finished goods storage

Table 2. Five S definition.

5S practice	English equivalence	Application
Seiri	Sort	Separate materials and tools that are needed in a work area from those that are not needed using a red tag. This minimizes the time it takes to look for things.
seiton	Set in order	Ensure there is one location to place those items remaining in the work area.
Seiso	Shine	Ensure all tools, equipment, and the work area are clean. Abnormal conditions such as oil leaks and rework are easier to see under these circumstances.
Seiketsu	Standardize	Ensure products are designed using standardized components and manufacturing processes, and that the work is performed the same way every time.
Shitsuke	Sustain	Form the habit of always following the first four Ss. Use control plans and deploy continuous improvement teams.

Table 3. Smooth flow of production processes improvements.

Area of improvement	Before	After
Transportation distance	322.6 m	189 m (reduction of 41%)
Backtracking	62%	7%
Unutilized plant space	16%	7%

3.5. Control Stage

In the Control stage we maintain the changes we made in order to sustain the improvements [8].

The control phase requires that policies and procedures be established to maintain the improvements created. Management commitment is essential to the implementation of this phase. Training and mentoring plans were put into place. The engineer responsible for the cutting process went through a training session on using the Smart-2D-Cutting ® software. Managers had several meetings with the operators explaining to them the new methods and procedures. Occasionally, operators resist changes, especially when it is imposed by upper management; however, since operators were brought in the early stage of this project, they welcomed these changes and implemented them. This indirectly affected the reckless and lack of motivation that existed among few workers. Elementary preventive maintenance program was implemented. In addition, examples of reports, data collection forms, and similar informational and control documents were prepared.

4. Conclusion

The DMAIC methodology was successfully used to implement lean manufacturing concepts in a plant specialized in the manufacturing of safety and fire resistance metal doors. The implementation included the use of the different quality and lean manufacturing tools resulting in improved operations such as value-added flow charts, C & E diagrams, and Pareto diagrams. The huge buy-in from the workers and operators in the shop floor to this improvement project was due to the fact that they were brought on board right from the beginning of the project and that their input and feedback were taken seriously by management. The upper management active involvement and open minded attitude encourage their workers to come up with ideas and solutions.

References

[1] Womack, J.P., Jones, D.T. and Roos, D. (2008) The Machine That Changed the World. Simon and Schuster, New York.

[2] Womack, J.P. and Jones, D.T. (2010) Lean Thinking: Banish Waste and Create Wealth in Your Corporation. Simon and Schuster, New York.

[3] Antony, J. (2009) Six Sigma vs. TQM: Some Perspectives from Leading Practitioners and Academics. *International Journal of Productivity and Performance Management*, **58**, 274-279. http://dx.doi.org/10.1108/17410400910938869

[4] Andersson, R., Eriksson, H. and Torstensson, H. (2006) Similarities and Differences between TQM, Six Sigma and Lean. *The TQM Magazine*, **18**, 282-296. http://dx.doi.org/10.1108/09544780610660004

[5] Munro, R.A. (2009) Lean Six Sigma for the Healthcare Practice: A Pocket Guide. ASQ Quality Press, Milwaukee.

[6]   Arafeh, M., *et al.* (2014) Six Sigma Applied to Reduce Patients' Waiting Time in a Cancer Pharmacy. *International Journal of Six Sigma and Competitive Advantage*, **8**, 105-124. http://dx.doi.org/10.1504/IJSSCA.2014.064256

[7]   Voehl, F., *et al.* (2013) The Lean Six Sigma Black Belt Handbook: Tools and Methods for Process Acceleration. CRC Press, Boca Raton. http://dx.doi.org/10.1201/b15163

[8]   Brue, G. and Formisano, R.A. (2002) Six Sigma for Managers. McGraw-Hill, New York.

[9]   Hambleton, L. (2007) Treasure Chest of Six Sigma Growth Methods, Tools, and Best Practices. Pearson Education, New York.

[10]   Karelahti, J. (2002) Solving the Cutting Stock Problem in the Steel Industry. Helsinki University of Technology, Helsinki.

[11]   Lunau, S. and John, A. (2008) Six Sigma+ Lean Toolset: Executing Improvement Projects Successfully. Springer, Berlin.

[12]   Carreira, B. (2005) Lean Manufacturing That Works: Powerful Tools for Dramatically Reducing Waste and Maximizing Profits. AMACOM Div American Management Association, New York.

[13]   Pyzdek, T. and Keller, P.A. (2003) The Six Sigma Handbook: A Complete Guide for Green Belts, Black Belts, and Managers at All Levels. McGraw-Hill, New York.

[14]   Pande, P.S. and Neuman, R.P. (2000) The Six Sigma Way: How GE, Motorola, and Other Top Companies Are Honing Their Performance. McGraw-Hill, New York.

[15]   Munro, R.A., *et al.* (2008) The Certified Six Sigma Green Belt Handbook. ASQ Quality Press, Milwaukee.

5

Multi-Objective Optimization of Time-Cost-Quality Using Hungarian Algorithm

Ventepaka Yadaiah[1], V. V. Haragopal[2]

[1]Department of Mathematics, Osmania University, Hyderabad, India
[2]Department of Statistics, Osmania University, Hyderabad, India
Email: v.yadaiah@gmail.com, haragopalvajjha@gmail.com

Abstract

In this paper, we propose an algorithm for solving multi-objective assignment problem (MOAP) through Hungarian Algorithm, and this approach emphasizes on optimal solution of each objective function by minimizing the resource. To illustrate the algorithm a numerical example (Sec. 4; Table 1) is presented.

Keywords

Assignment Problem, Hungarian Algorithm, Multi-Objectives

1. Introduction

General assignment problem includes "N" tasks that must assign to "N" workers where each worker has the competence to do all tasks. However, due to personal ability or other reasons, each worker may spend different amount of resource to finish different tasks. The objective is to assign each task to a proper worker so that the total resource that spends finishing all tasks can be minimized.

Many studies have been developed to solve the assignment problem [1]-[4], like time, cost, quality and risk in construction and development projects and investment has been taken into consideration and explanation [5], projects were completed by using the FDOT to establish a model to demonstrate the functional relationship between construction cost and time for the collected highway construction projects [6]. A Multi-Objective Ant Colony Optimization is developed to analyze the advanced time cost-quality trade-off problem [7], relationship between time, cost and quality management and the attainment of client objectives [8].

Most of the developed methods for the assignment problem consider only one-objective situation, such as (1) the minimum cost assignment problem, (2) the minimum finishing time assignment problem. The minimum cost

Table 1. Assigned cost mtrix (ACM). here cost unit: thousands, time unit: weeks, quality levels: 1, 3, 5, 7, 9.

Jobs/Machine	M_1	M_2	M_3	M_4	
J_1	9	7	4	6	$\leftarrow C_{ij}$
	2	1	8	2	$\leftarrow T_{ij}$
	1	1	1	5	$\leftarrow Q_{ij}$
J_2	12	5	5	8	C_{ij}
	9	9	1	8	T_{ij}
	7	5	5	9	Q_{ij}
J_3	9	9	9	11	C_{ij}
	8	9	5	6	T_{ij}
	1	7	5	7	Q_{ij}
J_4	2	7	11	8	C_{ij}
	1	5	4	9	T_{ij}
	1	3	5	3	Q_{ij}

assignment problem focuses on how to assign tasks to workers so that the total operation cost can be minimized. Such problems have been generally discussed and well developed in many operations researches. Geetha and Nair [9] provide a solution for an assignment problem that minimizes both time and cost. Tsai *et al.* [10] try to solve a multi-objective decision making problem associated with cost, time, and quality by fuzzy concept. Unfortunately, the provided approaches only deal with the 2-objective assignment problem.

2. Model Construction of Simple Assignment Problem

Assignment problem is one of the special cases of transportation problems. The goal of the assignment problem is to minimize the cost or time of completing a number of jobs by a number of persons. An important characteristic of the assignment problem is the number of sources is equal to the number of destinations. It is explained in the following way.

1) Only one job is assigned to person.
2) Each person is assigned with exactly one job.

Management has faced with problems whose structures are identical with assignment problems.

For example, a manager has five persons for five separate jobs and the cost of assigning each job to each person is given. His goal is to assign one and only job to each person in such a way that the total cost of assignment is minimized.

Balanced assignment problem: The number of persons is equal to the number of jobs.

Minimize (Maximize):

$$Z = \sum_{i=1}^{m}\sum_{i=1}^{n} C_{ij} X_{ij}$$

Subject to

$$\sum_{j=1}^{n} X_{ij} = 1; \text{ for } i = 1, 2, 3, \cdots, m$$

$$\sum_{i=1}^{m} X_{ij} = 1; \text{ for } j = 1, 2, 3, \cdots, n$$

where

$$X_{ij} = \begin{cases} 1, & \text{if the } i^{th} \text{ job is assigned to the } j^{th} \text{ machine.} \\ 0, & \text{if the } i^{th} \text{ job is not assigned to the } j^{th} \text{ machine.} \end{cases}$$

Problem definition: Consider a problem which consists a set of "n" machines $M = \{M_1, M_2, M_3, \cdots, M_n\}$. A set of "$m$" jobs $J = \{J_1, J_2, J_3, \cdots, J_m\}$ which is to be considered to assign for execution on "n" available machines, with an execution of cost $C_{ij}(,)$, time $T_{ij}(,)$, qality $Q_{ij}(,)$, where $i = 1, 2, \cdots, m$ and $j = 1, 2, \cdots, n$ are mentioned in the ACM of order $m \times n$, where $m = n$.

3. Scope Triangle

The triangle illustrates the relationship between three primary forces in an assignment. Time is available to deliver the assignment, cost represents the amount of money and quality represents fit for the purpose of assignment which should be a successful achievement.

3.1. Methodology

To determine the assignment of cost (C), time (T) and quality (Q) vs. machine (s) of an assignment problem for a set of "n" machines $M = \{M_1, M_2, M_3, \cdots, M_n\}$, which are to be considered as assigned for execution on "n" available machines with an execution cost C_{ij}, T_{ij}, Q_{ij}, where $i = 1, 2, \cdots, m$ and $j = 1, 2, \cdots, n$ are mentioned in the ACM of order, where $m = n$. First of all, we obtain the sum of cost, time, and quality in each job of the ACM (**Figure 1**). In this way we get single objective balanced assignment problem nature (**Table 2**). Now we apply the Hungarian algorithm approach [11]-[13] to obtain the exact optimum solution of balanced assignment problems. To solve this problem we follow the below algorithm.

3.2. Algorithm

Step 1: Consider "m" jobs on "n" machines costs given as a matrix (ACM), which is an balanced assignment problem, where $m = n$.

Step 2: Obtain the sum of cost, time, quality in each job of the ACM.

Step 3: If the total effectiveness of ACM is to be maximized, change the sign of each cost element in the effectiveness matrix and go to Step 4, otherwise go directly to Step 5 if ACM has the total value as minimum.

Step 4: If the minimum element in the ith row is not zero, then subtract this minimum element from each element in the row i ($i = 1, 2, 3, \cdots, m$).

Step 5: If the minimum element in the column j is not zero, then subtract this minimum element from each element in the column j ($j = 1, 2, 3, \cdots, m$).

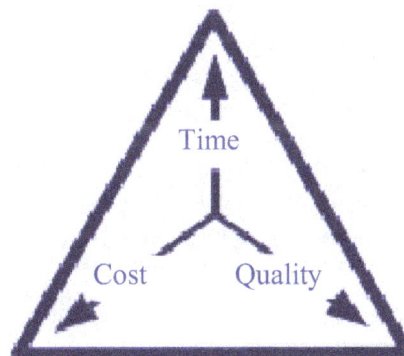

Figure 1. Scope-Triangle.

Table 2. Modified ACM (,).

	M_1	M_2	M_3	M_4
J_1	12	9	13	13
J_2	28	19	11	25
J_3	18	25	19	24
J_4	4	15	20	20

Step 6: Examine rows successively, beginning with row 1, for a row with exactly one unmarked zero. If at least one exists, mark this zero with the symbol (□) to denote an assignment. Cross out (X) the other zeros in the same column so that additional assignment will not be made to that column. Repeat the process until each row has no unmarked zeros or at least two unmarked zeros.

Step 7: Examine columns successively, beginning with column 1, for a column with exactly one unmarked zero. If at least one exists, mark this zero with the symbol (□) to denote an assignment. Cross out (X) the other zeros in the same row so additional assignment will not be made to that row. Repeat the process until each column has no unmarked zeros or at least two unmarked zeros.

Step 8: Repeat Steps 7 and 8 successively (if necessary) until one of the three things occurs:

Step 9: Every row has an assignment (□). Go to Step 16.

Step 10: There are at least two unmarked zeros in each row and each column. Go to Step 7.

Step 11: There are no zeros left unmarked and a complete assignment has not been made. Go to Step 10.

Step 12: Check (√) all rows for which assignment (□) has not been made.

Step 13: Check (√) columns not already checked which have a zero in checked rows.

Step 14: Check (√) rows not already checked which have assignments in the checked column.

Step 15: Repeat Steps 11 and 12 until the chain of checking ends.

Step 16: Draw lines through all unchecked rows and through all checked columns. This will necessarily give the minimum number of lines needed to cover each zero at least one time.

Step 17: Examine the elements that do not have at least one line through them. Select the smallest of these and subtract it from every element in each row that contains at least one uncovered element. Add the same element to every element in each column that has a vertical line through it. Return to Step 7.

Step 18: List the assignment cost and combination corresponding to sub problem.

Step 19: Add assignment cost of each sub problem to obtain the total assignment cost of the main problem, which shall be the optimal cost, and also rearrange the combinations.

Step 20: Stop.

4. Illustration

Solve the problem (**Table 1**) assuming that the objective is to minimize the total cost, time and quality.

Now obtain the sum of cost, time, quality in each job of the ACM and we get modified ACM (,) (**Table 2**).

Now apply the Hungarian method for Modified ACM (,), then the final optimal assignments of Modified ACM (,) (**Table 2**) as follows:

$$J_4 \rightarrow M_1, \ \ J_1 \rightarrow M_2, \ \ J_2 \rightarrow M_3, \ \ J_3 \rightarrow M_4$$

5. Conclusion

The above illustration was taken by the defined algorithm and implemented on several sizes of the problems to test the effectiveness of the algorithm. This approach was implemented on different sizes of multi-objective balanced assignment problems from the above. We noticed that by using standard Hungarian method we could get the optimum value. The time complexity verified and found that they were getting optimum in less time comparative to other methods.

References

[1]   Winston. L.W. (1991) Operations Research: Applications and Algorithm. Pws-Kent Puublishing Company, Boston.

[2]   Hillier. S.F. and Lieberman, J.G. (2001) Introduction to Operation Research. McGraw-Hill, Boston.

[3]   Taha, H.A. (1971) Operation Research: An Introduction. MacMillan Inc., New York.

[4]   Ravindran, A. and Ramaswami, V. (1977). On the Bottleneck Assignment Problem. *Journal of Optimization Theory and Applications*, **21**, 451-458. http://dx.doi.org/10.1007/BF00933089

[5]   Rezaian, A. (2011) Time-Cost-Quality-Risk of Construction and Development Projects or Investment. *Middle-East Journal of Scientific Research*, **10**, 218-223.

[6]   Shr, J.-F. and Chen, W.-T. (2006) Functional Model of Cost and Time for Highway Construction Projects. *Journal of Marine Science and Technology*, **14**, 127-138.

[7]   Afshar, A., Kaveh, A. and Shoghli, O.R. (2007) Multiobjective Optimization of Time-Cost-Quality Using Multi-Colony ant Algorithm. *Asian Journal of civil Engineering (Building and Housing)*, **8**, 113-124.

[8]   Bowen, P.A., Hall, K.A., Edwards, P.J., Pearl, R.G. and Cattell, K.S. (2002) Perceptions of Time, Cost, Quality Management on Building Projects. *The Australian Journal of Construction Economics and Building*, **2**, 48-56.

[9]   Geetha, S. and Nair, K.P.K. (1993) A Variation of the Assignment Problems. *European Journal of Operational Research*, **68**, 422-426. http://dx.doi.org/10.1016/0377-2217(93)90198-V

[10] Tsai, C., Wei, C.-C. and Cheng, C.-L. (1999) Multiobjective Fuzzy Deployment of Manpower. *International Journal of the Computer, the Internet and Management*, **7**,.

[11] Pandit, S.N.N. (1963) Some Quantitative Combinatorial Search Problems. PhD Thesis, IIT Khargpur.

[12] Ramesh, M. (1997) Lexi-Search Approach to Some Combinatorial Programming Problem. PhD Thesis, University of Hyderabad, Hyderabad.

[13] Gillett Billy, E. (2000) Introduction to Operations Research—A Computer Oriented Algorithm Approach. Tata Mc-Graw Hill, New Delhi.

6

Study on the Inventory Routing Problem of Refined Oil Distribution Based on Working Time Equilibrium

Zhenping Li, Zhiguo Wu

School of Information, Beijing Wuzi University, Beijing, China
Email: lizhenping66@163.com, 728432857@qq.com

Abstract

Taking the distribution route optimization of refined oil as background, this paper studies the inventory routing problem of refined oil distribution based on working time equilibrium. In consideration of the constraints of vehicle capacity, time window for unloading oil, service time and demand of each gas station, we take the working time equilibrium of each vehicle as goal and establish an integer programming model for the vehicle routing problem of refined oil distribution, the objective function of the model is to minimize the maximum working time of vehicles. To solve this model, a Lingo program was written and a heuristic algorithm was designed. We further use the random generation method to produce an example with 10 gas stations. The local optimal solution and approximate optimal solution are obtained by using Lingo software and heuristic algorithm respectively. By comparing the approximate optimal solution obtained by heuristic algorithm with the local optimal solution obtained by Lingo software, the feasibility of the model and the effectiveness of the heuristic algorithm are verified. The results of this paper provide a theoretical basis for the scheduling department to formulate the oil distribution plan.

Keywords

Working Time Equilibrium, Hard Time Window, Inventory Routing Problem, Mathematical Model, Heuristic Algorithm

1. Introduction

Inventory and transportation are the most important issues of the logistics system, which are the two main links of the logistics to obtain the "time value" and the "space value", and their consumption accounts for 2/3 of the total logistics cost [1]. The classical inventory routing problems mainly study one supplier providing distribution service to several customers, whose constraints are customers' demands, customers' delivery time windows and customers' inventory capacities; the objective function is to minimize the total cost. Many scholars have studied the inventory routing problem (IRP) and got fruitful theories. Clauclia *et al.* proposed the distribution problem of discrete time, which used the inventory and transportation cost minimization as the objective function [2]. Vansteenwegen and Mateo studied the single vehicle cycle inventory routing problem, considered the single vehicle cyclic distribution problem instead of the situation that unlimited vehicle could be used and took the total cost minimization as the main factor [3]. Li *et al.* studied the inventory routing problem of refined oil distribution under the assumption that each gas station could be served only once and obey the rule of maximum amount of replenishment. They built a mathematical model to minimize the travel time and designed a tabu search algorithm to solve this model [4]. In 2007, Li proposed the vehicle routing problem with time windows and random travel time, and designed a tabu search algorithm based on stochastic simulation [5]. When the vehicle routing problem with time windows was studied by Jiang, the VRPTW optimization model with penalty function was given, and the genetic algorithm was used to solve this problem [6]. Zhao *et al.* put forward the stochastic demand inventory-routing problem model with hard time window in 2014, and took the operating costs of system and the number of vehicles minimization as the objective function. A heuristic algorithm based on (s, S) inventory policy and modified C-W saving algorithm was given [7]. Milorad *et al.* proposed a mixed integer programming model to solve multi-product multi-period inventory routing problem and adopted a heuristic approach to observe the impact of fleet size costs on the solutions that were obtained [8]. Yan *et al.* presented a model for solving a multi objective vehicle routing problem with soft time-window constraints. The total transportation cost and the required fleet size were minimized in this model and a modified genetic algorithm was used to test the model [9]. In 2007, Raa *et al.* assumed that customer demand rates were deterministic constant when they studied the IRP problem. The objective of this model was to minimize average distribution and inventory costs without causing any stock-out at the customers and a heuristic solution approach was proposed for solving the model [10].

Most of the research on inventory routing optimization problem regarded the total distribution time or the total distribution cost minimum as the objective function, but few researchers considered the vehicle's working time equilibrium problem. In addition to the total distribution cost, another objective is to equilibrate each vehicle's working time as far as possible in actual arrangements for distribution planning. In this paper, we study the vehicle routing problem of refined oil distribution with hard time window, and formulate the problem into an integer programming model. The objective function of the model is to minimize the maximum working time of all vehicles. In the actual vehicle scheduling, it is often required to equilibrate the working time of the tankers as far as possible, so the problem of this paper has some practical significance. This model includes the constraints of vehicle capacity, demand of gas station and so on. We will design a heuristic algorithm to solve this model efficiently.

2. Problem Description and Mathematical Model

2.1. Problem Description

Taking the refined oil distribution logistics system into account which is managed by the gas station, the inventory routing problem of the refined oil distribution based on the working time equilibrium can be described as: an oil depot supplies a certain kind of refined oil to n gas stations, supposed that the oil depot has k tankers; When a tanker was filled with refined oil in the oil depot, it started from the oil depot and distributed refined oil to several gas stations, and then returned to oil depot after the distribution task was finished. Each gas station has a fixed unloading time window; the tanker must unload oil for the gas station at its time window. If the arrival time of tanker is earlier than the earliest unloading time of the gas station, the tanker must wait. If the arrival time of tanker is later than the latest time, the gas station will out of stock. The demand of a gas station can be carried out by multiple tankers; the tanker capacity, the demand of refined oil for each gas station are given the distance between gas stations, and distance between gas stations and oil depot are given; the service time and time window of each gas station are given. The working time of a tanker in the above problem refers to the total duration of the tanker from the depot to the end of the return to the oil depot. When all the working hours of the tankers are exactly same, the working time can reach a complete equilibrium state and the maximum working time of all tankers is the minimum. In practice, it is difficult to find an optimal routing scheme which makes the full equilibrium of the working time of each tanker because of the difference of tanker capacity and gas stations' service time window. When the operating time of each tanker is not exactly the same, the corresponding maximum working time of all vehicles will increased. According to this thought, a mathematical model for the inventory routing problem of refined oil distribution based on the working time equilibrium can be established.

2.2. Mathematical Model

Defines the following symbols and variables at first:

$I = \{0,1,2,\cdots,n,n+1\}$: set of oil depot and gas stations. In order to establish the model conveniently, we use two points to represent the oil depot, 0 represents the oil depot that tanker starts from, $n+1$ represents the oil depot that tanker returns to; $1,2,\cdots,n$ represent gas stations;

q_i: demand of gas station i;

$T = \{1,2,\cdots,K\}$: set of tankers;

p_k: capacity of tanker k;

d_{ik}: the amount of refined oil unload at gas station i by tanker k;

x_{ijk}: binary variable, $x_{ijk} = 1$, if tanker k passes through the path (v_i, v_j); $x_{ijk} = 0$, otherwise;

t_i^a: the earliest time of gas station i to receive service;

t_i^b: the latest time of gas station i to receive service;

r_{ik}: the time that tanker k starts to serve for gas station i;

r_{0k}: the time that tanker k sets out from oil depot;

$r_{n+1,k}$: the time that tanker k returns to oil depot $n + 1$;

s_i: the required time of gas station i for unloading oil;

t_{ij}: the time that a tanker travels from gas station i to j;

y: the maximum time that all the tankers return to the oil depot after the tasks is finished;

The mathematical model can be described as follows:

$$\min y \tag{1}$$

$$s.t. \sum_{k=1}^{K} \sum_{\substack{j=1, j\neq i}}^{n+1} x_{ijk} \geq 1 \quad i = 1,2,\cdots,n \tag{2}$$

$$\sum_{j=1}^{n} x_{0jk} = 1 \quad k = 1,2,\cdots,K \tag{3}$$

$$\sum_{i=1}^{n} x_{i,n+1,k} = 1 \quad k = 1,2,\cdots,K \tag{4}$$

$$\sum_{i=0}^{n} x_{ihk} - \sum_{j=1}^{n+1} x_{hjk} = 0 \quad k = 1,2,\cdots,K \tag{5}$$

$$\sum_{i=1}^{n} \sum_{j=1}^{n+1} d_{ik} x_{ijk} \leq p_k \quad k = 1,2,\cdots,K \tag{6}$$

$$r_{ik} + t_{ij} + s_i - r_{jk} \leq M\left(1 - x_{ijk}\right) \quad i = 1,2,\cdots,n, \ j = 1,2,\cdots,n, \ k = 1,2,\cdots,K \tag{7}$$

$$\sum_{i=0}^{n} x_{ijk} t_j^a \leq r_{jk} \leq \sum_{i=0}^{n} x_{ijk} t_j^b \quad j = 1,2,\cdots,n, \ k = 1,2,\cdots,K \tag{8}$$

$$\sum_{j=1}^{n+1} \sum_{k=1}^{K} d_{ik} x_{ijk} = q_i \quad i = 1,2,\cdots,n \tag{9}$$

$$r_{n+1,k} \leq y \quad k = 1,2,\cdots,K \tag{10}$$

$$r_{ik} \geq 0, t_{ij} \geq 0 \quad i = 1,2,\cdots,n, \ j = 1,2,\cdots,n+1, \ k = 1,2,\cdots,K \tag{11}$$

$$x_{ijk} \in \{0,1\} \quad i = 0,1,\cdots,n, \quad j = 1,2,\cdots,n+1, \quad k = 1,2,\cdots,K \tag{12}$$

The objective function (1) represents the minimization of the maximum working time of all tankers;

Constraint (2) indicates that each gas station is served by at least one tanker;

Constraints (3)-(4) indicate that the start point and the end point of each tanker's distribution route must be the oil depot;

Constraint (5) indicates that if a tanker drives into a gas station, it must leave the gas station when the task is completed;

Constraint (6) indicates that the total amount of the refined oil which is loaded in a tanker is no more than the capacity of the tanker;

Constraint (7) indicates the relationship of arrival time between two adjacent gas stations on the same distribution route;

Constraint (8) means that the time for tanker to arrive at a gas station must be in its time window;

Constraint (9) indicates that the amount of the refined oil of all tankers deliver to a certain gas station is equal to its demand;

Constraint (10) ensures that the time that all tankers return to the oil depot is no longer than the maximum working time;

Constraints (11)-(12) are the value constraints of variables.

3. Heuristic Algorithm

The mathematical model of inventory routing problem based on the working time equilibrium is an integer programming model. Lingo software can be used to solve small size problems directly, but it needs a long time to get the solution of the large scale problems, so this paper will design a heuristic algorithm to solve this model.

Several definitions are given as follows.

Set of feasible successor gas stations: the set of feasible successor gas stations that can be served when a tanker unload oil at a gas station. Every gas station has a fixed time window to accept the service, and tanker needs a certain time to drive from current gas station to the successor one. If the arrival time to the successor gas station just falls in its service time window, the gas station will be a feasible successor gas station. Since each gas station has a time window, it is impossible that each gas station becomes a successor gas station.

According to the following rules, we can calculate the set of feasible successor gas stations.

Because we consider the calculation of the set of feasible successor gas stations with hard time window, in order to avoid tankers waiting and gas station out of stock, it is necessary to calculate the feasible successor gas stations of oil depot and every gas station in the heuristic algorithm calculation process.

Assumed G_{ib} is the set of feasible successor gas stations of gas station i, G_{ob} indicates the set of feasible successor gas stations of oil depot. According to the time window calculation of gas station j, its **revised time window** corresponding to gas station i is:

$$t_{jr}^{a\prime} = t_j^a - t_{ij} - s_i, \quad t_{jr}^{b\prime} = t_j^b - t_{ij} - s_i.$$

If there is an overlap between the time window of gas station i and the revised time window of gas station j corresponding to the gas station i, gas station j can be a successor gas station of gas station i, then gas station j can be added into the set of feasible successor gas stations of gas station i, otherwise, gas station j is not allowed to join in the set of feasible successor gas stations of gas station i.

For every gas station j in the set of the feasible successor gas stations of gas station i, we calculate its transfer probability p_{ij} by the following equation:

$$p_{ij} = \frac{\zeta_{ij}}{\sum_j \zeta_{ij}}; \quad \zeta_{ij} = L\left(\left[t_{jr}^{a\prime}, t_{jr}^{b\prime}\right], \left[t_i^a, t_i^b\right]\right) \Big/ \left(t_i^b - t_i^a\right),$$

where $L\left(\left[t_{jb}^{a\prime}, t_{jb}^{b\prime}\right], \left[t_i^a, t_i^b\right]\right)$ means the length of the overlap between two time windows.

After finishing the service for gas station i, the tanker will select a gas station from the set of feasible successor gas stations based on the transfer probability. Firstly, the transfer probability is calculated, and the successor gas station is chosen by the roulette wheel method, and added to the tanker's route.

The heuristic algorithm can be described as follows:

Input: time window, demand and service time of each gas station, distance matrix and time matrix between gas stations and oil depot, the number of tankers, capacity of each tanker;

Step 0: Initialization

The departure time of tanker k from the oil depot is $t(k) = 0$, the current position of tanker k is oil depot $S(k) = 0$; sequence of gas stations that tanker k has served is $L(k) = \phi$, residual capacity of tanker k is $Q(k) = p_k$.

The set of gas stations that need to be served is $V = \{1, 2, \cdots, n\}$.

Step 1: Calculate the revised time window of each gas station j in the set of V to the gas station (or oil depot) where the tanker k is, and check whether there is an overlap between the time window of gas station (or oil depot) where tanker k is and the revised time window of gas station j. If there is an overlap, then gas station j will be added into the set $G_{s(k)b}$, which is the feasible successor gas stations of the gas station where tanker k is; otherwise, gas station j will not be added into $G_{s(k)b}$.

Step 2: Calculate the transfer probability of tanker k to each feasible successor gas station by equation

$$P_{s(k)j} = \frac{\varsigma_{s(k)j}}{\sum_j \varsigma_{s(k)j}}$$ and choose a successor gas station in $G_{s(k)b}$ by the roulette wheel method.

Suppose the successor gas station is j, update $L(k) = L(k) \cup \{j\}$, $S(k) = j$, go to Step 3.

Step 3: If the demand of gas station j is less than the residual capacity of tanker k, then the replenishment quantity for gas station j is q_j, update $V = V \setminus \{j\}$; otherwise, the replenishment quantity for gas station j equals to the residual capacity of the tanker k and update the demand of gas station j to $q_j = q_j - Q(k)$, go to Step 4.

Step 4: If the set $G_{s(k)b}$ is empty, the route of tanker k is determined, go to Step 5; otherwise, go to Step 1.

Step 5: If the set V is empty, terminate the algorithm; otherwise, $k = k + 1$, go to Step 6.

Step 6: If all tankers have been used up, terminate the algorithm; otherwise, go to Step 1.

Output: the distribution route and loading capacity of each tanker, the unloading capacity of the refined oil per tanker at each gas station, the working time of each tanker, and the maximum working time to complete the distribution task.

4. Simulations Analysis

Suppose there is an oil depot supplies refined oil to 10 gas stations.0 indicates oil depot; 1 - 10 indicate gas stations. There are 3 tankers in the oil depot, the driving speed of each tanker is 50 km/h, the capacity of each tanker is shown in **Table 1**; each gas station's demand, service time and hard time window are shown in **Table 2**, distance between gas stations, and distance between gas stations and oil pot are shown in **Table 3**; the tanker's running time between any pair of gas stations, gas station and oil depot is shown in **Table 4**.

Firstly, we use the Lingo software to solve the integer programming model which is established in this paper. When the solution option set to use global solver, Lingo cannot get the global optimal solution after 130 hours

Table 1. The capacity of tankers (ton).

Tanker	1	2	3
Capacity	52	48	54

Table 2. Related parameters of gas stations.

Gas station	1	2	3	4	5	6	7	8	9	10
q	14	17	15	13	16	14	20	12	15	17
TW	[0.4, 1.2]	[0.6, 1.5]	[0.8, 1.7]	[0.7, 1.6]	[1, 1.8]	[1, 1.6]	[0.4, 1.5]	[0.6, 2]	[1.4, 1.9]	[1.3, 2]
s	0.1	0.15	0.12	0.08	0.13	0.1	0.17	0.06	0.12	0.15

Table 3. The shortest distance between gas stations or oil depot (km).

Gas station	0	1	2	3	4	5	6	7	8	9	10
0	-	20	25	24	28	27	22	23	20	21	26
1		-	16	23	18	28	24	19	34	18	11
2			-	9	16	22	27	30	17	13	24
3				-	30	27	39	31	12	18	24
4					-	18	24	21	33	12	26
5						-	26	20	14	30	27
6							-	34	26	18	27
7								-	21	32	25
8									-	26	35
9										-	25
10											-

Table 4. Driving time between gas stations or oil depot (hour).

Gas station	0	1	2	3	4	5	6	7	8	9	10
0	-	0.4	0.5	0.48	0.56	0.54	0.44	0.46	0.40	0.42	0.52
1		-	0.32	0.46	0.36	0.56	0.48	0.38	0.68	0.36	0.22
2			-	0.18	0.32	0.44	0.54	0.60	0.34	0.26	0.48
3				-	0.60	0.54	0.78	0.62	0.24	0.36	0.48
4					-	0.36	0.48	0.42	0.66	0.24	0.52
5						-	0.52	0.40	0.28	0.60	0.54
6							-	0.68	0.52	0.36	0.54
7								-	0.42	0.64	0.50
8									-	0.52	0.70
9										-	0.50
10											-

operating. Therefore, the solution option set to the local solver. After 70 hours of operation, the local optimal solution of the model is obtained by Lingo, and the results are as follows:

The distribution routes of the 3 tankers are respectively: 0-1-4-6-9-0; 0-8-7-5-0; 0-1-2-3-10-0.

The arrival time of each tanker to the gas stations is shown in **Table 5**.

From **Table 5**, we can learn: the working time of tanker1 is 2.42 hours, that of tanker 2 is 2.42 hours and that of tanker 3 is 2.42 hours. The maximum working time is 2.42 hours, which means the optimal solution obtained by Lingo is 2.42 hours.

The amount of refined oil distributed for each gas station is shown in **Table 6**.

Next, according to the heuristic algorithm coded by Matlab software, we obtain an approximate optimal distribution route for 3 tankers: 0-1-3-2-8-0; 0-1-4-6-9-0; 0-7-5-10-0.

The arrival time of each tanker to the gas station is shown in **Table 7**.

From **Table 7**, we can learn that the working time of tanker 1 is 2.21 hours, which of tanker 2 is 2.42 hours and that of tanker 3 is 2.37 hours. The maximum working time is 2.42 hours, which means the optimal solution obtained by heuristic algorithm is 2.42 hours.

The amount of refined oil per tanker distributed for each gas station is shown in **Table 8**.

By comparison, we find that the approximate optimal solution obtained by the heuristic algorithm is approximately the same as the local optimal solution obtained by Lingo. But the local optimal solution of Lingo is better in terms of eliminating the working time difference of each tanker. But it needs a long time to obtain local optimal solution by Lingo, which cannot satisfy the requirement to obtain the optimal solution in short time. The operation time is greatly shorter by using heuristic algorithm coded by Matlab than using Lingo software to solve the integer programming model.

Table 5. The arrival time of each tanker to the gas stations (unit: hour).

k \ r \ i	0	1	2	3	4	5	6	7	8	9	10	0
1	0	0.4	0	0	0.86	0	1.42	0	0	1.88	0	2.42
2	0	0	0	0	0	1.75	0	1.18	0.6	0	0	2.42
3	0	0.4	0.82	1.15	0	0	0	0	0	0	1.75	2.42

i indicates the gas station; r represents arrival time; k refers to the tanker.

Table 6. The amount of refined oil distributed for each gas station (ton).

k \ d \ i	0	1	2	3	4	5	6	7	8	9	10	0
1	0	9	0	0	13	0	14	0	0	15	0	0
2	0	0	0	0	0	16	0	20	12	0	0	0
3	0	5	17	15	0	0	0	0	0	0	17	0

i indicates the gas station; d represents distribution quantity; r refers to the tanker.

Table 7. The arrival time of each tanker to the gas stations (hour).

k \ r \ i	0	1	2	3	4	5	6	7	8	9	10	0
1	0	0.4	1.26	0.96	0	0	0	0	1.75	0	0	2.21
2	0	0.4	0	0	0.86	0	1.42	0	0	1.88	0	2.42
3	0	0	0	0	0	1.03	0	0.46	0	0	1.70	2.37

i indicates the gas station; r represents arrival time; k refers to the tanker.

Table 8. The amount of refined oil distributed for each gas station (ton).

k \ d \ i	0	1	2	3	4	5	6	7	8	9	10	0
1	0	8	17	15	0	0	0	0	12	0	0	0
2	0	6	0	0	13	0	14	0	0	15	0	0
3	0	0	0	0	0	16	0	20	0	0	17	0

i indicates the gas station; d represents distribution quantity; k refers to the tanker.

5. Conclusions

The inventory routing problem is the key problem in making the distribution plan of the refined oil. The working time equilibrium of each tanker often needs to be considered in the actual arrangement of the oil delivery scheme. In this paper, we study the inventory routing problem, which is to make the working time equilibrium of the tanker as far as possible. The mathematical model of the problem is established and Lingo program is written for solving the model. We further design a heuristic algorithm to solve the problem quickly. The model and algorithm in this paper provide a theoretical basis for the formulation of the oil distribution plan.

This paper only considers the inventory routing problem of single refined oil distribution and assumes that the demand of each gas station is determined, and a tanker is allowed to unload at multiple gas stations. In practice, the demand of gas stations is usually a random variable and the tanker for distribution of the refined oil usually has a plurality of compartments of different volumes. In order to facilitate the measurement of gas stations, the oil in one compartment must be unloaded to one gas station, in other words, the oil in one compartment can't be unloaded to multiple gas stations. We will consider a variety of refined oil distribution problem in the future research and add the vehicle compartment constraints and other conditions in order to get the results that are more suitable to the actual distribution plan.

Acknowledgements

This work was supported by the National Natural Science Foundation of China (11131009, 71540028, F012408), the Funding Project for Academic Human Resources Development in Institutions of Higher Learning under the Jurisdiction of Beijing Municipality (CIT & TCD20130327), and Major Research Project of Beijing Wuzi University. Funding Project for Technology Key Project of Municipal Education Commission of Beijing (ID: TSJHG 201310037036); Funding Project for Beijing Key Laboratory of Intelligent Logistics System; Funding Project of Construction of Innovative Teams and Teacher Career Development for Universities and Colleges Under Beijing Municipality (ID: IDHT20130517); Funding Project for Beijing Philosophy and Social Science Research Base Specially Commissioned Project Planning (ID: 13JDJGD013).

References

[1] Herer, Y.T. and Levy, R. (1997) The Metered Inventory Routing Problem, an Integrative Heuristic Algorithm. *International Journal of Production Economics*, **51**, 69-81. http://dx.doi.org/10.1016/S0925-5273(97)00059-5

[2] Clauclia, A., Nicola, B., StafanIrnich, M. and Grazia, S. (2014) Formulations for an Inventory Routing Problem. *International Transactions in Operational Research*, **21**, 353-374.

[3] Vansteenwegen, P. and Mateo, M. (2014) An Iterated Local Search Algorithm for the Single-Vehicle Cyclic Inventory Routing Problem. *European Journal of Operational Research*, **237**, 802-813. http://dx.doi.org/10.1016/j.ejor.2014.02.020

[4] Li, K.P., Chen, B., Sirakumar, A. and Wu, Y. (2013) An Inventory-Routing Problem with the Objective of Travel Time Minimization. *European Journal of Operational Research*, **236**, 936-945. http://dx.doi.org/10.1016/j.ejor.2013.07.034

[5] Li, X. (2007) Research on Model and Algorithm of Vehicle Routing Problem. Shanghai Jiao Tong University, Shanghai, 91-105. (In Chinese)

[6] Jiang, B. (2010) Research on Vehicle Routing Problem with Time Windows Based on Genetic Algorithm. Beijing Jiaotong University, Beijing, 8-44. (In Chinese)

[7] Zhao, D., Li, J., Ma, D. and Li, Y. (2014) Optimization Algorithm for Solving Stochastic Demand Inventory Routing Problem with Hard Time Window Constraints. *Operations Research and Management Science*, **23**, 27-37. (In Chinese)

[8] Milorad, V., Dražen, P. and Branislava, R. (2014) Mixed Integer and Heuristics Model for the Inventory Routing Problem in Fuel Delivery. *International Journal of Production Economics*, **147**, 593-604.

[9] Yan, Q.Y., Zhang, Q. and Torres, D.F.M. (2015) The Optimization of Transportation Costs in Logistics Enterprises with Time-Window Constraints. *Discrete Dynamics in Nature and Society*, **2015**, Article ID: 365367.

[10] Raa, B. and Aghezzaf, E.-H. (2009) A Practical Solution Approach for the Cyclic Inventory Routing Problem. *European Journal of Operational Research*, **192**, 429-441. http://dx.doi.org/10.1016/j.ejor.2007.09.032

7

A Newsvendor with Priority Classes and Shortage Cost

Bo Li[1], Pillaiboothamgudi Sundararaghavan[2], Udayan Nandkeolyar[2]*

[1]Ashland University, Ashland, OH, USA
[2]University of Toledo, Toledo, OH, USA
Email: bli@ashland.edu, p.sundararaghavan@utoledo.edu

Abstract

We consider an extension of the standard newsvendor problem by allowing for multiple classes of customers. The product is first sold to customers with the highest priority, and the remaining units (if any) are sold at a discounted price to customers in decreasing order of priority until all classes of customers have been served, limited only by the available stock. Unsold items, if any, have a salvage value. The demands of different priority customers are independent random variables with known probability distributions. The problem is to find the purchase quantity that maximizes the expected profit. We show that this problem actually reduces to the standard newsvendor problem with the demand distribution being a mixture of the input demand distributions. Since this mixture of distributions is typically hard to handle analytically, we propose a simple general heuristic which can be implemented using different types of distributions. Some of these implementations produce near optimal solutions. We tested these implementations for the case of two demand classes of customers and found that they outperform previously published heuristics in almost all instances. We suggest applications for this model in the Chinese pharmaceutical industry, apparel industry, and perishable goods among others. We also propose an extension involving shortage cost.

Keywords

Newsvendor Problem, Priority Demand Classes, Mixture Distribution, Heuristics

1. Introduction

Arrow *et al.* [1] and Morse and Kimball [2] were the first who formulated and solved the single-period, single-

---

*The author has passed away.

item stochastic inventory problem also known as the newsvendor or newsboy problem. Given one period as a planning horizon, relevant unit parameters such as purchasing cost c, selling price p, salvage value s, and the cumulative distribution function (CDF) F of random demand, the problem is to find the order quantity that maximizes the expected profit. This optimal quantity was found to be the smallest q for which $(q) \geq \dfrac{p-c}{p-s}$.

There have been many extensions to the newsvendor problem [3]. The extension studied in this paper assumes n classes of customers, and is in the spirit of the model introduced by Şen and Zhang [4]. In general, the product is sold for p_i to the ith class of customers before the remaining units (if any) are sold at a discounted price p_{i+1} to the $(i + 1)^{th}$ class of customers. Thus, the newsvendor sells newspapers starting at the highest price location, and will move to the next highest price location after meeting the realized demand at the current location. For example, the price charged in the morning is higher than that in the afternoon. The model we examine is common, for example, in the apparel industry and retailing of perishable goods where discounts are used to sell excess inventory; see [5] [6]. We have also found an interesting potential application from our study of the Chinese pharmaceutical supply chain. In this supply chain, there are two classes of customers: hospitals and pharmacies. The demand of hospitals is expected to be met first, and the remaining units are sold to pharmacies at a lower price. Drugs may be considered similar to perishable products since distributors generally like to sell out each batch rather than carry forward inventory because of the proximity of expiry dates. Also items such as flu shots which are supplied only once a year are never carried forward.

The main theoretical result of this paper is that the newsvendor problem with multiple demand classes characterized by decreasing prices actually reduces to the standard newsvendor problem with the demand distribution being a mixture of the input distributions, the selling price p_1, the unit cost c, and the salvage value s. Since typically this mixture distribution does not have a closed analytical form, we propose a general heuristic that replaces it by a more tractable distribution with the same first two moments. We demonstrate that some implementations of this general heuristic produce near optimal solutions, and they significantly outperform the heuristics reported in the literature. We also present some extensions of the studied problem that involve shortage penalties and the case when only the means and the standard deviations of random demands are known [7].

The paper is organized as follows. In Section 2, we provide some additional motivations related to the main problem addressed in this paper, and in Section 3 we formulate the problem and show its reduction to the standard newsvendor problem. In Section 4, we describe our proposed heuristics; their performance is empirically examined in Section 5. Section 6 indicates some extensions to the problem, while Section 7 presents some concluding remarks.

2. Motivation

First, we discuss the motivation generated by the practices we gleaned through our recent survey of the Chinese Pharmaceutical industry in Central China. Generally, the customers of a pharmaceutical distributor in China are classified into two types: Hospitals (including clinics) and retail pharmaceutical franchisees (Pharmacies). In China, hospitals and pharmacies are separate and both play important roles in the Pharmaceutical Supply Chain [8] [9]. Hospitals are the high-end customers to the Distributor, and the pharmacies represent the low-end segment. The State Drug & Food Administration holds an annual bidding conference for matching distributors in a region with medicines and the hospital network. At the end of the conference, every high volume medicine is assigned to a distributor exclusively for supplying all hospitals in the region at a fixed price for the entire year. Typical profit margin for the distributor selling to hospitals is around 15%. Each distributor is free to purchase any medicine from manufacturers or wholesalers, but they cannot supply to hospitals unless they have exclusive rights earned in the annual bidding process. The supply chain from manufacturer to distributor to hospitals is tightly regulated and has higher profit margin and monopolistic pattern for each medicine. However, the supply chain from manufacturer to wholesaler to distributors to pharmacies is loosely organized with both supply and price flexibility at the purchasing and selling end for each player. Typically, pharmacies offer lower profit margins, around 8%, because of significant competition from the supply and demand sides. These are governed by market forces. There are pharmaceuticals which are produced only once a year due to their seasonal nature of demand (such as flu shots) or due to constraints on expiry dates. From the perspective of the pharmaceutical distributor, this situation can be modeled as a problem of finding the order quantity given two types of customers with fixed prices (p_1 and p_2) and $p_1 > p_2 > c$, where c is the unit cost.

This model is also applicable for annual order placements by US manufactures to their Asian contract manufactures for apparels and fashion items, as well as for the general purchase of perishable items by the organizations which have multiple classes of customers with fixed known prices.

3. Main Theoretical Result

The general newsvendor problem with decreasing priority demand classes of customers can be formulated as follows. A product with a unit cost of c is sold to n classes of customers in a sequential order. It is sold first to the first class of costumers at a price of p_1, next (if any units left) to the second class of customers at a price of p_2, and so on until the last class characterized by the price p_n can be served. Any unsold units have a salvage values, and $p_1 \geq p_2 \geq \cdots \geq p_n > c > s$ is assumed. For $j = 1, 2, \cdots, n$, let X_j be the random demand of the jth class of customers and F_j be its CDF. The random variables $X_1, X_2, \cdots, X_n$ are independent and, without loss of generality, they are assumed to be continuous. If Q_j and Q_{n+1} are the quantities sold at p_j and s, then for a given purchase quantity q, we have

$$Q_1 = \min(q, X_1)$$

$$Q_j = \min\left(q, \sum_{i=1}^{j} X_i\right) - \min\left(q, \sum_{i=1}^{j-1} X_i\right) \text{ for } j = 2, \cdots, n; \text{ and } Q_{n+1} = q - \min\left(q, \sum_{j=1}^{n} X_j\right)$$

Therefore, letting $p_{n+1} = s$, the profit is

$$\Pi(q) = \sum_{j=1}^{n+1} p_j Q_j - cq$$

$$= p_1 \min(q, X_1) + \sum_{j=2}^{n} p_j \left[\min\left(q, \sum_{i=1}^{j} X_i\right) - \min\left(q, \sum_{i=1}^{j-1} X_i\right)\right]$$

$$+ s\left[q - \min\left(q, \sum_{j=1}^{n} X_j\right)\right] - cq$$

$$= \sum_{j=1}^{n} (p_j - p_{j+1}) \min\left(q, \sum_{i=1}^{j} X_i\right) - (c - s)q.$$

Since for any random variable Y whose CDF G has a finite mean,

$$E[\min(q, Y)] = \int_{-\infty}^{\infty} \min(q, y) dG(y) = q - \int_{-\infty}^{q} G(y) dy \text{ holds, the expected profit is expressed by}$$

$$\pi(q) = E[\Pi(q)] = \sum_{j=1}^{n} (p_j - p_{j+1})\left(q - \int_{-\infty}^{q} G_j(y) dy\right) - (c - s)q$$

$$= (p_1 - c)q - \sum_{j=1}^{n} (p_j - p_{j+1}) \int_{-\infty}^{q} G_j(y) dy, \tag{1}$$

where $G_j = F_1 \circ F_2 \circ \cdots \circ F_j$ is the CDF of $Y_j = \sum_{i=1}^{j} X_i$. Recall here that the CDF of $X_1 + X_2$ is the convolution

$$(F_1 \circ F_2)(x) = P(X_1 + X_2 \leq x) = \int_{-\infty}^{\infty} F_2(x - x_1) dF_1(x_1).$$

The first derivative of $\pi(q)$ is

$$\frac{d\pi(q)}{dq} = (p_1 - c) - \sum_{j=1}^{n} (p_j - p_{j+1}) G_j(q),$$

and the second derivative is obviously non-positive. Therefore, the expected profit $\pi(q)$ is concave in q, and the optimal purchase quantity q^* is a solution to the equation $\dfrac{d\pi(q)}{dq} = 0$, that is, q^* satisfies

$$\sum_{j=1}^{n} (p_j - p_{j+1}) G_j(q^*) = p_1 - c. \tag{2}$$

Letting now $w_j = \dfrac{p_j - p_{j+1}}{p_1 - s}$ for $j = 1, 2, \cdots, n$, and $w = \dfrac{p_1 - c}{p_1 - s}$, we observe that (2) is equivalent to

$$\sum_{j=1}^{n} w_j G_j(q^*) = w. \tag{3}$$

Since $\sum_{j=1}^{n} w_j = 1$ and $0 \leq w_j, w \leq 1$, one can define the mixture distribution whose CDF is

$G = \sum_{j=1}^{n} w_j G_j$. The optimal purchase quantity is then $q^* = G^{-1}(w) = G^{-1}\left(\dfrac{p_1 - c}{p_1 - s}\right)$. Thus, we have shown that

the newsvendor problem with n decreasing priority demand classes of customers actually reduces to the standard newsvendor problem with the demand CDF G, the unit selling price p_1, the unit purchase cost c, and the unit salvage value s. Note here that all prices p_j and all CDFs G_j are imbedded in the CDF G.

4. Heuristics

Şen and Zhang [4] proposed the following two heuristics, whose idea is to replace the original problem by some standard newsvendor problems:

H1. Define the standard newsvendor problem with the demand $Y_n = \sum_{j=1}^{n} X_j$, the selling price

$\bar{p} = \dfrac{\sum_{j=1}^{n} \mu_j p_j}{\sum_{j=1}^{n} \mu_j}$, where μ_j denotes the mean of X_j, and assume $q^{H1} = G_n^{-1}\left(\dfrac{\bar{p} - c}{\bar{p} - s}\right)$.

H2. Solve separately n standard newsvendor problems with the demands X_j, the selling prices p_j, and sum

up the obtained purchase quantities. Thus, $q^{H2} = \sum_{j=1}^{n} F_j^{-1}\left(\dfrac{p_j - c}{p_j - s}\right)$.

We have shown above that the problem under study can be reduced to a specific standard newsvendor problem, and the optimal purchase quantity is $q^* = G^{-1}\left(\dfrac{p_1 - c}{p_1 - s}\right)$, where $G = \sum_{j=1}^{n} w_j G_j$. However, the mixture

CDF G cannot be assumed to have a closed analytical form, and hence its inverse G^{-1} cannot be easily determined. Also, the equation $\sum_{j=1}^{n} w_j G_j(q^*) = w$ requires the use of numerical methods for estimating q^*. Therefore, the use of heuristics is fully justified. Although G is hard to handle, its moments (around zero) are easily determined. Clearly, if $E[X_j] = \mu_j$ and $E[X_j^2] = m_j$ for $j = 1, 2, \cdots, n$, then the first two moments of G are:

$$\mu_G = \sum_{j=1}^{n} w_j E[Y_j] = \sum_{j=1}^{n} w_j \sum_{i=1}^{j} \mu_i, \tag{4}$$

and

$$m_G = \sum_{j=1}^{n} w_j E[Y_j^2] = \sum_{j=1}^{n} w_j E\left[\left(\sum_{i=1}^{j} X_i\right)^2\right]$$
$$= \sum_{j=1}^{n} w_j \sum_{i=1}^{j} m_i + 2\sum_{j=2}^{n} w_j \sum_{h=1}^{j-1} \mu_h \sum_{i=h+1}^{j} \mu_i. \tag{5}$$

We propose a general heuristic named H3, which can be implemented by replacing the mixture CDF G by a more tractable CDF $\tilde{G}$ with the same first two moments as G. In other words, the means and the standard deviations of G and $\tilde{G}$ are assumed to be the same. Consequently, the optimal purchase quantity $q^* = G^{-1}\left(\dfrac{p_1 - c}{p_1 - s}\right)$ is approximated by $q^{H3} = \tilde{G}^{-1}\left(\dfrac{p_1 - c}{p_1 - s}\right)$. In our preliminary search for the best CDF $\tilde{G}$, we developed four implementations of H3, named H3N, H3L, H3G and H3W, in which $\tilde{G}$ is assumed to be normal, lognormal, gamma, and Wei bull, respectively.

5. Computational Results

For a given heuristic H that yields the purchase quantity q^H, we are interested in the relative percentage error induced by H:

$$RPE(\mathrm{H})=\frac{\pi\left(q^{*}\right)-\pi\left(q^{\mathrm{H}}\right)}{\pi\left(q^{*}\right)}\times100\%.$$

For every conducted experiment, we computed the average and maximum relative percentage errors *ARPE* (H) and *MRPE* (H).

We have limited our experiments to the case $n = 2$, though they are easily extendable to any value of n. The demands X_1 and X_2 were assumed to follow normal, uniform or exponential distributions.

All numerical computations were performed using MS Excel. The optimal purchase quantity q^{*} was found by Excel Solver applied on the equation:

$$G\left(q^{*}\right)=w_{1}F_{1}\left(q^{*}\right)+w_{2}\left(F_{1}\circ F_{2}\right)\left(q^{*}\right)=w$$

to which (3) is reduced for $n = 2$. Excel Solver was also used to determine q^{H1} from $G_{2}\left(q^{\mathrm{H1}}\right)=\dfrac{\overline{p}-c}{\overline{p}-s}$ in the

case of exponentially distributed demands. For normally distributed demands, we employed Excel function NORM.INV to find both q^{H1} and q^{H2}. In the case of uniformly distributed demands, the two purchase quantities could be analytically determined.

Whenever the expected profit,

$$\pi(q)=\left(p_{1}-c\right)q-\left(p_{1}-p_{2}\right)\int_{-\infty}^{q}F_{1}(x)\mathrm{d}x-\left(p_{2}-s\right)\int_{-\infty}^{q}\left(F_{1}\circ F_{2}\right)(x)\mathrm{d}x$$

(assume $n = 2$ in (1)) could not be analytically determined, we applied the Simpson method for approximating the integrals $\int_{-\infty}^{q}F_{1}(x)\mathrm{d}x$ and $\int_{-\infty}^{q}\left(F_{1}\circ F_{2}\right)(x)\mathrm{d}x$.

5.1. Instances of Şen and Zhang [4]

We reconsidered the 240 instances defined in Şen and Zhang [4]. In their work, the demands X_1 and X_2 were assumed to be normally distributed, $X_j \sim N\left(\mu_j,\sigma_j\right)$, with means $\mu_1 = 1$, $\mu_2 \in \{0.5,1,2\}$, and constant coefficients of variation *CV*, that is, $\sigma_1/\mu_1 = \sigma_2/\mu_2$. They also assumed $CV \in \{0.1,0.2,0.3,0.4,0.5\}$, $c = 1$, $s = 0$, $p_1 \in \{1.2,2,3,5\}$, and $p_2 = rp_1$, where $r \in \{0.2,0.4,0.6,0.8\}$. We applied the two heuristics proposed by them and the four implementations of our general heuristics to that dataset. The obtained results are shown in **Table 1**.

It should be added here that the presented errors are strongly biased upward by some rather unrealistic instances for which $p_1 > c > p_2$. Actually such instances should be ignored because our general heuristic H is valid when $p_1 \geq p_2 > c$.

We found that in general the heuristics H3N and H3G with normal and gamma distributions for $\tilde{G}$ outperform heuristics H3L and H3W,which respectively use lognormal and Weibull distributions. Hence for brevity, in the rest of the paper, we present only the results concerning the performance of H1, H2, H3N, and H3G. Moreover, only the realistic instances for which $p_1 \geq p_2 > c$ are considered.

Table 1. The results on Şen and Zhang [4] instances.

Heuristics	ARPE	MRPE
H1	22.91%	100.00%
H2	2.91%	36.84%
H3N	2.00%	28.65%
H3G	1.71%	29.89%
H3L	2.03%	38.96%
H3W	3.48%	49.48%

5.2. Simulation Experiments

In all of our simulation experiments, it is assumed that there are only two classes of customers and $p_1 = 1 > p_2 > c > s = 0$. Three datasets were defined by different ways of generating p_2 and c. Dataset 1 was generated by setting $p_2 = \max(x, y)$ and $c = \min(x, y)$, where x and y were drawn from U((0, 1)), that is, the uniform distribution on the interval (0,1). Thus, the average values of p_2 and c were 2/3 and 1/3, respectively. This data set is characterized by a relatively low purchase cost and evenly spread importance of the two classes of customers. In order to evaluate the heuristics over wider combinations of p_2 and c, we generated two additional data sets. For Dataset 2, the average values of p_2 and c were secured to be 3/4 and 1/2, while for Dataset 3, they were 4/7 and 3/7. Thus, Dataset 2 may represent products with high purchase cost and evenly spread importance of the two classes of customers. On the other hand, Dataset 3 may represent products with moderate purchase cost, but with dominant higher priority customers. For each of the three data sets, we assigned 100 randomly generated values of p_2 and c.

In order to have comparable results, we used the same 100 mean demands μ_1 and μ_2 across all three data sets; they were randomly generated from U((500, 1500)). Furthermore, these 100 pairs of means were used for the different types of the demand distributions tested. In the case of the assumed normal and uniform distributions, we considered three cases related to different coefficients of variation, while in the case of exponentially distributed demands only one case could be assumed. Thus, to analyze the impact of the type of demand distributions, 7 different cases were considered. Consequently, we designed $3 \times 7 = 21$ experiments, each included 100 randomly generated instances. For each of these experiments, we tested the statistical significance of the difference between the results produced by the four heuristics H1, H2, H3N, and H3G.

Let $\mu_{RPE\,(H)}$ be the mean relative percentage error induced by heuristic H for a particular simulation experiment conducted on 100 problem instances. We are interested in testing the following null and alternative hypotheses:

$$H_0 : \mu_{RPE(H1)} = \mu_{RPE(H2)} = \mu_{RPE(H3N)} = \mu_{RPE(H3G)}$$

H_1 : not all population means are the same.

Since all of the heuristics are applied on the same instances, we have a dependent (matched) sampling. We tried to apply the ANOVA test for a randomized block design. Unfortunately, we were unable to verify the needed normality assumptions in any of the simulation experiments. Consequently, we turned to the non-parametric Friedman test followed by the non-parametric HSD (honestly significance difference) Tukey's test. For each simulation experiment, the null hypothesis stated above was rejected by Friedman's test with a p-value of virtually zero in all of the 21 experiments. Therefore, below we present only the results of Tukey's multiple comparison test conducted at a 5% significance level. For example, the notation $\{H3N, H3G\} \prec \{H1, H2\}$ means that both H3N and H3 Goutper form H1 and H2, in terms of the mean relative percentage errors, but no dominance relation could be establish between H3N and H3G, and between H1 and H2. The tables presented below also include the obtained average and maximum relative percentage errors, $ARPE$ (H) and $MRPE$ (H), for the four examined heuristics.

First, we consider normally distributed demands $X_j \sim N(\mu_j, \sigma_j)$ for $j = 1, 2$. To reduce the number of randomly generated parameters, X_1 and X_2 were assumed to have the same coefficient of variation $CV \in \{1/4, 1/3, 1/2\}$. Thus, for given μ_1 and μ_2, the standard deviations were $\sigma_i = CV \cdot \mu_i$ for the three considered values of CV. The obtained results are reported in **Tables 2-4**. It may be noted that H3N dominates all other heuristics for all datasets and CV values reported. H3G performs well for many scenarios, though in some instances one of the other heuristics may do just as well.

Next, we consider uniformly distributed demands X_1 and X_2 with means μ_1 and μ_2, and the same range $2d$.

Since the coefficient of variation for the uniform distribution on the interval $[\mu - d, \mu + d]$ is $\dfrac{d}{\mu\sqrt{3}}$, the average coefficient for X_1 and X_2 is $\dfrac{d}{2\sqrt{3}}\left(\dfrac{\mu_1 + \mu_2}{\mu_1 \mu_2}\right)$. Therefore, to secure the comparison with the results for normally distributed demands, we assumed $d = \dfrac{2\sqrt{3}\mu_1\mu_2}{\mu_1 + \mu_2}CV$, where $CV \in \{1/4, 1/3, 1/2\}$. The obtained results

Table 2. The results on Dataset 1 for normally distributed demands.

Heuristic	CV = 1/2		CV = 1/3		CV = 1/4	
	ARPE	*MRPE*	*ARPE*	*MRPE*	*ARPE*	*MRPE*
H1	0.94%	8.67%	0.93%	6.62%	0.79%	6.08%
H2	1.41%	21.86%	0.59%	2.60%	0.42%	1.28%
H3N	0.08%	1.16%	0.11%	0.95%	0.29%	2.62%
H3G	0.23%	1.36%	0.36%	2.89%	0.63%	5.02%
Tukey test	{H3N,H3G} ≺ {H1,H2}		{H3N,H3G} ≺ {H2} ≺ {H1}		{H3N,H2} ≺ {H1,H3G}	

Table 3. The results on Dataset 2 for normally distributed demands.

Heuristic	CV = 1/2		CV = 1/3		CV = 1/4	
	ARPE	*MRPE*	*ARPE*	*MRPE*	*ARPE*	*MRPE*
H1	1.19%	7.27%	1.60%	8.27%	1.48%	6.41%
H2	5.25%	49.87%	0.91%	5.88%	0.39%	1.94%
H3N	0.21%	5.06%	0.12%	1.72%	0.43%	2.20%
H3G	0.22%	1.41%	0.33%	2.00%	0.80%	3.68%
Tukey test	{H1,H3N,H3G} ≺ {H2}		{H3N,H3G} ≺ {H2} ≺ {H1}		{H2,H3N} ≺ {H3G} ≺ {H1}	

Table 4. The results on Dataset 3 for normally distributed demands.

Heuristic	CV = 1/2		CV = 1/3		CV = 1/4	
	ARPE	*MRPE*	*ARPE*	*MRPE*	*ARPE*	*MRPE*
H1	2.65%	9.09%	3.29%	9.21%	3.03%	7.65%
H2	7.83%	72.43%	1.23%	18.53%	0.52%	5.71%
H3N	0.42%	4.54%	0.20%	1.87%	0.32%	1.59%
H3G	0.21%	2.51%	0.22%	2.02%	0.53%	3.89%
Tukey test	{H1,H3N,H3G} ≺ {H2}		{H3N,H3G} ≺ {H2} ≺ {H1}		{H2,H3N,H3G} ≺ {H1}	

are presented in **Tables 5-7**. It may be noted that H3N dominates all other heuristics for all datasets and *CV* values reported. H3G falls in the best performing heuristics group in all but three cases.

Finally, we consider exponentially distributed demands X_1 and X_2 with means μ_1 and μ_2. For the exponential distribution, the standard deviation equals the mean, so $CV = 1$. The results are reported in **Table 8**. Under this rather extreme assumption about demand distributions, H3G has evidently the lowest *APRE* and *MPRE* and together with H1 it dominates H2 and H3N. It is not a surprise that H3N performs poorly in comparison with H3G because the exponential distribution is a special case of the gamma distribution.

6. Extensions

In the standard newsvendor problem, no shortage cost is assumed if the purchased quantity is less than the demand. Although this cost might be difficult to define in practice, the authors of [7] and [10] assumed a known unit lost sales (shortage) cost of ℓ. The optimal purchase quantity q^* is then defined as $q^* = F^{-1}\left(\dfrac{p+\ell-c}{p+\ell-s}\right)$.

If in the newsvendor problem with n decreasing priority demand classes, ℓ_j denotes the unit lost sale cost for the jth class, the expected profit is:

$$\pi(q) = \sum_{j=1}^{n}\left(p_j + \ell_j - p_{j+1}\right)E\left[\min\left(q, \sum_{i=1}^{j}X_j\right)\right] - \sum_{j=1}^{n}\ell_j\sum_{i=1}^{j}\mu_i - (c-s)q$$

$$= \left(p_1 + \sum_{j=1}^{n}\ell_j - c\right)q - \sum_{j=1}^{n}\left(p_j + \ell_j - p_{j+1}\right)\int_{-\infty}^{q}G_j(y)\,\mathrm{d}y - \sum_{j=1}^{n}\ell_j\sum_{i=1}^{j}\mu_i.$$

Table 5. The results on Dataset 1 for uniformly distributed demands.

Heuristic	CV = 1/2		CV = 1/3		CV = 1/4	
	ARPE	MRPE	ARPE	MRPE	ARPE	MRPE
H1	1.00%	7.44%	1.01%	7.24%	0.84%	6.78%
H2	2.24%	14.49%	1.05%	3.34%	0.74%	2.09%
H3N	0.07%	0.51%	0.16%	2.05%	0.37%	4.17%
H3G	0.27%	2.02%	0.40%	4.30%	0.71%	6.87%
Tukeytest	{H3N,H3G} ≺ {H1} ≺ {H2}		{H3N,H3G} ≺ {H1,H2}		{H3N} ≺ {H1,H2,H3G}	

Table 6. The results on Dataset 2 for uniformly distributed demands.

Heuristic	CV = 1/2		CV = 1/3		CV = 1/4	
	ARPE	MRPE	ARPE	MRPE	ARPE	MRPE
H1	0.93%	5.81%	1.59%	7.07%	1.54%	6.01%
H2	6.66%	20.03%	1.80%	7.35%	0.85%	4.37%
H3N	0.10%	0.41%	0.14%	1.91%	0.50%	5.62%
H3G	0.22%	0.79%	0.33%	2.40%	0.85%	5.70%
Tukeytest	{H1,H3N,H3G} ≺ {H2}		{H3N,H3G} ≺ {H1,H2}		{H2,H3N,H3G} ≺ {H1}	

Table 7. The results on Dataset 3 for uniformly distributed demands.

Heuristic	CV = 1/2		CV = 1/3		CV = 1/4	
	ARPE	MRPE	ARPE	MRPE	ARPE	MRPE
H1	2.42%	8.38%	3.50%	9.16%	3.21%	7.27%
H2	5.11%	23.44%	1.29%	7.35%	0.81%	3.55%
H3N	0.15%	0.92%	0.23%	1.12%	0.47%	3.26%
H3G	0.27%	0.82%	0.24%	2.65%	0.64%	4.80%
Tukeytest	{H3N,H3G} ≺ {H1} ≺ {H2}		{H3N,H3G} ≺ {H2} ≺ {H1}		{H2,H3N,H3G} ≺ {H1}	

Table 8. The results on Datasets 1, 2 and 3 for exponentially distributed demands.

Heuristic	Dataset 1		Dataset 2		Dataset 3	
	ARPE	MRPE	ARPE	MRPE	ARPE	MRPE
H1	0.27%	5.16%	0.25%	4.15%	0.56%	5.17%
H2	4.32%	32.19%	16.38%	41.45%	13.42%	47.46%
H3N	2.35%	10.40%	3.53%	10.83%	4.63%	15.64%
H3G	0.01%	0.22%	0.01%	0.18%	0.01%	0.13%
Tukeytest	{H1,H3G} ≺ {H3N} ≺ {H2}		{H1,H3G} ≺ {H3N} ≺ {H2}		{H1,H3G} ≺ {H3N} ≺ {H2}	

Consequently, the optimal purchase quantity q^* satisfies:

$$\sum_{j=1}^{n}\left(p_j+\ell_j-p_{j+1}\right)G_j\left(q^*\right)=p_1+\ell-c,\tag{6}$$

where $\ell=\sum_{j=1}^{n}\ell_j$. Letting now $v_j=\dfrac{p_j+\ell_j-p_{j+1}}{p_1+\ell-s}$ for $j=1,2,\cdots,n$, and $v=\dfrac{p_1+\ell-c}{p_1+\ell-s}$, we observe that

equation (6) is equivalent to $\sum_{j=1}^{n}v_jG_j\left(q^*\right)=v$. Since $\sum_{j=1}^{n}v_j=1$ and $0\le v_j,v\le 1$, the optimal purchase

quantity is $q^*=G^{-1}\left(v\right)=G^{-1}\left(\dfrac{p_1+\ell-c}{p_1+\ell-s}\right)$, where $G=\sum_{j=1}^{n}v_jG_j$. Thus, the newsvendor problem with n de-

creasing priority demand classes and the shortage penalties reduces to the standard newsvendor problem with the demand CDF G, the selling price $p_1+\ell$, the unit purchase cost c, and the salvage value s. Evidently, this result is an extension of that shown in Section 3.

We assumed so far that the demands X_j of the n classes of customers are independent random variables with known CDFs F_j. Since these CDFs might be difficult to determine in practice, suppose that only their means μ_j and the standard deviations σ_j are available. Thus, we consider the problem under incomplete probabilistic information. To solve it, let $\mathcal{G}$ denote the family of all CDFs with the mean μ_G and the standard deviation $\sigma_G=\sqrt{m_G-\mu_G^2}$ defined by (4) and (5), where second moment of X_j is $m_j=E\left[X_j^2\right]=\mu_j^2+\sigma_j^2$. Recall that the CDFs $\underline{G}$ and $\overline{G}$ bound the family $\mathcal{G}$ in the sense of increasing concave order if for every

$G\in\mathcal{G}$ and q, $\int_{-\infty}^{q}\underline{G}(x)\mathrm{d}x\ge\int_{-\infty}^{q}G(x)\mathrm{d}x\ge\int_{-\infty}^{q}\overline{G}(x)\mathrm{d}x$. The sharp bounds are: $\underline{G}(x)=\dfrac{1}{2}\left[1+\dfrac{x-\mu_G}{\sqrt{\left(x-\mu_G\right)^2+\sigma_G^2}}\right]$,

and $\overline{G}(x)=0$ for $x<\mu_G$ and $\overline{G}(x)=1$ for $x\ge\mu_G$ [11]. The corresponding purchase quantities

$\underline{q}=\underline{G}^{-1}\left(\dfrac{p_1+\ell-c}{p_1+\ell-s}\right)=\mu_G+\dfrac{\sigma_G\left(p_1+\ell+s-2c\right)}{2\sqrt{\left(p_1+\ell-c\right)\left(c-s\right)}}$ and $\overline{q}=\mu_G$ yield sharp lower and upper bounds,

$\left(p_1-c\right)\mu_G-\sigma_G\sqrt{\left(p_1+\ell-c\right)\left(c-s\right)}$ and $\left(p_1-c\right)\mu_G$, on the optimal expected profit defined over all $G\in\mathcal{G}$. The quantities $\underline{q}$ and $\overline{q}$ should be regarded as those found under the worst-case and best-case scenarios, respectively. The quantity $\overline{q}=\mu_G$ can be also verified by the use of Jensen's inequality. On the other hand,

$\underline{q}=\mu_G+\dfrac{\sigma_G\left(p_1+\ell+s-2c\right)}{2\sqrt{\left(p_1+\ell-c\right)\left(c-s\right)}}$ is actually a generalization of the well-known Scarf formula,

$q_{\text{Scarf}}=\mu+\dfrac{\sigma\left(p+s-2c\right)}{2\sqrt{\left(p-c\right)\left(c-s\right)}}$, derived for the standard $n=1$ case; see [7] [12] [13].

7. Conclusions

We reconsidered the Şen and Zhang [4] extension of the standard newsvendor problem with multiple classes of demand with decreasing selling prices. We showed that this extension actually reduces to the standard newsvendor problem with the demand distribution being a mixture of the input distributions, the selling price p_1, the unit cost c, and the salvage value s. Since the mixture distribution is typically hard to handle analytically, we developed a simple general heuristic. This was implemented by replacing the mixture distribution with other distributions having the same mean and standard deviation which led to the two final heuristics proposed in this paper (H3N and H3G).

The two heuristics along with the two from previous work [4] were tested across different demand distributions and price/cost structures. This resulted in a total of 2100 problem instances that were solved using an exact algorithm and the four heuristics. In general, at least one of our heuristics produced a near optimal solution and was statistically proved to outperform the two heuristics proposed in Şen and Zhang [4]. Additional studies are needed to provide strict guidelines on the types of demand distributions for which a particular implementation of our heuristic (H3N or H3G) performs better. One can see a direct application of this model in determining order quantity for some Chinese pharmaceuticals which are ordered once a year or other applications such as annual orders for winter jackets and one time orders for perishables.

The reduction of the newsvendor problem with decreasing priority demand classes to the standard newsvendor problem revealed in this paper, remains valid when penalties are imposed for not meeting the demands. This reduction is also very useful in the case of incomplete probabilistic information about the demand distributions. In particular, we showed extensions of Scarf's ordering rule when the means and standard deviations of random demands are the only parameters available. Additional studies are needed to consider different assumptions concerning the demands whose distributions cannot be fully specified. For example, the work of [10] seems to provide new opportunities in this matter.

Acknowledgements

We would like to thank Prof. Jerzy Kamburowski for his significant contributions to Sections 3 and 6.

References

[1] Arrow, K.J., Harris, T. and Marschak, J. (1951) Optimal Inventory Policy. *Econometrica*, **19**, 250-272. http://dx.doi.org/10.2307/1906813

[2] Morse, M.P. and Kimball, G.E. (1951) Methods of Operations Research. M.I.T. Press, Cambridge, MA.

[3] Qin, Y., Wang, R., Vakharia, A.J., Chen, Y. and Seref, M.M.H. (2011) The Newsvendor Problem: Review and Directions for Future Research. *European Journal of Operational Research*, **213**, 361-374. http://dx.doi.org/10.1016/j.ejor.2010.11.024

[4] Şen, A. and Zhang, A.X. (1999) The Newsboy Problem with Multiple Demand Classes. *IIE Transactions*, **31**, 431-444. http://dx.doi.org/10.1080/07408179908969846

[5] Khouja, M. (1995) The Newsboy Problem under Progressive Multiple Discounts. *European Journal of Operational Research*, **84**, 458-466. http://dx.doi.org/10.1016/0377-2217(94)00053-F

[6] Chung, K.-J., Ting, P.-S. and Hou, K.-L. (2009) A Simple Cost Minimization Procedure for the (Q,r) Inventory System with a Specified Fixed Cost per Stockout Occasion. *Applied Mathematical Modelling*, **33**, 2538-2543. http://dx.doi.org/10.1016/j.apm.2008.08.023

[7] Alfares, H.K. and Elmorra, H.H. (2005) The Distribution-Free Newsboy Problem: Extensions to the Shortage Penalty Case. *International Journal of Production Economics*, **93-94**, 465-477. http://dx.doi.org/10.1016/j.ijpe.2004.06.043

[8] Hu, J., Dai, Y. and Gu, K. (2010) Pharmaceutical Supply Chain in China: Challenges and Opportunities. CAPS Research: Institute for Supply Management and Carey School of Business at Arizona State University, 81.

[9] Meng, Q., Cheng, G., Silver, L., Sun, X., Rehnberg, C. and Tomson, G. (2005) The Impact of China's Retail Drug Price Control Policy on Hospital Expenditures: A Case Study in Two Shandong Hospitals. *Health Policy and Planning*, **20**, 185-196. http://dx.doi.org/10.1093/heapol/czi018

[10] Perakis, G. and Roels, G. (2008) Regret in the Newsvendor Model with Partial Information. *Operations Research*, **56**, 188-203. http://dx.doi.org/10.1287/opre.1070.0486

[11] Müller, A. and Stoyan, D. (2002) Comparison Methods for Stochastic Models and Risks. John Wiley & Sons, Hoboken.

[12] Scarf, H. (1958) A Min-Max Solution of an Inventory Problem. In: *Studies in the Mathematical Theory of Inventory and Production*, Stanford University Press, Redwood City, CA, 201-209.

[13] Gallego, G. and Moon, I. (1993) The Distribution Free Newsboy Problem: Review and Extensions. *Journal of the Operational Research Society*, 825-834. http://dx.doi.org/10.1057/jors.1993.141

8

Using DEA and AHP for Multiplicative Aggregation of Hierarchical Indicators

Mohammad Sadegh Pakkar

Faculty of Management, Laurentian University, Sudbury, Canada
Email: ms_pakkar@laurentian.ca

Abstract

The author [Pakkar, M.S. (2014) Using Data Envelopment Analysis and Analytic Hierarchy Process to Construct Composite Indicators. *Journal of Applied Operational Research*, 6(3), 174-187.] recently proposed a multiplicative approach using Data Envelopment Analysis (DEA) and Analytic Hierarchy Process (AHP) to reflect the priority weights of indicators in constructing composite indicators (CIs). Nonetheless, this approach is limited to the situations with a single level hierarchy which might not satisfy the needs of a multiple level hierarchy. Therefore, the current paper extends this approach to the situations in which the indicators of similar characteristics can be grouped into sub-categories and further linked into categories to form a three-level hierarchical structure. An illustrative example of road safety performance for a set of European countries highlights the usefulness of the proposed "extended approach".

Keywords

Data Envelopment Analysis, Analytic Hierarchy Process, Composite Indicators, Multiplicative Aggregation, Hierarchical Structures

1. Introduction

A composite indicator (CI) is a mathematical tool to aggregate a set of multidimensional indicators in order to produce a single measure of performance. In a recent paper, Pakkar [1] proposed a multiplicative approach using DEA and AHP to reflect the priority weights of indicators in constructing CIs. This approach can be organized into the following steps:

1) Using a multiplicative DEA based-CI model to compute the composite value of each Decision Making Unit (DMU). The computed composite values are used in the next step.

2) Using a minimax distance model to obtain the optimal weights of indicators for each DMU (minimum composite loss).

3) Using the minimax distance model bounded by AHP to obtain the priority weights of indicators for each DMU (maximum composite loss).

4) Using a parameter goal programming model to assess the performance of each DMU in terms of its relative closeness to the priority weights of indicators.

In the basic multiplication DEA based-CI models, all indicators are simply treated to be at the same level of hierarchy [2]. Nonetheless, these indicators might also belong to different sub-categories and further be linked to one another constituting a three-level hierarchical structure. To overcome this limitation, we integrate AHP to a three-level DEA-based CI model in a multiplicative context. A three-level DEA based-CI model can reflect the characteristics of the generalized multi-level DEA based-CI model developed by [3].

2. Methodology

2.1. DEA-Based CI Model

A DEA-based CI model can be formulated similar to a multiplicative DEA model without explicit inputs [2]. In the following, and in line with the more common CI terminology, we will often refer to outputs as "indicators". In order to eliminate the scale differences between all (output) indicators, and moreover, to ensure that all of them are in the same direction of change the normalized counterparts of indicators, using a min-max method, are computed as follows [4]:

$$y_{rj} = \left(\frac{\hat{y}_{rj} - \hat{y}_{r(\min)}}{\hat{y}_{r(\max)} - \hat{y}_{r(\min)}} \right) + x, \quad \hat{y}_{r(\max)} = \max\left\{ \hat{y}_{r1}, \hat{y}_{r2}, \cdots, \hat{y}_{rn} \right\} \text{ for desirable indicators,} \tag{1}$$

$$y_{rj} = \left(\frac{\hat{y}_{r(\max)} - \hat{y}_{rj}}{\hat{y}_{r(\max)} - \hat{y}_{r(\min)}} \right) + x, \quad \hat{y}_{r(\min)} = \min\left\{ \hat{y}_{r1}, \hat{y}_{r2}, \cdots, \hat{y}_{rn} \right\} \text{ for undesirable indicators.} \tag{2}$$

Here, y_{rj} is the normalized value of (output) indicator $r(r = 1, 2, \cdots, s)$ for DMU $j(j = 1, 2, \cdots, n)$. Since in a multiplicative aggregation, the value of each indicator must always be larger than 1, we add a positive constant x to the normalized values of each indicator. We choose x so that $\left(y_{r(\min)} + x \right)$ turns to 1.01 while $y_{r(\min)}$ is the minimum normalized value of indicator r for all DMUs. Although the model used in this paper does not satisfy the desirable unit invariant property, it is very robust to changes in the measurement units [2]. Therefore, this would only slightly change the composite values without making a significant change in DMUs' rankings. Then a multiplicative optimization model in the construction of a composite indicator can be formulated as

$CI_k = \max \prod_{r=1}^{s} y_{rk}^{u_r}$, subject to $\prod_{r=1}^{s} y_{rj}^{u_r} \le e$ with $u_r \ge 0$, where CI_k is the composite value of DMU$_k$

$(k = 1, 2, \cdots, n)$ or the DMU under assessment, u_r is the weight of indicator $r(r = 1, 2, \cdots, s)$ and e is the base of the natural logarithm. Taking logarithms with base e, the multiplicative model can be converted to the following log-linear programming model:

$$\text{Max} \quad \tilde{CI}_k = \sum_{r=1}^{s} u_r \tilde{y}_{rk} \tag{3}$$

s.t.

$$\sum_{r=1}^{s} u_r \tilde{y}_{rj} \le 1 \quad \forall j, \tag{4}$$

$$u_r \ge 0 \quad \forall r, \tag{5}$$

where the tilde symbol (~) denotes natural logarithms. The combination of (3)-(5) forms a single level DEA-based CI model in a log-linear context that looks like an output-oriented DEA model without explicit inputs. This model is theoretically similar to the log-linear DEA model for efficiency analysis introduced in [5].

2.2. Three-Level DEA-Based CI Model

We develop our formulation based on a generalized distance model (for example, see [6]) in such a way that the hierarchical structures of indicators, using a weighted-average approach, are taken into consideration [3]. Let $\tilde{y}_{ll'rj} = \ln\left(y_{ll'rj}\right)$ be the value of indicator $r(r = 1, 2, \cdots, s)$ of sub-category $l'(l' = 1, 2, \cdots, S')$ of category $l(l = 1, 2, \cdots, S)$ for DMU $j(j = 1, 2, \cdots, n)$ after normalizing the original data. Let $u_{ll'r}$ be the internal weight of indicator r of sub-category l' of category l while $\sum_{r=1}^{s} u_{ll'r} = 1$. Then the value of sub-category l' of category l for the DMU j is defined as $\tilde{y}_{ll'j} = \sum_{r=1}^{s} u_{ll'r}\tilde{y}_{ll'rj}$. Let $p_{ll'}$ be the internal weight of sub-category l' of category l while $\sum_{l'=1}^{S'} p_{ll'} = 1$. Then the value of category l is defined as $\tilde{y}_{lj} = \sum_{l'=1}^{S'} p_{ll'}\tilde{y}_{ll'j}$. Let p_l be the weight of category l.

To develop a linear model, the new multiplier of indicator r of sub-category l' of category l is defined as: $u'_{ll'r} = p_l p_{ll'} u_{ll'r}$. Similarly, the new multiplier of sub-category l' of category l is defined as: $p'_{ll'} = p_l p_{ll'}$. Let $\tilde{CI}_k^*(k = 1, 2, \cdots, n)$ be the best attainable composite value for the DMU under assessment, calculated from the one-level DEA-based CI model. We want the composite value $\tilde{CI}_k(u'_{ll'r})$, calculated from the set of weights $u'_{ll'r}$ to be *closest* to $\tilde{CI}_k^*$. Our definition of *closest* is that the largest distance is at its minimum. Hence we choose the form of the minimax model: $\min_{u'_{ll'r}} \max_k \left\{\tilde{CI}_k^* - \tilde{CI}_k(u'_{ll'r})\right\}$ to minimize a single deviation which is equivalent to the following linear model:

$$\text{Min } \theta \tag{6}$$

s.t.

$$\tilde{CI}_k^* - \sum_{l=1}^{S}\sum_{l'=1}^{S'}\sum_{r=1}^{s} u'_{ll'r}\tilde{y}_{ll'rk} \leq \theta, \tag{7}$$

$$\sum_{l=1}^{S}\sum_{l'=1}^{S'}\sum_{r=1}^{s} u'_{ll'r}\tilde{y}_{ll'rj} \leq \tilde{CI}_j^* \qquad \forall j, \tag{8}$$

$$\sum_{l'=1}^{S'}\sum_{r=1}^{s} u'_{ll'r} = p_l \qquad \forall l, \tag{9}$$

$$\sum_{r=1}^{s} u'_{ll'r} = p'_{ll'} \qquad \forall l, l', \tag{10}$$

$$u'_{ll'r}, p'_{ll'}, p_l > 0 \qquad \forall l, l', r, \tag{11}$$

$$\theta \geq 0. \tag{12}$$

The combination of (6)-(12) forms a three-level DEA based-CI model in the log-linear context that identifies the minimum composite loss $\theta_{\min}$ needed to arrive at an optimal set of weights. Constraint (7) ensures that each DMU loses no more than θ of its best attainable composite value, $\tilde{CI}_k^*$. The second set of constraints (8) satisfies that the composite values of all DMUs are less than or equal to their upper bound of $\tilde{CI}_j^*$. Two sets of constraints (9) and (10) are added to the model. This implies that the sum of weights under each (sub-)sub-category equals to the weight of that (sub-)sub-category. It should be noted that the original (or internal) weights used for calculating the weighted averages are obtained as $u_{ll'r} = u'_{ll'r}/p'_{ll'}$ and $p_{ll'} = p'_{ll'}/p_l$.

2.3. Prioritizing Indicator Weights Using AHP

The three-level DEA based-CI model identifies the minimum composite loss $\theta_{\min}$ needed to arrive at a set of weights of indicators by the internal mechanism of DEA. On the other hand, the priority weights of indicators, and the corresponding (sub-)categories are defined out of the internal mechanism of DEA by AHP.

In order to more clearly demonstrate how AHP is integrated into the three-level DEA-based CI model, this research presents an analytical process in which indicator weights are bounded by the AHP method. The AHP procedure for imposing weight bounds may be broken down into the following steps:

Step 1: A decision maker makes a pairwise comparison matrix of different criteria, denoted by A, with the entries of $a_{lq}(l = q = 1, 2, \cdots, S)$. The comparative importance of criteria is provided by the decision maker using a rating scale. Saaty [7] recommends using a 1-9 scale.

Step 2: The AHP method obtains the priority weights of criteria by computing the eigenvector of matrix A (Equation (13)), $w = (w_1, w_2, \cdots, w_S)^{\mathrm{T}}$, which is related to the largest eigenvalue, $\lambda_{\max}$.

$$Aw = \lambda_{\max} w . \tag{13}$$

To determine whether or not the inconsistency in a comparison matrix is reasonable the random consistency ratio, *C.R.*, can be computed by the following equation:

$$C.R. = \frac{\lambda_{\max} - N}{(N-1) R.I.} \tag{14}$$

where *R.I.* is the average random consistency index and N is the size of a comparison matrix. In a similar way, the priority weights of (sub-)sub-criteria under each (sub-)criterion can be computed. To obtain the weight bounds for indicator weights in the three-level DEA-based CI model, this study aggregates the priority weights of three different levels in AHP as follows:

$$\bar{u}_{ll'r} = w_l e_{ll'} f_{ll'r}, \quad \sum_{l=1}^{S} w_l = 1, \quad \sum_{l'=1}^{S'} e_{ll'} = 1 \text{ and } \sum_{r=1}^{s} f_{ll'r} = 1 \tag{15}$$

where w_l is the priority weight of criterion $l(l = 1, \cdots, S)$ in AHP and $e_{ll'}$ is the priority weight of sub-criterion $l'(l' = 1, 2, \cdots, S')$ under criterion l and $f_{ll'r}$ is sub-sub-criterion $r(r = 1, \cdots, s)$ under sub-criterion l'.

In order to estimate the maximum composite loss $\theta_{\max}$ necessary to achieve the priority weights of indicators for each DMU the following set of constraints is added to the three-level DEA-based CI model:

$$u'_{ll'r} = \alpha \bar{u}_{ll'r} \quad \forall l, l', r, \text{ while } \alpha > 0. \tag{16}$$

The set of constraints (16) changes the AHP computed weights to weights for the new system by means of a scaling factor α. The scaling factor α is added to avoid the possibility of contradicting constraints leading to infeasibility or underestimating the relative composite scores of DMUs [8].

2.4. Parametric Goal Programming Model

In this stage we develop a parametric goal programming model that can be solved repeatedly to generate the various sets of weights for the discrete values of the parameter θ, such that $\theta_{\min} \leq \theta \leq \theta_{\max}$. The purpose of the model is to minimize the total deviations from the priority weights of indicators with a city block distance measure. Choosing such a distance measure, each deviation is being equally weighted subject to the following constraints:

$$\text{Min } Z_k(\theta) = \sum_{l=1}^{S} \sum_{l'=1}^{S'} \sum_{r=1}^{s} \left(d^+_{ll'r} + d^-_{ll'r} \right) \tag{17}$$

s.t.

$$u'_{ll'r} - d^+_{ll'r} + d^-_{ll'r} = \alpha \bar{u}_{ll'r} \quad \forall l, l', r, \tag{18}$$

$$d^+_{ll'r}, d^-_{ll'r} \geq 0 \quad \forall l, l', r, \tag{19}$$

and constraints (7)-(12).

Here, $d^+_{ll'r}$ and $d^-_{ll'r}$ are the positive and negative deviations from the priority weight of indicator r under sub-category l' of category l, for DMU$_k$. The set of equations (18) indicates the goal equations whose right-hand sides are the priority weights of hierarchical indicators adjusted by the obtained value of the scaling variable in (16).Because the range of deviations computed by the objective function is different for each DMU, it is necessary to normalize it by using relative deviations rather than absolute ones. Hence, the normalized deviations can be computed by:

$$\Delta_k(\theta) = \frac{Z_k^*(\theta_{\min}) - Z_k^*(\theta)}{Z_k^*(\theta_{\min})}, \tag{20}$$

where $Z_k^*(\theta)$ is the optimal value of objective function (17) for $\theta_{\min} \leq \theta \leq \theta_{\max}$. We define $\Delta_k(\theta)$ as a *measure of closeness* which represents the relative closeness of each DMU to the weights obtained from the three-level DEA based-CI model in the range [0, 1] after adding the set of constraints (16) to it. Increasing the parameter (θ), we improve the deviations between the two systems of weights obtained from the three-level DEA based-CI model before and after adding the set of constraints (16). This may lead to different ranking positions for each DMU in comparison to the other DMUs. It should be noted that in a special case where the parameter $\theta = \theta_{\max} = 0$, we assume $\Delta_k(\theta) = 1$.

3. Numerical Example

In this section we present the application of the proposed approach to assess the road safety performance of a set of European countries (or DMUs). The data for eight hierarchical indicators that compose road safety performance indicators (SPIs) for 11 European countries has been adopted from [9]. **Table 1** presents the normalized data, using (1) and (2), for SPIs on a logarithmic scale.

The notations in **Table 1** are as follows: $\tilde{y}_1$ = Road user behavior, $\tilde{y}_2$ = Vehicle, $\tilde{y}_{11}$ = Alcohol, $\tilde{y}_{12}$ = Seat belt, $\tilde{y}_{111}$ = Roadside police alcohol tests per 1000 population in 2008, $\tilde{y}_{112}$ = The percentage of drivers above legal alcohol limit in roadside checks in 2008, $\tilde{y}_{121}$ = Daytime seat belt wearing rates on front seats aggregated of cars in 2009, $\tilde{y}_{122}$ = Daytime wearing rates of seat belts on rear seats of cars in 2009, $\tilde{y}_{21} = \tilde{y}_{211}$ = The average percentage of occupant protection score for new cars sold in 2008, $\tilde{y}_{22} = \tilde{y}_{221}$ = The average percentage of pedestrian protection score for new cars sold in 2008, $\tilde{y}_{23} = \tilde{y}_{231}$ = Renewal rate of passenger cars in 2007, $\tilde{y}_{24} = \tilde{y}_{241}$ = Median age of passenger cars in 2008. AT = Austria, BE = Belgium, BG = Bulgaria, CY = Cyprus, CZ = Czech Republic, DK = Denmark, EE = Estonia, FI = Finland, FR = France, DE = Germany, EL = Greece.

The results of the AHP model for prioritizing hierarchical SPIs as constructed by the author in Expert Choice software are presented in **Table 2**. One can argue that the priority weights of SPIs must be judged by road safety experts. However, since the aim of this section is just to show the application of the proposed approach on numerical data, we see no problem to use our judgment alone.

Solving the three-level DEA based-CI model for the country under assessment, we obtain an optimal set of weights with minimum composite loss $(\theta_{\min})$. It should be noted that the composite value of all countries calculated from the three-level DEA based-CI model is identical to that calculated from the one-level DEA based-CI model. Therefore, the minimum composite loss for the country under assessment is $\theta_{\min} = 0$ (**Table 3**). This implies that the measure of relative closeness to the AHP weights for the country under assessment is $\Delta_k(\theta_{\min}) = 0$. On the other hand, solving the three-level DEA based-CI model for the country under assessment

Table 1. Normalized data for hierarchical SPIs on a logarithmic scale.

| Country | $\tilde{y}_1$ | | | | $\tilde{y}_2$ | | | |
| | $\tilde{y}_{11}$ | | $\tilde{y}_{12}$ | | $\tilde{y}_{21}$ | $\tilde{y}_{22}$ | $\tilde{y}_{23}$ | $\tilde{y}_{24}$ |
	$\tilde{y}_{111}$	$\tilde{y}_{112}$	$\tilde{y}_{121}$	$\tilde{y}_{122}$	$\tilde{y}_{211}$	$\tilde{y}_{221}$	$\tilde{y}_{231}$	$\tilde{y}_{241}$
AT	0.579	0.523	0.517	0.505	0.558	0.412	0.448	0.558
BE	0.555	0.553	0.382	0.469	0.587	0.313	0.698	0.570
BG	0.654	0.625	0.480	0.352	0.395	0.364	0.010	0.249
CY	0.430	0.519	0.382	0.010	0.698	0.698	0.641	0.435
CZ	0.539	0.666	0.552	0.372	0.463	0.545	0.209	0.233
DK	0.650	0.423	0.603	0.557	0.496	0.503	0.517	0.558
EE	0.567	0.688	0.517	0.487	0.587	0.503	0.375	0.233
FI	0.010	0.682	0.603	0.683	0.644	0.545	0.295	0.415
FR	0.417	0.614	0.698	0.646	0.587	0.412	0.448	0.511
DE	0.497	0.362	0.683	0.691	0.587	0.313	0.517	0.511
EL	0.506	0.621	0.272	0.039	0.463	0.503	0.448	0.375

Table 2. The AHP hierarchical model for SPIs.

Objective level	Criteria level	Sub-criteria level	Sub-sub-criteria level
Prioritizing road safety performance indicators	Road user behavior $w_1 = 0.65$	Alcohol $e_{11} = 0.60$	Roadside police alcohol tests $f_{111} = 0.40$
			Driving above legal alcohol limit $f_{112} = 0.60$
		Seat belt $e_{12} = 0.40$	Seat belt wearing in front seats $f_{121} = 0.70$
			Seat belt wearing in rear seats $f_{122} = 0.30$
	Vehicle $w_2 = 0.35$	Occupant protection for cars $e_{21} = 0.30$	Occupant protection for cars $f_{211} = 1.00$
		Pedestrian protection for cars $e_{22} = 0.20$	Pedestrian protection for cars $f_{221} = 1.00$
		Renewal rate of passenger cars $e_{23} = 0.40$	Renewal rate of passenger cars $f_{231} = 1.00$
		Age of passengercars $e_{24} = 0.10$	Age of passenger cars, $f_{241} = 1.00$

Table 3. Minimum and maximum losses in composite values for each country.

Countries	$\tilde{CI}_k^*$	$\theta_{\min}$	$\theta_{\max}$
AT	1.000	0.000	0.1516
BE	1.000	0.000	0.1594
BG	1.000	0.000	0.3328
CY	1.000	0.000	0.1942
CZ	1.000	0.000	0.1626
DK	1.000	0.000	0.1621
EE	1.000	0.000	0.0462
FI	1.000	0.000	0.1142
FR	1.000	0.000	0
DE	1.000	0.000	0.2289
EL	1.000	0.000	0.3466

after adding the set of constraints (16), we adjust the priority weights of hierarchical SPIs obtained from AHP in such a way that they become compatible with the weights' structure in the three level DEA-based CI model. **Table 4** presents the optimal weights of hierarchical SPIs as well as its scaling factor for all countries. It should be noted that the priority weights of AHP used for incorporating weight bounds on indicator weights after adding (16) to the three-level model are obtained as $\bar{u}_{ll'r} = \dfrac{u'_{ll'r}}{\alpha}$. Similarly, the priority weights of AHP at criteria level can be obtained as $w_l = \dfrac{p_l}{\alpha}$ while $\sum_{l'=1}^{S'}\sum_{r=1}^{s} u'_{ll'r} = p_l$ and $\sum_{r=1}^{s} u'_{ll'r} = p'_{ll'}$.

In addition, the priority weights of AHP at sub-criteria and sub-sub-criteria levels can be obtained as $e_{ll'} = p'_{ll'}/p_l$ and $f_{ll'r} = u'_{ll'r}/p'_{ll'}$, respectively.

The maximum composite loss for each country to achieve the corresponding weights in the three-level DEA-based CI model after adding (16) is equal to $\theta_{\max}$ (**Table 3**). As a result, the measure of relative closeness to the priority weights of SPIs for the country under assessment is $\Delta_k(\theta_{\max}) = 1$.

Going one step further to the solution process of the parametric goal programming model, we proceed to the estimation of total deviations from the AHP weights for each country while the parameter θ is $0 \le \theta \le \theta_{\max}$. **Table 5** represents the ranking position of each country based on the minimum deviation from the priority weights of indicators for $\theta = 0$. It should be noted that in a special case where the parameter $\theta = \theta_{\max} = 0$ we

Table 4. Optimal weights of hierarchical SPIs obtained from three-level DEA based-CI model bounded by AHP weights for all countries.

Weights of categories	Weights of sub-categories	Weights of sub-sub-categories
$p_1 = 1.2322$	$p'_{11} = 0.7393$	$u'_{111} = 0.2957$
		$u'_{112} = 0.4436$
	$p'_{12} = 0.4929$	$u'_{121} = 0.3450$
		$u'_{122} = 0.1479$
$p_2 = 0.6636$	$p'_{21} = 1.991$	$u'_{211} = 0.1991$
	$p'_{22} = 0.1327$	$u'_{221} = 0.1327$
	$p'_{23} = 0.2654$	$u'_{231} = 0.2654$
	$p'_{24} = 0.0664$	$u'_{241} = 0.0664$
$\alpha = 1.8957$		

Table 5. The ranking position of each country based on the minimum distance to priority weights of hierarchical SPIs.

Countries	$Z^*(\theta_{min})$	Rank
AT	0.9288	9
BE	0.4619	5
BG	1.2375	10
CY	0.3956	4
CZ	0.8410	8
DK	0.5135	6
EE	0.1307	2
FI	0.3775	3
FR	0.0000	1
DE	0.7205	7
EL	1.5300	11

assume $\Delta_k(\theta) = 1$. **Table 5** shows that France (FR) is the best performer in terms of the $\tilde{CI}$ value and the relative closeness to the priority weights of indicators in comparison to the other countries. Nevertheless, increasing the value of θ from 0 to θ_{max} has two main effects on the performance of the other countries: improving the degree of deviations and reducing the value of composite indicator. This, of course, is a phenomenon, one expects to observe frequently. The graph of $\Delta(\theta)$ versus θ, as shown in **Figure 1**, is used to describe the relation between the relative closeness to the priority weights of indicators and composite loss for each country. This may result in different ranking positions for each country in comparison to the other countries (**Table A1**).

4. Conclusion

We develop a multiplicative (or log-linear) aggregation approach based on DEA and AHP methodologies to construct CIs for hierarchical indicators. We define two sets of weights of hierarchical indicators in a three-level DEA framework. All indicators are treated as benefit type which satisfy the property of "the larger the better". The first set represents the weights of indicators with minimum composite loss. The second set represents

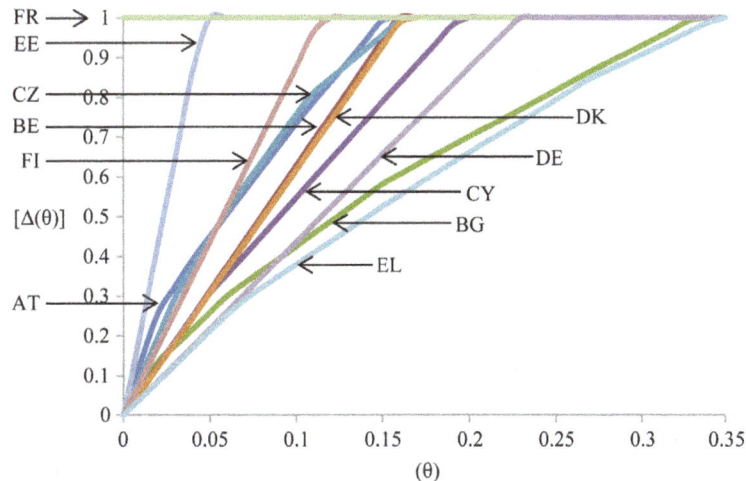

Figure 1. The relative closeness to the priority weights of hierarchical indicators [$\Delta(\theta)$], versus composite loss (θ) for each country.

the corresponding priority weights of hierarchical indicators, using AHP, with maximum composite loss. We assess the performance of each DMU in comparison to the other DMUs based on the relative closeness of the first set of weights to the second set of weights. Improving the measure of relative closeness in a defined range of composite loss, we explore the various ranking positions for the DMU under assessment in comparison to the other DMUs. To demonstrate the effectiveness of the proposed approach, we apply it to construct a composite road safety performance index for eight hierarchical indicators that compose SPIs for 11 European countries.

References

[1]   Pakkar, M.S. (2014) Using Data Envelopment Analysis and Analytic Hierarchy Process to Construct Composite Indicators. *Journal of Applied Operational Research*, **6**, 174-187.

[2]   Zhou, P., Ang, B.W. and Zhou, D.Q. (2010) Weighting and Aggregation in Composite Indicator Construction: A Multiplicative Optimization Approach. *Social Indicators Research*, **96**, 169-181. http://dx.doi.org/10.1007/s11205-009-9472-3

[3]   Shen, Y., Hermans, E., Brijs, T. and Wets, G. (2013) Data Envelopment Analysis for Composite Indicators: A Multiple Layer Model. *Social Indicators Research*, **114**, 739-756. http://dx.doi.org/10.1007/s11205-012-0171-0

[4]   OECD (2008) Handbook on Constructing Composite Indicators: Methodology and User Guide. OECD Publishing.

[5]   Charnes, A., Cooper, W.W., Seiford, L. and Stutz, J. (1982) A Multiplicative Model for Efficiency Analysis. *Socio-Economic Planning Sciences*, **16**, 223-224. http://dx.doi.org/10.1016/0038-0121(82)90029-5

[6]   Hashimoto, A. and Wu, D.A. (2004) A DEA-Compromise Programming Model for Comprehensive Ranking. *Journal of the Operation Research Society of Japan*, **47**, 73-81.

[7]   Saaty, T.S. (1980) The Analytic Hierarchy Process. McGraw-Hill, New York.

[8]   Podinovski, V.V. (2004) Suitability and Redundancy of Non-Homogeneous Weight Restrictions for Measuring the Relative Efficiency in DEA. *European Journal of Operational Research*, **154**, 380-395. http://dx.doi.org/10.1016/S0377-2217(03)00176-0

[9]   Bax, C., Wesemann, P., Gitelman, V., Shen, Y., Goldenbeld, C., Hermans, E., Doveh, E., Hakkert, S., Wegman, F. and Aarts, L. (2012) Developing a Road Safety Index. Deliverable 4.9 of the EC FP7 Project DaCoTA.

Appendix

Table A1. The measure of relative closeness to the priority weights of hierarchical SPIs [$\Delta_k(\theta)$] vs. composite loss [θ] for each country.

θ	AT	BE	BG	CY	CZ	DK	EE	FI	FR	DE	EL
0	0.0000	0.0000	0.0000	0.0000	0.0000	0.0000	0.0000	0.0000	1.0000	0.0000	0.0000
Rank	N/A	N/A	N/A	N/A	N/A	N/A	N/A	N/A	1	N/A	N/A
0.01	0.1373	0.0627	0.0695	0.0627	0.0993	0.0617	0.2163	0.0875	1.0000	0.0437	0.0428
Rank	3	7	6	8	4	9	2	5	1	10	11
0.02	0.2576	0.1255	0.1358	0.1253	0.1987	0.1234	0.4326	0.1751	1.0000	0.0874	0.0856
Rank	3	7	6	8	4	9	2	5	1	10	11
0.03	0.3261	0.1882	0.1771	0.1880	0.2980	0.1850	0.6489	0.2626	1.0000	0.1311	0.1283
Rank	3	6	9	7	4	8	2	5	1	10	11
0.04	0.3942	0.2510	0.2184	0.2506	0.3742	0.2467	0.8652	0.3502	1.0000	0.1747	0.1711
Rank	3	6	9	7	4	8	2	5	1	10	11
0.05	0.4508	0.3137	0.2597	0.3076	0.4454	0.3084	1.0000	0.4377	1.0000	0.2184	0.2131
Rank	3	6	9	8	4	7	1	5	1	10	11
0.06	0.5075	0.3765	0.3009	0.3556	0.5166	0.3701	1.0000	0.5253	1.0000	0.2621	0.2542
Rank	5	6	9	8	4	7	1	3	1	10	11
0.07	0.5642	0.4392	0.3329	0.4036	0.5775	0.4318	1.0000	0.6128	1.0000	0.3058	0.2906
Rank	5	6	9	8	4	7	1	3	1	10	11
0.08	0.6209	0.5020	0.3639	0.4516	0.6373	0.4935	1.0000	0.7004	1.0000	0.3495	0.3212
Rank	5	6	9	8	4	7	1	3	1	10	11
0.09	0.6776	0.5647	0.3950	0.4997	0.6970	0.5551	1.0000	0.7879	1.0000	0.3932	0.3509
Rank	5	6	9	8	4	7	1	3	1	10	11
0.1	0.7342	0.6275	0.4260	0.5477	0.7568	0.6168	1.0000	0.8755	1.0000	0.4368	0.3799
Rank	5	6	10	8	4	7	1	3	1	9	11
0.11	0.7874	0.6902	0.4571	0.5957	0.8103	0.6785	1.0000	0.9630	1.0000	0.4805	0.4089
Rank	5	6	10	8	4	7	1	3	1	9	11
0.12	0.8385	0.7530	0.4881	0.6437	0.8477	0.7402	1.0000	1.0000	1.0000	0.5242	0.4379
Rank	5	6	10	8	4	7	1	1	1	9	11
0.13	0.8897	0.8157	0.5191	0.6917	0.8835	0.8019	1.0000	1.0000	1.0000	0.5679	0.4669
Rank	4	6	10	8	5	7	1	1	1	9	11
0.14	0.9408	0.8785	0.5502	0.7397	0.9192	0.8635	1.0000	1.0000	1.0000	0.6116	0.4959
Rank	4	6	10	8	5	7	1	1	1	9	11
0.15	0.9919	0.9412	0.5811	0.7877	0.9550	0.9252	1.0000	1.0000	1.0000	0.6553	0.5249
Rank	4	6	10	8	5	7	1	1	1	9	11
0.16	1.0000	1.0000	0.6058	0.8357	0.9908	0.9869	1.0000	1.0000	1.0000	0.6990	0.5524
Rank	1	1	10	8	6	7	1	1	1	9	11
0.17	1.0000	1.0000	0.6305	0.8837	1.0000	1.0000	1.0000	1.0000	1.0000	0.7426	0.5797
Rank	1	1	10	8	1	1	1	1	1	9	11

Continued

0.18	1.0000	1.0000	0.6542	0.9317	1.0000	1.0000	1.0000	1.0000	1.0000	0.7863	0.6071
Rank	1	1	10	8	1	1	1	1	1	9	11
0.19	1.0000	1.0000	0.6774	0.9798	1.0000	1.0000	1.0000	1.0000	1.0000	0.8300	0.6343
Rank	1	1	10	8	1	1	1	1	1	9	11
0.2	1.0000	1.0000	0.7006	1.0000	1.0000	1.0000	1.0000	1.0000	1.0000	0.8737	0.6607
Rank	1	1	10	1	1	1	1	1	1	9	11
0.21	1.0000	1.0000	0.7238	1.0000	1.0000	1.0000	1.0000	1.0000	1.0000	0.9174	0.6871
Rank	1	1	10	1	1	1	1	1	1	9	11
0.22	1.0000	1.0000	0.7470	1.0000	1.0000	1.0000	1.0000	1.0000	1.0000	0.9611	0.7135
Rank	1	1	10	1	1	1	1	1	1	9	11
0.23	1.0000	1.0000	0.7703	1.0000	1.0000	1.0000	1.0000	1.0000	1.0000	1.0000	0.7398
Rank	1	1	10	1	1	1	1	1	1	1	11
0.24	1.0000	1.0000	0.7935	1.0000	1.0000	1.0000	1.0000	1.0000	1.0000	1.0000	0.7662
Rank	1	1	10	1	1	1	1	1	1	1	11
0.25	1.0000	1.0000	0.8167	1.0000	1.0000	1.0000	1.0000	1.0000	1.0000	1.0000	0.7926
Rank	1	1	10	1	1	1	1	1	1	1	11
0.26	1.0000	1.0000	0.8399	1.0000	1.0000	1.0000	1.0000	1.0000	1.0000	1.0000	0.8190
Rank	1	1	10	1	1	1	1	1	1	1	11
0.27	1.0000	1.0000	0.8631	1.0000	1.0000	1.0000	1.0000	1.0000	1.0000	1.0000	0.8454
Rank	1	1	10	1	1	1	1	1	1	1	11
0.28	1.0000	1.0000	0.8864	1.0000	1.0000	1.0000	1.0000	1.0000	1.0000	1.0000	0.8668
Rank	1	1	10	1	1	1	1	1	1	1	11
0.29	1.0000	1.0000	0.9081	1.0000	1.0000	1.0000	1.0000	1.0000	1.0000	1.0000	0.8868
Rank	1	1	10	1	1	1	1	1	1	1	11
0.3	1.0000	1.0000	0.9296	1.0000	1.0000	1.0000	1.0000	1.0000	1.0000	1.0000	0.9068
Rank	1	1	10	1	1	1	1	1	1	1	11
0.31	1.0000	1.0000	0.9510	1.0000	1.0000	1.0000	1.0000	1.0000	1.0000	1.0000	0.9268
Rank	1	1	10	1	1	1	1	1	1	1	11
0.32	1.0000	1.0000	0.9725	1.0000	1.0000	1.0000	1.0000	1.0000	1.0000	1.0000	0.9468
Rank	1	1	10	1	1	1	1	1	1	1	11
0.33	1.0000	1.0000	0.9939	1.0000	1.0000	1.0000	1.0000	1.0000	1.0000	1.0000	0.9668
Rank	1	1	10	1	1	1	1	1	1	1	11
0.34	1.0000	1.0000	1.0000	1.0000	1.0000	1.0000	1.0000	1.0000	1.0000	1.0000	0.9868
Rank	1	1	1	1	1	1	1	1	1	1	11
0.35	1.0000	1.0000	1.0000	1.0000	1.0000	1.0000	1.0000	1.0000	1.0000	1.0000	1.0000
Rank	1	1	1	1	1	1	1	1	1	1	1

9

Necessary Optimality Conditions for Multi-Objective Semi-Infinite Variational Problem

Bharti Sharma[1], Promila Kumar[2]

[1]Department of Mathematics, University of Delhi, New Delhi, India
[2]Department of Mathematics, Gargi College, University of Delhi, New Delhi, India
Email: bharti.sharma3135@yahoo.in, kumar.promila@gmail.com

Abstract

In this paper, necessary optimality conditions for a class of Semi-infinite Variational Problems are established which are further generalized to a class of Multi-objective Semi-Infinite Variational Problems. These conditions are responsible for the development of duality theory which is an extremely important feature for any class of problems, but the literature available so far lacks these necessary optimality conditions for the stated problem. A lemma is also proved to find the topological dual of $\left(L^2 \left(I, \mathbb{R} \right) \right)^N$ as it is required to prove the desired result.

Keywords

Semi-Infinite, Variational Problem, Efficient Solution, Necessary Optimality Conditions

1. Introduction

A Semi-infinite Programming Problem (SIP) [1]-[3] is an optimization problem in which the index set of inequality constraints is an arbitrary and not necessarily finite set. It has wide variety of applications in various fields like economics, engineering, mathematical physics and robotics. While browsing the literature, we observe that much attention has been paid to SIP which is static in nature in the sense that time does not enter into consideration. Whereas in practical problems we come across situations where time plays an important role and hence cannot be neglected.

Semi-infinite Programming Problem is tightly interwoven with Variational Problem [4]-[9]. Both these subjects have undergone independent development, hence mutual adaptation of ideas and techniques have always been appreciated.

In this article, we propose Semi-infinite Variational Problem for which necessary optimality conditions are established. These optimality conditions are further extended to Multi-objective Semi-infinite Variational Problem (MSVP). We also clarify, with proper reasoning, certain points which were left for later validation in [9].

Necessary optimality conditions are important because these conditions lay down foundation for many computational techniques in optimization problems as they indicate when a feasible point is not optimal. At the same time these conditions are useful in the development of numerical algorithms for solving certain optimization problems. Further, these conditions are also responsible for the development of duality theory on which there exists an extensive literature and a substantial use of which (duality theory) has been made in theoretical as well as computational applications in many diverse fields. While browsing the literature, we found that necessary optimality conditions were not proved for the class of semi-infinite variational problems.

The paper is organized as follows: In section 2 some basic definitions and preliminaries are given. Section 3 deals with necessary optimality conditions for semi-infinite variational problem; single objective as well as multi-objective. In section 4, we prove a lemma which is required to prove necessary optimality conditions of section 3, for semi-infinite variational problem.

2. Definitions and Preliminaries

Let E be a topological vector space over the field of real numbers and E' denotes the topological dual space of E. For a set $C \subset E$, the topological polar cone C' of C is $C' = \{\xi \in E' \mid \xi(c) \geq 0, \text{ for all } c \in C\}$. Let r and n be two positive integers. For a given real interval $I = [a,b]$, let $x : I \to \mathbb{R}^n$ be a piecewise smooth state function with its derivative $\dot{x}$. For notational convenience we write $x, \dot{x}$ in place of $x(t), \dot{x}(t)$. Let $M = \{1, 2, \cdots, r\}$, $f^i : I \times \mathbb{R}^n \times \mathbb{R}^n \to \mathbb{R}, i \in M$ and $g^j : I \times \mathbb{R}^n \times \mathbb{R}^n \to \mathbb{R}, j \in \mathbb{N}$ be continuously differentiable functions with respect to each of their argument. We also denote the partial derivative of $f^i, i \in M$ with respect to t, x and $\dot{x}$ by $f_t^i, f_x^i, f_{\dot{x}}^i$ respectively. Analogously, we write the partial derivative of $g^i, i \in \mathbb{N}$. For the sake of notational convenience we write $f_x^j(t)$ for $f_x^j(t, x(t), \dot{x}(t))_{\text{T}}$ and $f_{\dot{x}}^j(t)$ for $f_{\dot{x}}^j(t, x(t), \dot{x}(t))$ for $j = 1, 2, \cdots, r$.

For any $x = (x^1, x^2, \cdots, x^n)^{\text{T}}$, $y = (y^1, y^2, \cdots, y^n)^{\text{T}}$ in n-dimensional Euclidean space $\mathbb{R}^n$,

1) $x = y \Leftrightarrow x^i = y^i$ for all $i = 1, 2, \cdots, n$.
2) $x < y \Leftrightarrow x^i < y^i$ for all $i = 1, 2, \cdots, n$.
3) $x \leqq y \Leftrightarrow x^i \leq y^i$ for all $i = 1, 2, \cdots, n$.
4) $x \leq y \Leftrightarrow x \leqq y$ and $x \neq y$.

Let $\mathbb{R}_+^n$ and $\text{int} \mathbb{R}_+^n$ denote the non negative and positive orthant of $\mathbb{R}^n$ respectively. Let X be the space of piecewise smooth state functions $x : I \to \mathbb{R}^n$ which equipped with the norm $\|x\| = \|x\|_\infty + \|Dx\|_\infty$, where the differential operator D is given by

$$u = Dx \Leftrightarrow x(t) = x(a) + \int_a^t u(s) \, ds. \quad \text{Therefore,} \quad D = \frac{d}{dt} \quad \text{except at discontinuities.}$$

Consider the following Multi-objective Semi-infinite Variational Problem (MSVP):

$$(\text{MSVP}) \, \text{Minimize} \left(\int_a^b f^1(t, x, \dot{x}) \, dt, \cdots, \int_a^b f^r(t, x, \dot{x}) \, dt \right)$$

subject to

$$g^i(t, x, \dot{x}) \leqq 0, i \in \mathbb{N}, t \in I, x \in X, \quad (1)$$

$$x(a) = 0, x(b) = 0. \quad (2)$$

$\overline{X} = \{x \in X \mid x(a) = 0, x(b) = 0\}$ with the norm defined as above is a Banach space.

Let $X_0 = \{x \in X \mid g^i(t, x, \dot{x}) \leqq 0, i \in \mathbb{N}, t \in I, x(a) = 0, x(b) = 0\}$ be the set of all feasible solutions of (MSVP).

Definition 1 A point $\overline{x} \in X_0$ is said to be an efficient solution for (MSVP) if there is no other $x \in X_0$ such that

$$\int_a^b f^i(t, x, \dot{x}) \, dt \leq \int_a^b f^i(t, \overline{x}, \dot{\overline{x}}) \, dt, \quad \text{for all } i \in M \text{ and,}$$

$$\int_a^b f^j(t, x, \dot{x}) \, dt < \int_a^b f^j(t, \overline{x}, \dot{\overline{x}}) \, dt, \quad \text{for at least one } j \in M.$$

3. Necessary Optimality Conditions

Let us first prove necessary optimality conditions for the following single objective Semi-infinite Variational Problem (SVP):

$$(\text{SVP}) \text{ Minimize} \int_a^b \phi(t,x,\dot{x})\,dt$$

subject to

$$g^i(t,x,\dot{x}) \leq 0, i \in \mathbb{N}, t \in I, x \in X, \tag{3}$$

$$x(a) = 0, x(b) = 0. \tag{4}$$

where $\phi : I \times \mathbb{R}^n \times \mathbb{R}^n \to \mathbb{R}$ is continuously differentiable function with respect to each of its argument. The problem (SVP) may be rewritten as Cone Constrained Problem (CCP):

$$(\text{CCP}) \text{ Minimize } \Phi(x)$$

subject to

$$-G(x) \in K, x \in \overline{X}. \tag{5}$$

where $\Phi : \overline{X} \to \mathbb{R}$ is defined as

$$\Phi(x) = \int_a^b \{\phi(t,x,\dot{x})\}\,dt, x \in \overline{X}. \tag{6}$$

$$L^2(I,\mathbb{R}) = \left\{ f : I \to \mathbb{R} \text{ such that } f \text{ is measurable and } \int_a^b |f(t)|^2\,d\mu(t) < \infty \right\}$$

$$= \left\{ f : I \to \mathbb{R} \text{ such that } f \text{ is measurable and } \int_a^b |f(t)|^2\,dt < \infty \right\},$$

where μ is Lebesgue measure.

$$\left(L^2(I,\mathbb{R})\right)^{\mathbb{N}} = \left\{ f = (f^i)_{i \in \mathbb{N}} \mid f^i \in L^2(I,\mathbb{R}), i \in \mathbb{N} \right\},$$

$$K = \left\{ (\gamma^i)_{i \in \mathbb{N}} = \gamma \in \left(L^2(I,\mathbb{R})\right)^{\mathbb{N}} \mid \gamma^i(t) \geq 0, i \in \mathbb{N}, t \in I \right\};$$

$G : \overline{X} \to \left(L^2(I,\mathbb{R})\right)^{\mathbb{N}}$ is defined as

$$(G(x))^i(t) = g^i(t,x,\dot{x}), i \in \mathbb{N}, t \in I, x \in \overline{X}. \tag{7}$$

Theorem 2 Let $\overline{x}$ be an optimal solution of (SVP). Then there exist $\overline{\tau} \in \mathbb{R}_+$ and piecewise smooth functions $\overline{\lambda}^i : I \to \mathbb{R}, \overline{\lambda}^i \neq 0$ for finitely many $i \in \mathbb{N}$ such that

$$\overline{\tau}\phi_{\overline{x}}(t) + \sum_{i \in \mathbb{N}} \overline{\lambda}^i(t) g_{\overline{x}}^i(t) = \frac{d}{dt}\left[\overline{\tau}\phi_{\dot{\overline{x}}}(t) + \sum_{i \in \mathbb{N}} \overline{\lambda}^i(t) g_{\dot{\overline{x}}}^i(t) \right], t \in I, \tag{8}$$

$$\int_a^b \sum_{i \in \mathbb{N}} \overline{\lambda}^i(t) g^i(t,\overline{x},\dot{\overline{x}})\,dt = 0, \tag{9}$$

$$\left(\tau, \overline{\lambda} = (\overline{\lambda}^i)_{i \in \mathbb{N}} \right) \neq 0, \overline{\lambda}^i(t) \geq 0 \quad a.e., t \in I, i \in \mathbb{N}. \tag{10}$$

Proof. Since $\overline{x}$ is an optimal solution of (SVP), so is of (CCP). Therefore there exist $\overline{\tau} \in \mathbb{R}_+$ and $\overline{y} \in K'$ (topological polar cone of K) [10] such that

$$\overline{\tau}\Phi'(\overline{x}) + \overline{y}G'(\overline{x}) = 0, \tag{11}$$

$$\overline{y}G(\overline{x}) = 0. \tag{12}$$

where $\Phi'(\overline{x})$ and $G'(\overline{x})$ are Frechet derivatives of Φ and G at $\overline{x}$.

Also for every $v \in \overline{X}$,

$$(\Phi)'(\overline{x})(v) = \int_a^b \left\{ \phi_{\overline{x}}(t) v(t) + \phi_{\dot{\overline{x}}}(t) \dot{v}(t) \right\} dt \tag{13}$$

$$\left(G'(\overline{x}) \right)^i (v)(t) = g_{\overline{x}}^i(t) v(t) + g_{\dot{\overline{x}}}^i(t) \dot{v}(t), i \in \mathbb{N}, t \in I. \tag{14}$$

Since $\overline{y} \in K'$,

$$\Rightarrow \overline{y} \in \left(\left(L^2(I, \mathbb{R}) \right)^{\mathbb{N}} \right)'.$$

By Lemma 1 (proved in Section 4)

$$\overline{y} = \left(\overline{y}^i \right)_{i \in \mathbb{N}}, \overline{y}^i \in L^2(I, \mathbb{R}), \overline{y}^i \neq 0, \text{ for finitely many } i \in \mathbb{N}.$$

Let $S = \left\{ i \in \mathbb{N} \mid \overline{y}^i \neq 0 \right\}$. For $i \in S, \overline{y}^i \in L^2(I, \mathbb{R})$, by Riesz representation theorem [11] there exist $\overline{\lambda}^i \in L^2(I, \mathbb{R})$ such that

$$\overline{y}^i(h) = \int_a^b \overline{\lambda}^i(t) h(t) dt, \text{ for all } h \in L^2(I, \mathbb{R}), \tag{15}$$

for $i \notin S$, choose $\overline{\lambda}^i = 0$, therefore for any $v \in \left(L^2(I, \mathbb{R}) \right)^{\mathbb{N}}$,

$$\overline{y}(v) = \int_a^b \sum_{i \in \mathbb{N}} \overline{\lambda}^i(t) v^i(t) dt. \tag{16}$$

Substituting $v = G(\overline{x})$ in (16) and using (12), we arrive at (9).

Now it follows from (11)

$$\overline{\tau} \Phi'(\overline{x})(h) + \overline{y} G'(\overline{x})(h) = 0, \text{ for all } h \in \overline{X}. \tag{17}$$

(13) along with (16) implies

$$\int_a^b \left[\overline{\tau} \phi_{\overline{x}}(t) h(t) + \overline{\tau} \phi_{\dot{\overline{x}}}(t) \dot{h}(t) + \sum_{i \in \mathbb{N}} \overline{\lambda}^i(t) G'(\overline{x})^i(h)(t) \right] dt = 0, \text{ for all } h \in \overline{X}. \tag{18}$$

On using (14), we get

$$\int_a^b \left[\overline{\tau} \phi_{\overline{x}}(t) h(t) + \overline{\tau} \phi_{\dot{\overline{x}}}(t) \dot{h}(t) + \sum_{i \in \mathbb{N}} \overline{\lambda}^i(t) g_{\overline{x}}^i(t) h(t) + \sum_{i \in \mathbb{N}} \overline{\lambda}^i(t) g_{\dot{\overline{x}}}^i(t) \dot{h}(t) \right] dt = 0, \text{ for all } h \in \overline{X}. \tag{19}$$

$$\Rightarrow \int_a^b \left[\left(\overline{\tau} \phi_{\overline{x}}(t) + \sum_{i \in \mathbb{N}} \overline{\lambda}^i(t) g_{\overline{x}}^i(t) \right) h(t) + \left(\overline{\tau} \phi_{\dot{\overline{x}}}(t) + \sum_{i \in \mathbb{N}} \overline{\lambda}^i(t) g_{\dot{\overline{x}}}^i(t) \right) \dot{h}(t) \right] dt = 0, \text{ for all } h \in \overline{X}. \tag{20}$$

Integrating by parts the following function and using boundary condition of h,

$$\int_a^b \left[\left(\overline{\tau} \phi_{\dot{\overline{x}}}(t) + \sum_{i \in \mathbb{N}} \overline{\lambda}^i(t) g_{\dot{\overline{x}}}^i(t) \right) \dot{h}(t) \right] dt = \int_a^b -\frac{d}{dt} \left(\overline{\tau} \phi_{\dot{\overline{x}}}(t) + \sum_{i \in \mathbb{N}} \overline{y}^i(t) g_{\dot{\overline{x}}}^i(t) \right) h(t) dt. \tag{21}$$

Using above equation in (20), we get

$$\int_a^b \left[\left(\overline{\tau} \phi_{\overline{x}}(t) + \sum_{i \in \mathbb{N}} \overline{\lambda}^i(t) g_{\overline{x}}^i(t) \right) - \frac{d}{dt} \left(\overline{\tau} \phi_{\dot{\overline{x}}}(t) + \sum_{i \in \mathbb{N}} \overline{y}^i(t) g_{\dot{\overline{x}}}^i(t) \right) \right] h(t) dt = 0, \text{ for all } h \in \overline{X}. \tag{22}$$

By fundamental theorem of calculus of variation [12]

$$\left(\overline{\tau} \phi_{\overline{x}}(t) + \sum_{i \in \mathbb{N}} \overline{\lambda}^i(t) g_{\overline{x}}^i(t) \right) = \frac{d}{dt} \left(\overline{\tau} \phi_{\dot{\overline{x}}}(t) + \sum_{i \in \mathbb{N}} \overline{y}^i(t) g_{\dot{\overline{x}}}^i(t) \right), t \in I. \tag{23}$$

Claim 1: $\bar{\lambda}^i(t) \geq 0$ a.e., $t \in I$, $i \in S$.

Without loss of generality assume that $S = \{1, 2, \cdots, s\}$.

Since $\bar{y} \in K'$,

$$\Rightarrow \bar{y}(y) \geq 0, \text{ for all } y \in K, \tag{24}$$

$$\Rightarrow \int_a^b \sum_{i \in S} \bar{\lambda}^i(t) y^i(t) \, dt \geq 0, \text{ for all } y^i \in L^2(I, \mathbb{R}) \text{ such that } y^i(t) \geq 0, t \in I, i \in S. \tag{25}$$

In particular

$$\int_a^b \bar{\lambda}^1(t) y^1(t) \, dt \geq 0, \text{for all } y^1 \in L^2(I, \mathbb{R}) \text{ such that } y^1(t) \geq 0, \, t \in I. \tag{26}$$

Claim 2: $\mu(A) = 0$, where $A = \{t \in I \mid \bar{\lambda}^1(t) < 0\}$.

Let if possible $\mu(A) > 0$, then $\int_A \bar{\lambda}^1(t) \, dt < 0$.

Define $\hat{y}^1(t) = \begin{cases} 1 & t \in A \\ 0 & t \notin A. \end{cases}$

Then $\hat{y}^1(t) \geq 0, t \in I$, but $\int_a^b \bar{\lambda}^1(t) \hat{y}^1(t) \, dt = \int_A \bar{\lambda}^1(t) \, dt < 0$, a contradiction.

Hence Claim 2 holds, that is, $\bar{\lambda}^1(t) \geq 0$ a.e., $t \in I$.

Using the same argument $\bar{\lambda}^i(t) \geq 0$ a.e., $t \in I$, $i = 2, \cdots, s$. Hence claim 1 also holds.

The relations (16) are generally valid only if $\bar{\lambda}(\cdot) = (\bar{\lambda}^i(\cdot))_{i \in \mathbb{N}}$ is Schwarz distribution. Condition (8) is a linear first order differential equation for $\bar{\lambda}(\cdot)$, therefore for given $\bar{x}$, equation (8) is solvable for piecewise smooth function $\bar{\lambda}(\cdot)$ [9] [13].

Theorem 3 (Necessary Optimality Conditions) Let $\bar{x}$ be a normal efficient solution for (MSVP). Then there exist $(\bar{\tau}^1, \bar{\tau}^2, \cdots, \bar{\tau}^r) \in \mathbb{R}^r$ and piecewise smooth functions $\bar{\lambda}^i : I \to \mathbb{R}, \bar{\lambda}^i \neq 0$ for finitely many $i \in \mathbb{N}$ such that the following conditions hold:

$$\sum_{i=1}^r \bar{\tau}^i f_{\bar{x}}^i(t) + \sum_{i \in \mathbb{N}} \bar{\lambda}^i(t) g_{\bar{x}}^i(t) = \frac{d}{dt} \left[\sum_{i=1}^r \bar{\tau}^i f_{\dot{\bar{x}}}^i(t) + \sum_{i \in \mathbb{N}} \bar{\lambda}^i(t) g_{\dot{\bar{x}}}^i(t) \right], t \in I, \tag{27}$$

$$\int_a^b \sum_{i \in \mathbb{N}} \bar{\lambda}^i(t) g^i(t, \bar{x}, \dot{\bar{x}}) \, dt = 0, \tag{28}$$

$$\bar{\tau} \geq 0, \sum_{i=1}^r \bar{\tau}^i = 1, \bar{\lambda}^i(t) \geq 0 \text{ a.e., } t \in I, i \in \mathbb{N}. \tag{29}$$

Proof. This theorem can be proved by using Theorem 2 and proceeding on the similar lines of ([14], Theorem 3.4).

The following example illustrates the validity of Theorem 3.

Example 4 Consider *the problem* (P1):

$$(P1) \text{ Minimize} \left(\int_0^1 \{t - 2x(t)\}^2 \, dt, \int_0^1 \{2 - 4\dot{x}(t)\}^2 \, dt \right)$$

Subject to

$$-x(t) + \dot{x}(t) \leq \frac{k}{2}, k \in \mathbb{N}, t \in I = [0, 1], \tag{30}$$

$$x(0) = 0, \, x(1) = \frac{1}{2}. \tag{31}$$

where $x : I \to \mathbb{R}$ is a piecewise smooth state function. It is trivial that $\bar{x}(t) = \frac{t}{2}, t \in I$ is a normal efficient solution for (P1). It can be verified that there exist $\bar{\tau} = \left(\dfrac{1}{2}, \dfrac{1}{2}\right) \in \mathbb{R}^2$ and smooth functions

$\bar{\lambda}^k : I \to \mathbb{R}, \bar{\lambda}^k(t) = 0$, for $k \in \mathbb{N}, t \in I$ such that (27), (28) and (29) hold.

The following example illustrates that a feasible solution of (MSVP) fails to be a normal efficient solution if it does not satisfy any one of the necessary optimality conditions (27), (28) or (29).

Example 5 Consider the problem (P2):

$$(P2) \text{ Minimize} \left(\int_0^1 \{x(t) + \dot{x}(t)\} \, dt, \int_0^1 \{x(t) + t\} \, dt \right)$$

Subject to

$$x^2(t) + \dot{x}^2(t) \le k^2, \ k \in \mathbb{N}, \ t \in I = [0,1], \tag{32}$$

$$x(0) = 0, x(1) = 0. \tag{33}$$

where $x : I \to \mathbb{R}$ is a piecewise smooth state function. Then $\bar{x}(t) = 0, t \in I$ is feasible solution for (P2). But not a normal efficient solution, since it not satisfied condition (27) for any $(\tau^1, \tau^2) \in \mathbb{R}_+^2$ and for any piecewise smooth functions $\bar{\lambda}^i : I \to \mathbb{R}, \bar{\lambda}^i \ne 0$ for finitely many $i \in \mathbb{N}, t \in I$.

4. Topological Dual of $\left(L^2(I, \mathbb{R})\right)^{\mathbb{N}}$

Let us summarizes some basic concepts and tools to find topological dual of $\left(L^2(I, \mathbb{R})\right)^{\mathbb{N}}$.

1) $\left(L^2(I, \mathbb{R})\right)^{\mathbb{N}}$ is a Riesz space ([15], p. 313) as it is partially ordered by the pointwise ordering $f \ge g$ in $\left(L^2(I, \mathbb{R})\right)^{\mathbb{N}}$ if and only if $f^i \ge g^i$ in $L^2(I, \mathbb{R})$, for each $i \in \mathbb{N}$. Its lattice operations are given pointwise

$$(f \vee g)(t) = \left(\max\{f^1(t), g^1(t)\}, \max\{f^2(t), g^2(t)\}, \cdots \right) \tag{34}$$

$$(f \wedge g)(t) = \left(\min\{f^1(t), g^1(t)\}, \min\{f^2(t), g^2(t)\}, \cdots \right). \tag{35}$$

2) $D = \left\{ f = \left(f^i\right)_{i \in \mathbb{N}} \in \left(L^2(I, \mathbb{R})\right)^{\mathbb{N}} \mid f^i \ne 0, \text{ for finitely many } i \in \mathbb{N} \right\}$ is also a Riesz space.

3) Order dual of $\left(L^2(I, \mathbb{R})\right)^{\mathbb{N}}$ is a Riesz space ([15], Theorem 8.24).

4) $\left(L^2(I, \mathbb{R})\right)$ is a Frechet lattice, as it is Banach lattice ([15], p. 348). Since countable cartesian product of Frechet lattice is Frechet lattice ([16], Theorem 5.18) which imply $\left(L^2(I, \mathbb{R})\right)^{\mathbb{N}}$ is Frechet lattice equipped with the product topology.

5) Given $k \in \left(L^2(I, \mathbb{R})\right)^{\mathbb{N}}$ define the n-tail of $k = \left(k^i\right)_{i \in \mathbb{N}}$ by

$$k^{(n)} = \left(0, \cdots, 0, k^{n+1}, k^{n+2}, \cdots\right). \tag{36}$$

Motivated by the topological dual of $\mathbb{R}^{\mathbb{N}}$ ([15], Theorem 16.3), we now find the topological dual of $\left(L^2(I, \mathbb{R})\right)^{\mathbb{N}}$ in the following lemma.

Lemma 1 The topological dual of $\left(L^2(I, \mathbb{R})\right)^{\mathbb{N}}$ is

$$D = \left\{ f = \left(f^i\right)_{i \in \mathbb{N}} \in \left(L^2(I, \mathbb{R})\right)^{\mathbb{N}} \mid f^i \ne 0, \text{ for finitely many } i \in \mathbb{N} \right\}.$$

Proof. For any $h \in D$, define,

$$\Psi_h(k) = \langle k, h \rangle = \int_a^b \sum_{i \in \mathbb{N}} k^i(t) h^i(t) \, dt. \tag{37}$$

Clearly Ψ_h is a continuous linear functional on $\left(L^2(I, \mathbb{R})\right)^{\mathbb{N}}$.

For the converse, assume that $\Psi : \left(L^2(I, \mathbb{R})\right)^{\mathbb{N}} \to \mathbb{R}$ is continuous linear functional. The continuity of Ψ at zero element of $\left(L^2(I, \mathbb{R})\right)^{\mathbb{N}}$ guarantees that there exist $\delta > 0$ and $l > 0$ such that $k \in \left(L^2(I, \mathbb{R})\right)^{\mathbb{N}}$ and $\|k^i\| < \delta$ for $i = 1, 2, \cdots, l$ imply $|\Psi(k)| < 1$.

So for each $k \in \left(L^2(I, \mathbb{R})\right)^{\mathbb{N}}, n \left|\Psi\left(k^{(l)}\right)\right| = \left|\Psi\left(nk^{(l)}\right)\right| < 1$ for each n, hence $\left|\Psi\left(k^{(l)}\right)\right| = 0$.

For each $i \in \mathbb{N}$ define $e^i : L^2(I, \mathbb{R}) \to \left(L^2(I, \mathbb{R})\right)^{\mathbb{N}}$ as

$$e^i(f) = \left(0, 0, \cdots, \underbrace{f}_{i\text{th place}}, \cdots, 0, 0\right). \tag{38}$$

Then $\Psi_o e^i : L^2(I, \mathbb{R}) \to \mathbb{R}$ is a continuous linear functional.
By Riesz representation theorem, for $i = 1, 2, \cdots, l$ there exist $\lambda^i \in L^2(I, \mathbb{R})$ such that

$$\Psi_o e^i(\alpha) = \int_a^b \alpha(t) \lambda^i(t) \, dt, \text{ for all } \alpha \in L^2(I, \mathbb{R}). \tag{39}$$

Now let $h = \left(\lambda^1, \lambda^2, \lambda^3, \cdots, \lambda^l, 0, \cdots, 0\right) \in D$,

note that $\Psi(k) = \sum_{i=1}^{l} \Psi_o e^i \left(k^i\right) = \sum_{i=1}^{l} \int_a^b k^i(t) \lambda^i(t) \, dt = \langle k, h \rangle$, for each $k \in \left(L^2(I, \mathbb{R})\right)^{\mathbb{N}}$.

That is $\Psi = \Psi_h$ and h is uniquely determined.
Now, if $h \geq 0$, then $\Psi_h(k) \geq 0$, for all $k \geq 0$
Conversely, proceeding similarly as in claim 1 of Theorem 2, it can be shown that
if $\Psi_h(k) \geq 0$, for all $k \geq 0$ then $h \geq 0$.

This infers $h \to \Psi_h$ is a lattice isomorphism from D onto $\left(\left(L^2(I, \mathbb{R})\right)^{\mathbb{N}}\right)'$.

Hence $D = \left(\left(L^2(I, \mathbb{R})\right)^{\mathbb{N}}\right)'$ ([15], Theorem 9.11).

5. Conclusion

In this paper, we have developed necessary optimality conditions for a Semi-Infinite Variational Problem. These optimality conditions are further extended to Multi-objective Semi-infinite Variational Problem (MSVP) as Theorem 3. The results proved in this article are significant for the growth of optimality and duality theory for the class of semi-infinite variational problems. An example is presented to demonstrate the validity of the theorem proved. Another example illustrates that a feasible solution of (MSVP) fails to be a normal efficient solution if it does not satisfy any one of the necessary optimality conditions stated in the theorem. Vital part of the result depends on the topological dual of $\left(L^2(I, \mathbb{R})\right)^{\mathbb{N}}$ which was proved as a lemma in the last section.

Acknowledgements

We thank the Editor and the referee for their comments. The first author was supported by Council of Scientific and Industrial Research, Junior Research Fellowship, India (Grant no 09/045(1350)/2014-EMR-1).

References

[1] Lopez, M. and Still, G. (2007) Semi-Infinite Programming. *European Journal of Operational Research*, **180**, 491-518. http://dx.doi.org/10.1016/j.ejor.2006.08.045

[2] Kanzi, N. and Nobakhtian, S. (2010) Optimality Conditions for Non-Smooth Semi-Infinite Programming. *Optimization*, **59**, 717-727. http://dx.doi.org/10.1080/02331930802434823

[3] Mishra, S.K., Jaiswal, M. and An, L.T.H. (2012) Duality for Nonsmooth Semi-Infinite Programming Problems. *Optimization Letters*, **6**, 261-271. http://dx.doi.org/10.1007/s11590-010-0240-8

[4] Mond, B. and Hanson, M.A. (1967) Duality for Variational Problems. *Journal of Mathematical Analysis and Applications*, **18**, 355-364. http://dx.doi.org/10.1016/0022-247x(67)90063-7

[5] Hanson, M.A. (1964) Bounds for Functionally Convex Optimal Control Problems. *Journal of Mathematical Analysis and Applications*, **8**, 84-89. http://dx.doi.org/10.1016/0022-247x(64)90086-1

[6] Bhatia, D. and Kumar, P. (1996) Duality for Variational Problems with B-Vex Functions. *Optimization*, **36**, 347-360. http://dx.doi.org/10.1080/02331939608844189

[7] Bhatia, D. and Mehra, A. (1999) Optimality Conditions and Duality for Multiobjective Variational Problems with Generalized B-Invexity. *Journal of Mathematical Analysis and Applications*, **234**, 341-360.

http://dx.doi.org/10.1006/jmaa.1998.6256

[8]   Antczak, T. (2014) On Efficiency and Mixed Duality for a New Class of Nonconvex Multiobjective Variational Control Problems. *Journal of Global optimization*, **59**, 757-785. http://dx.doi.org/10.1007/s10898-013-0092-8

[9]   Husain, I. and Jabeen, Z. (2005) On Variational Problems Involving Higher Order Derivatives. *Journal of Applied Mathematics and Computing*, **17**, 433-455. http://dx.doi.org/10.1007/BF02936067

[10] Luc, D.T. (1989) Theory of Vector Optimization, Lecture Notes in Economics and Mathematical Systems, Vol. 319. Springer-Verlag, New York, 75.

[11] Kreyszig, E. (1989) Introductory Functional Analysis with Applications. Wiley Classics Library, Wiley, New York, 188.

[12] Gelfand, I.M. and Fomin, S.V. (1963) Calculas of Variations. Prentice-Hall, Inc., Englewood Cliffs, 34-35.

[13] Chandra, S., Craven, B.D. and Husain, I. (1985) A Class of Non-Differentiable Continuous Programming Problems. *Journal of Mathematical Analysis and Applications*, **107**, 122-131. http://dx.doi.org/10.1016/0022-247X(85)90357-9

[14] Mititelu, S. and Stancu-Minasian, I.M. (2009) Efficiency and Duality for Multiobjective Fractional Variational Problems with (ρ,b)-Quasiinvexity. *Yugoslav Journal of Operations Research*, **19**, 85-99. http://dx.doi.org/10.2298/yjor0901085m

[15] Aliprantis, C.D. and Border, K.C. (1999) Infinite Dimensional Analysis, A Hitchhiker's Guide. Springer-Verlag, Berlin Herdelberg, 313, 324, 348, 352, 528. http://dx.doi.org/10.1007/978-3-662-03961-8

[16] Aliprantis, C.D. and Burkinshaw, O. (2003) Locally Solid Riesz Spaces with Applications to Economics. American Mathematical Society, Providence, 127. http://dx.doi.org/10.1090/surv/105

10

Using Harmonic Mean to Solve Multi-Objective Linear Programming Problems

Nejmaddin A. Sulaiman, Rebaz B. Mustafa*

Department of Mathematics, College of Education, University of Salahaddin, Erbil, Iraq
Email: *bahramrebaz@yahoo.com

Abstract

In this paper, we have suggested a new technique to transform multi-objective linear programming problem (MOLPP) to the single objective linear programming problem by using Harmonic mean for values of function and an algorithm is suggested for its solution, the computer application of algorithm has been demonstrated by solving some numerical examples. We have used some other techniques, such as (sen, arithmetic mean, median) to solve the same problems, the results in Table 3 indicate that the new technique in general is promising.

Keywords

MOLPP, Harmonic Mean

1. Introduction

Linear programming is a relatively new mathematical discipline, dating from the invention of the simplex method by G. B. Dantzig in 1947. He proposed the simplex algorithm as an efficient method to solve a linear programming problem.

A multi-objective linear programming problem is introduced by Chandra Sen [1] and suggests an approach to construct the multi-objective function under the limitation that the optimum value of individual problem was greater than zero. [2] studied the multi-objective function by solving the multi-objective programming problem, using mean and mean value. [3] solved the multi objective fractional programming problem by Chandra Sen's technique. In order to extend this work, we have defined a multi-objective linear programming problem and in-

---

*Corresponding author.

vestigated the algorithm to solve linear programming problem for multi-objective functions. By new technique, we use harmonic mean (HM) of the values of objective functions. The computer application of our algorithm has also been discussed by solving some numerical examples. Finally we have showed results and comparison among the new technique and Chandra Sen's approach [1] and Sulaiman's approach [2].

2. Mathematical Definition of Multi-Objective Programming Problems (MOPP)

A deterministic (MOPP) model is usually formulated to maximize and/or minimize several objectives simultaneously subject to a constraint set with "$\geq$" and/or "$\leq$" relationships the equality constraints may be expressed as a combination of both of inequality constraints.

Mathematically, the MOPP problems can be defined as:

$$\left. \begin{array}{ll} \text{Max. } f_i = c_i X + \alpha_i & i = 1, \cdots, r \\ \text{Min. } f_i = c_i X + \alpha_i & i = r+1, \cdots, s \end{array} \right\} \tag{1.1}$$

subject to:

$$AX \begin{bmatrix} \geq \\ = \\ \leq \end{bmatrix} B \tag{1.2}$$

$$X \geq 0 \tag{1.3}$$

where x is an n-dimensional vector of decision variables c is n-dimensional vector of constants, B is m-dimensional vector of constants, r is the number of objective function to be maximized, s the number of objective function to maximized plus minimized, $(s-r)$ is the number of objective that is to be minimized, A is a $(m \times n)$ matrix of coefficients all vectors are assumed to be column vectors unless transposed, $\alpha_i \ (i = 1, \cdots, s)$ are scalar constants, $c_i X + \alpha_i \ (i = 1, \cdots, s)$ are linear factors for all feasible solutions [3].

If $\alpha_i = 0$; for all $i = 1, \cdots, s$, then the mathematical form become:

$$\left. \begin{array}{ll} \text{Max. } f_i = c_i X + \alpha_i & i = 1, \cdots, r \\ \text{Min. } f_i = c_i X + \alpha_i & i = r+1, \cdots, s \end{array} \right\} \tag{2.1}$$

subject to:

$$AX = B \tag{2.2}$$

$$X \geq 0 \tag{2.3}$$

The problem said to be multi-objective linear programming problem (MOLPP) if all the objective functions and constraint functions are linear, and all the variables are continuous variables.

3. The New Technique for Solving MOLPP by Using Harmonic Mean

Before solving MOLPP, and preface an algorithm to it, we will need to define Harmonic Mean.

Harmonic Mean [4]

Harmonic mean of a set of observations is defined as the reciprocal of the arithmetic average of the reciprocal of the given values. If $x_1, x_2, \cdots, x_n$ are n observations, $HM = \dfrac{n}{\sum_{i=1}^{n} \left(\dfrac{1}{x_i} \right)}$

4. Multi-Objective Functions Formulation

Suppose we optimize (maximize or minimize) all the objective functions individually in (2.1), (2.2) and (2.3) and obtain the values as follows.

$$\left.\begin{array}{l} \text{Max. } f_1 = \Psi_1 \\ \text{Max. } f_2 = \Psi_2 \\ \quad\vdots \\ \text{Max. } f_r = \Psi_r \\ \text{Min. } f_{r+1} = \Psi_{r+1} \\ \quad\vdots \\ \text{Min. } f_s = \Psi_s \end{array}\right\} \tag{3.1}$$

where $\Psi_i, (i = 1, 2, \cdots, s)$ are the values of objective functions.

We require the common set of decision variable to be the best compromising optimal solution [5]. Here we can determine the common set of decision variables from the following combined objective function.

Formulate the multi-objective linear programming problem given in (1.1) can be translated by our technique to:

$$\text{Max. } F = \sum_{k=1}^{r} \left(\text{Max. } f_k\right)\big/ Hm_1 - \sum_{k=r+1}^{s} \left(\text{Min. } f_k\right)\big/ Hm_2 \tag{3.2}$$

where

$$Hm_1 = r\big/\left(\sum_{i=1}^{r} 1/\Psi_i\right), \quad Hm_2 = (s-r)\big/\left(\sum_{i=r+1}^{s} 1/\Psi_i\right) \tag{3.3}$$

And $\Psi_i \neq 0$; $(i = 1, 2, \cdots, s)$ subject to the same constraints (1.2), (1.3) and the optimum value of the functions Ψ_k $(k = 1, 2, \cdots, s)$ may be positive or negative, Hm_1 the value of harmonic mean of maximized Ψ_k $(k = 1, 2, \cdots, r)$ and Hm_2 the value of harmonic mean of minimized Ψ_k $(k = r+1, \cdots, s)$. $s - r \geq 2$, if $s - r = 1$ then the combined formula (3.2) becomes.

$$\text{Max. } F = \sum_{k=1}^{r} \left(\text{Max. } f_k\right)\big/ Hm_1 - \left(\text{Min. } f_{r+1}\right)\big/ \left|\Psi_{r+1}\right|$$

If $s - r = 0$ then the function $\text{Max. } F = \sum_{k=1}^{r} \left(\text{Max. } f_k\right)\big/ Hm_1$. We can solve this (MOLPP) by Chandra Sen's approach [1]-[3] by using mean and median and algorithms in above researches for solving MOLPP as explained in [1]-[3].

4.1. Algorithm

This algorithm is to obtain the optimal solution for the MOLPP defined previously can be summarized as follows.

Step 1: Assign arbitrary values to each of the individual objective functions that to be maximized and minimized.

Step 2: Solve the first objective function { $\text{Max. } f_1$ subject to constraints (1.2) and (1.3)} by simplex method.

Step 3: Check the feasibility of the solution in step 2, if it is feasible then go to step 4, otherwise, use dual simplex method to remove infeasibility.

Step 4: Assign a name to the optimum value of the first objective function f1 say Ψ_1

Step 5: Repeat step 2, for $i = 1, \cdots, r$ for the kth objective function, for all $i = i + 1, \cdots, s$

Step 6: Determine Harmonic Mean Hm1 for $\Psi_i, i = 1, \cdots, r$ and Hm_2 for $i = r+1, \cdots, s$

Step 7: Optimize the combined objective function $\text{Max. } F = \sum_{k=1}^{r} \left(\text{Max. } f_k\right)\big/ Hm_1 - \sum_{k=r+1}^{s} \left(\text{Min. } f_k\right)\big/ Hm_2$ under the same constraints (1.2) and (1.3) by repeating Steps 2-4.

4.2. Used Notation

The following notations were used in our algorithm:

$$HA_i = \left|\Psi A_i\right|, \quad \forall i = 1, 2, \cdots, r$$

$$HL_i = \left|\Psi L_i\right|, \quad \forall i = r+1, \cdots, s$$

where ΨA_i = The value of objective function which is to be maximized, and
$\quad\quad \Psi L_i$ = The value of objective function which is to be minimized.
$\quad\quad Hm_1$ = The value of Harmonic mean of maximized $\left(Hm_1 = r \big/ \left(\sum_{i=1}^{r} 1/\Psi_i \right) \right)$.

$\quad\quad Hm_2$ = The value of Harmonic mean of minimized $\left(Hm_2 = (s-r) \big/ \left(\sum_{i=r+1}^{s} 1/\Psi_i \right) \right)$.

$$SL = \sum_{i=1}^{r} \text{Max. } f_i; \quad SS = \sum_{i=r+1}^{s} \text{Min. } f_i$$

$$\text{Max. } F = SL/Hm_1 - SS/Hm_2$$

4.3. Numerical Examples

Ex. (1)

$$\left.\begin{array}{l} \text{Max. } f_1 = x_1 + 2x_2 \\ \text{Max. } f_2 = 3x_1 \\ \text{Min. } f_3 = -2x_1 - 3x_2 \\ \text{Min. } f_4 = -x_2 \end{array}\right\} \tag{4.1}$$

s.to:

$$\left.\begin{array}{l} 6x_1 + 8x_2 \leq 48 \\ x_1 + x_2 \geq 3 \\ x_1 \leq 4 \\ x_2 \leq 3 \\ x_1, x_2 \geq 0 \end{array}\right\} \tag{4.2}$$

Solution:
After finding the value of each of individual objective function by simplex method the results as below in (**Table 1**): $\{\Psi_1 = 10, \Psi_2 = 4, \Psi_3 = -17, \Psi_4 = -3\}$ by using Harmonic Mean technique (3.3) we get $Hm_1 = 80/14$ and $Hm_2 = 51/10$
After that using equation (3.2) for transform we get:

$$\left.\begin{array}{l} \text{Max. } F = 0.7421x_1 + 1.1343x_2 \\ \text{Subject to given constraints } (5.2): \end{array}\right\} \tag{4.3}$$

Solving (4.3) by simplex method we get:

$$\text{Max. } F = 6.3715 \quad x_1 = 4, \ x_2 = 3$$

Solve (4.1) by:
1. Using Chandra Sen's approach, [1]: after convert the MOLPP to the single objective problem we get Max. $F = 0.4676x_1 + 0.7098x_2$ subject to the same constraints (4.2) by simplex method it is optimal solution Max. $F = 3.9999$ $x_1 = 4, x_2 = 3$.

Table 1. Results of example (1).

I	Ψ_i	X_i	HA_i	HL_i	Hm_1	Hm_2
1	10	(4,3)	10			
2	4	(4,3)	4		80/14	
3	−17	(4,3)		17		
4	−3	(4,3)		3		51/10

2. Using modified approach, [2]:

A-using Mean: after convert the MOLPP to the single objective problem we get   Max. $F = 0.4857x_1 + 0.7857x_2$ subject to the same constraints (4.2) by simplex method it is optimal solution   Max.$F = 4.2999$ $x_1 = 4, x_2 = 3$

B-using Median: after convert the MOLPP to the single objective problem we get

Max. $F = 0.4857x_1 + 0.7857x_2$ subject to the same constraints (4.2) by simplex method it is optimal solution Max. $F = 4.2999$ $x_1 = 4, x_2 = 3$.

Ex. (2)

$$\left.\begin{aligned} &\text{Max. } f_1 = x_1 \\ &\text{Max. } f_2 = 2 + x_1 + 2x_2 \\ &\text{Max. } f_3 = 3 + x_2 \\ &\text{Min. } f_4 = -3x_2 \\ &\text{Min. } f_5 = -x_1 - 3x_2 \end{aligned}\right\} \tag{5.1}$$

Subject to:

$$\left.\begin{aligned} &2x_1 + 3x_2 \le 6 \\ &x_1 \le 4 \\ &x_1 + 2x_2 \le 2 \\ &x_1, x_2 \ge 0 \end{aligned}\right\} \tag{5.2}$$

Solution:

After finding the value of each of individual objective function by simplex method the results as below in (**Table 2**): $\{\Psi_1 = 2, \Psi_2 = 4, \Psi_3 = 4, \Psi_4 = -3 \text{ and } \Psi_5 = -3\}$ by using Harmonic Mean technique (3.3) we get $Hm_1 = 3$ and $Hm_2 = 3$.

After that using equation (3.2) for transform we get:-

$$\left.\begin{aligned} &\text{Max. } F = 5 + 3.5x_1 + 12x_2 \\ &\text{Subject to given constraints} (6.2): \end{aligned}\right\} \tag{5.3}$$

Solving (5.3) by simplex method we get:

$$\text{Max. } F = 4.6 \quad x_1 = 0, \ x_2 = 1$$

Solve (5.1) by:

1. using Chandra Sen approach, [1]: after convert the MOLPP to the single objective problem we get Max. $F = 1.25 + 1.08333x_1 + 2.75x_2$ subject to the same constraints (5.2) by simplex method it is optimal solution Max. $F = 4$ $x_1 = 0, x_2 = 1$.

2. using modified approach, [2]:

A-using Mean: after convert the MOLPP to the single objective problem we get

Max. $F = 1.5 + 0.93333x_1 + 2.9x_2$ subject to the same constraints (5.2) by simplex method it is optimal solution Max. $F = 4.4$ $x_1 = 0, x_2 = 1$.

B-using Median: after convert the MOLPP to the single objective problem we get

Table 2. Results of example (2).

I	Ψ_i	X_i	HA_i	HL_i	Hm_1	Hm_2
1	2	(2, 0)	2			
2	4	(0, 1)	4		3	
3	4	(0, 1)	4			
4	−3	(0, 1)		3		3
5	−3	(0, 1)		3		

Table 3. Comparison between results obtained by different numerical approaches.

Examples	Chandra Sen's approach	Modified approach		New approach using harmonic mean
		Using mean	Using median	
Example (1)	Max. $F = 3.9999$ $x_1 = 4, x_2 = 3$	Max. $F = 4.2999$ $x_1 = 4, x_2 = 3$	Max. $F = 4.2999$ $x_1 = 4, x_2 = 3$	Max. $F = 6.3715$ $x_1 = 4, x_2 = 3$
Example (2)	Max. $F = 4$ $x_1 = 0, x_2 = 1$	Max. $F = 4.4$ $x_1 = 0, x_2 = 1$	Max. $F = 4$ $x_1 = 0, x_2 = 1$	Max. $F = 4.6$ $x_1 = 0, x_2 = 1$

Max. $F = 1.25 + 0.83333x_1 + 2.75x_2$ subject to the same constraints (6.2) by simplex method it is optimal solution Max. $F = 4$ $x_1 = 0, x_2 = 1$.

5. Conclusions

Our aim was to develop an approach for solving multi-objective programming problem (MOLPP) and to suggest a new algorithm to convert the MOLPP into a single LPP by using harmonic mean of the values of objective functions and its computer application by using programming mathematical language (Matlab Programming). Moreover, we used different methods to solve the problems, and applied our technique and the other methods to the same examples in order to compare the results.

From this comparison, we observed that our technique gave identical results with that obtained by the other methods, for this see **Table 3**. So we conclude that this method is better than other methods considered in solving MOLP problems.

References

[1]   Chandra, S. (1983) A New Approach Objective Planning. *The Indian Economic Journal*, **30**, 91-96.

[2]   Sulaiman, N.A. and Sadiq Gulnar, W. (2006) Solving the Multi-Objective Programming Problem; Using Mean and Median Value. *Ref. J of Comp & Math's*, **3**.

[3]   Abdul-Kadir, M.S. and Sulaiman, N.A. (1993) An Approach for Multi-Objective Fractional Programming Problem. *Journal of College of Education, University of Salahaddin*, **3**, 1-5.

[4]   Jothikumar, J. (2004) STATISTICS. 60 G.S.M Paper, p.100.

[5]   Azapagic, A. and Clift, R. (1999) Life Cycle Assessment and Multi Objective Optimization. *Journal of Cleaner Production*, **7**, 135-143. http://dx.doi.org/10.1016/S0959-6526(98)00051-1

11

System Dynamics Approach in LCA for PET-Renewable Raw Materials Impact

Pablo Aarón Anistro Jiménez, Carlos E. Escobar Toledo

Department of Chemical Engineering, National University of Mexico (UNAM), Mexico City, Mexico
Email: carloset@unam.mx

Abstract

It has been shown that system dynamics and life-cycle analysis together provide an analytical tool to study different impacts between variables in decision-making process. This paper presents an application of this approach toward to a plastic material of quotidian use, considering both, a production chain using renewable resources and petrochemical production of Polyethylene Terephthalate, used as a beverage bottle application, and showing its impacts to the system when production replacement of non-renewable resources is allowed. We have already studied the substitution between different materials: glass and aluminum.

Keywords

System Dynamics, Life Cycle Analysis (LCA), Dynamical Modeling, Polyethylene Terephthalate (PET)

1. Introduction

The life cycle analysis (LCA) of a plastic resin, can respond to complex behaviors of market interventions and technological change, as well as, to social requirements expressed during governmental policies formulation.

System dynamics is a discipline that provides distinctive concepts and tools to allow the study of accumulation and feedback effects in complex systems. These effects can make a system to behave in complicated forms very different from the expected behavior compared with static or linear models [1]. The system dynamics models focus on the structure and behavior of systems composed of interacting feedback loops. System dynamics models are different from other optimization and equilibrium models as it focuses on the disequilibrium dynamics and complexities, through decision rules, stocks and flows.

The strength of system dynamics consists in its ability to examine how the system structure influences decisions and how systems react to these decisions over time.

As said early, systems dynamics models allow exploration of futures and justify the development of possible scenarios.

In the present study, the relevant technological routes are: 1) PET production from renewable resources and its counterpart, production from oil and gas, non-renewable resources, and, 2) recovered and recycled material coming from the packaging industry.

System dynamics models are particularly useful in LCA studies, in order to estimate the environmental impacts of future changes in existing products [2]-[4]. They are also support to the probability of success estimation to product changes and assist in identifying the uncertainty sources in order to find robust solutions. Essentially, system dynamics brings temporality features and feedback to quantitative indicators from LCA.

2. Plastics Dynamics

Man-made polymers, obtained from natural sources, were displaced by those of petrochemical origin considering the petrochemical industry growth, mainly in the 1930's. It is from the 1980's and particularly the 1990's, biopolymers, and natural fibers, have increased their participation and significance in a growing variety of applications. Raw materials coming from biological processes are employed in the production of biodegradable materials, as well as non-biodegradable materials. Bio-materials superiority in terms of environmental impact has been the main driving force that has generated this increase, and it is expecting to be maintained in the future.

Reviews to different LCA studies applied to polymers obtained from biological and petrochemical resources, [5]; Patel [6], reveals questionable assumptions and uncertainties, as for example, the framework used to perform LCA, in which the commonly used approach is known as cradle to factory gate. This excludes relevant stages of materials flows to get to its final disposal. Those stages include the consideration of various options for waste treatment and recycling, due to high impact in final result.

Between the relevant results of LCA studies reviewed by Patel to different materials, materials produced from biological sources, clearly contribute to energy saving and greenhouse gases (GHG) reduction; however, at the same time it is clear that performance of these materials will not exceed its synthetic counterpart.

On the other hand, Pilz [7] has considered some of the weaknesses identified in previous studies, highlighting the approach, now from the cradle to the grave, to quantify plastics products impacts throughout the full life cycle; the study does not intent to set a premise of superiority of the plastics over other materials, since , all materials have characteristics that make them well suited to a greater or lesser extent of applications, in many cases, solution to a specific application , in terms of efficiency, is a combination of different materials.

In this approach, material selection will depend on factors such as the deposition impact or a suitable waste recovery system. In these cases, the solution is dependent on the country and is attached to the number or proportion of applications for the material. However this study does not address plastics from renewable sources because today market low impact recognizes the future role of the materials from renewable sources on plastics industry.

The study makes distinction of two production categories for renewable sources plastics. First, the monomer production to polymer manufacture, where the commercial challenge is competing against high-volume production, in terms of production economics and processing equipment adaptation. Second, monomer high volume production from renewable sources, provide raw material to polymerization plants. In both cases, chemical mechanisms are tested and presented important developments at an increasing rate of growth [8]. In the present work is considered this second category which presents an important technological development.

Methods

The present study is constituted as a contribution to the methodology developed by Escobar [9], in order to incorporate system dynamics approach [10], to the analysis of plastic products systems **Figure 1**, prior to exergetic and multicriteria analysis proposed by Mejia [11]. Major interest has been developed in plastics life cycleanalysis and research, to minimize their deposition and therefore its environmental impact. The proposed model is focused to bottle PET resin behavior.

Using multicriteria decision making we have also studied the possibility of substitution between glass and aluminum. The criteria used in this case are the following:
  1) Total irreversibility (MJ/fu).

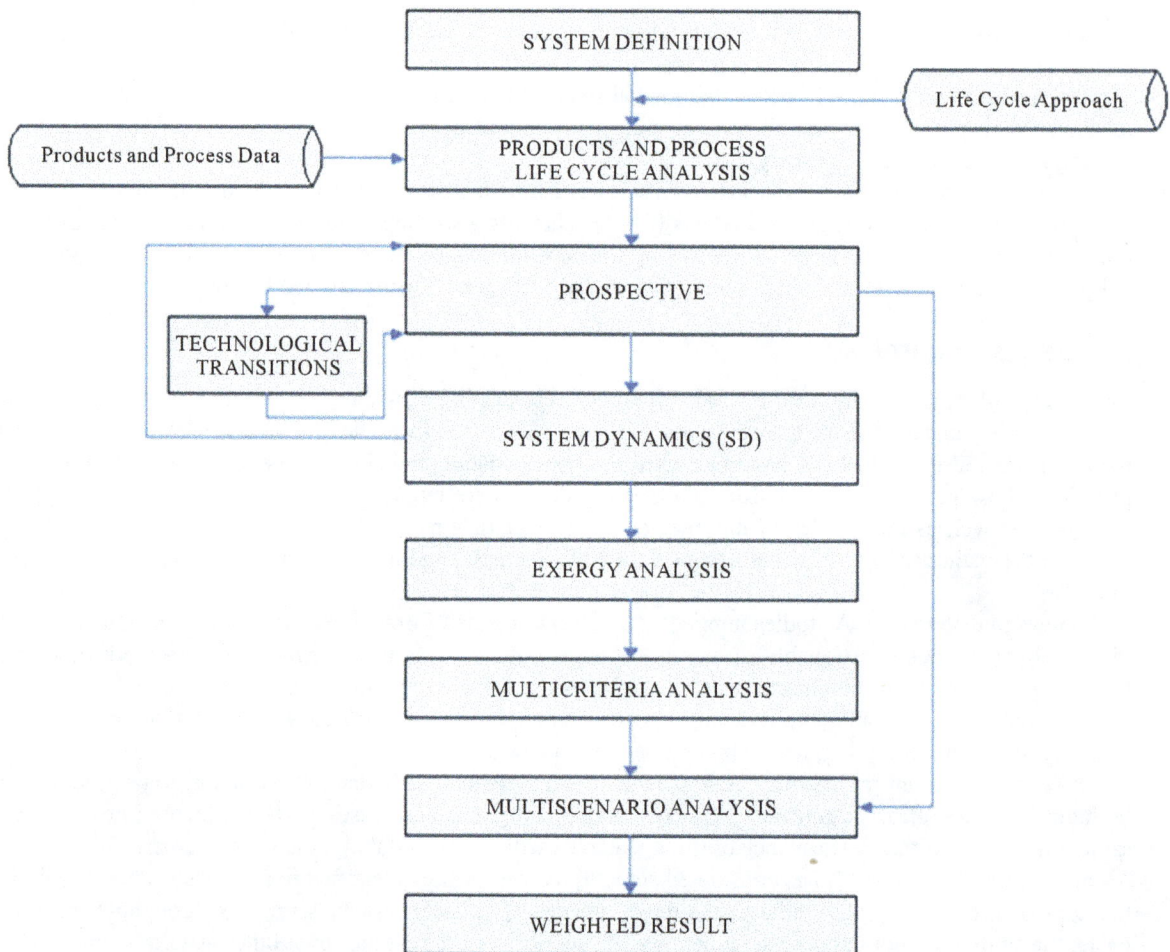

Figure 1. General description of methodology with system dynamics incorporation.

2) Energy consumption (MJ/fu) to be interpreted also as the measure of natural resources.

3) GHG emissions (Ton. CO_{2eq}).

4) Profit ($/fu).

The alternatives studied are divided into two groups: bottles made from primary materials (virgin) and secondary materials (recycled) for each of the three types of packaging materials considered.

Figure 2 shows the resulting multicriteria ranking that has been analyzed with Visual PROMETHEE.

3. Present Research

The present research, focus on both: the impact of PET production via renewable resources to the PET life cycle [12], and the growth of recycled PET on its market. This market presents a strong material recovery rate and an increasing proportion of this material reincorporates to bottle production [13].

Within the system there are a lot of relationships that influence the variables already described. Udo de Hæs [14] emphasizes that attempt to take into account the dynamic effects of all processes involved in the life cycle, makes modeling work very complex and data requirements very extensive. Therefore modeling a small amount of key variables of the system is suggested as better solution.

The present study is an example of the approach mentioned above, where the model is built using first, the influence relationship between the key variables through structural analyses [15]. The model incorporates three dynamic processes, the first corresponds to PET production via hydrocarbons; the second, to the PET production via renewable resources, and the third, material recycling from PET waste deposition [16] [17].

The projection horizon is a long term one, considering 50 years.

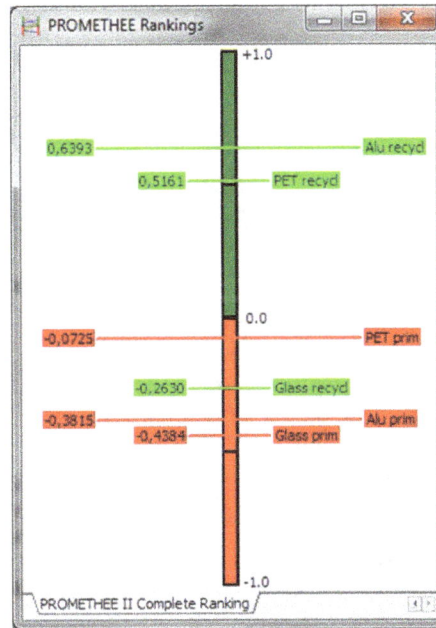

Figure 2. PROMETHEE ranking.

The data was obtained from NAPCOR, National Association for PET Container Resources. See reference [13].

The model assumes that:

1) We will study the influence of the rate of recycled material for bottles production, Equation (1), on the consumption annual growth rate for PET bottles, Equation (2), are function of the simulation time, Equation (3), according to NAPCOR's postconsumer PET reports data.

2) The presence of delays in the system due to renewable sources and recovery technologies.

$$BUR = 0.049442x^{0.541451} \tag{1}$$

$$AGR = \frac{0.17972}{x^{0.508789}} \tag{2}$$

$$x = time + TIMESTEP \tag{3}$$

where *time* corresponds to continuous standard time and *TIMESTEP* corresponds to integration method time step model selected, avoiding indetermination problems.

Figure 3 displays consumption annual growth rate behavior. It is clearly an initial boom on PET bottles consumption growth. Nevertheless, this growth reduced, mainly due to optimization of bottles design, recovery and recycled materials. On the other hand, PET bottle utilization rate, shows constant growth consumption due to increasingly interest and benefits on recycled and reformed material and the technologies involved.

To perform the calculations, a stock-and-flow model was built using VENSIM PLE® (version 5.11A) modeling software. Model structure is presented in **Figure 4**.

We should consider that material recovery feedback loop is turn off, then defining "recovery efficiency = 0". Then, if recycling loop is active, "recovery efficiency" is considered 30.0%.

On the other hand virgin material production process from renewable sources is subject to the definition of participation percentage or desired objective, into total virgin material production, allows disabling this process, if other scenario is required. In this study the renewable resources participation is 10%.

The application to the present research which focus on the impact of PET production via renewable resources on the PET life cycle and the growth of recycled PET on its market. This market presents a strong material recovery rate and an increasing proportion of this material reincorporates to bottle production. Then the results and its application are based on **Figure 4**, stock and flow model.

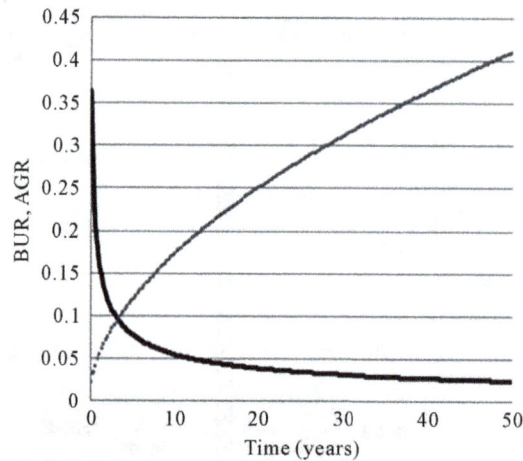

Figure 3. Annual growth rate and bottle utilization rate functions.

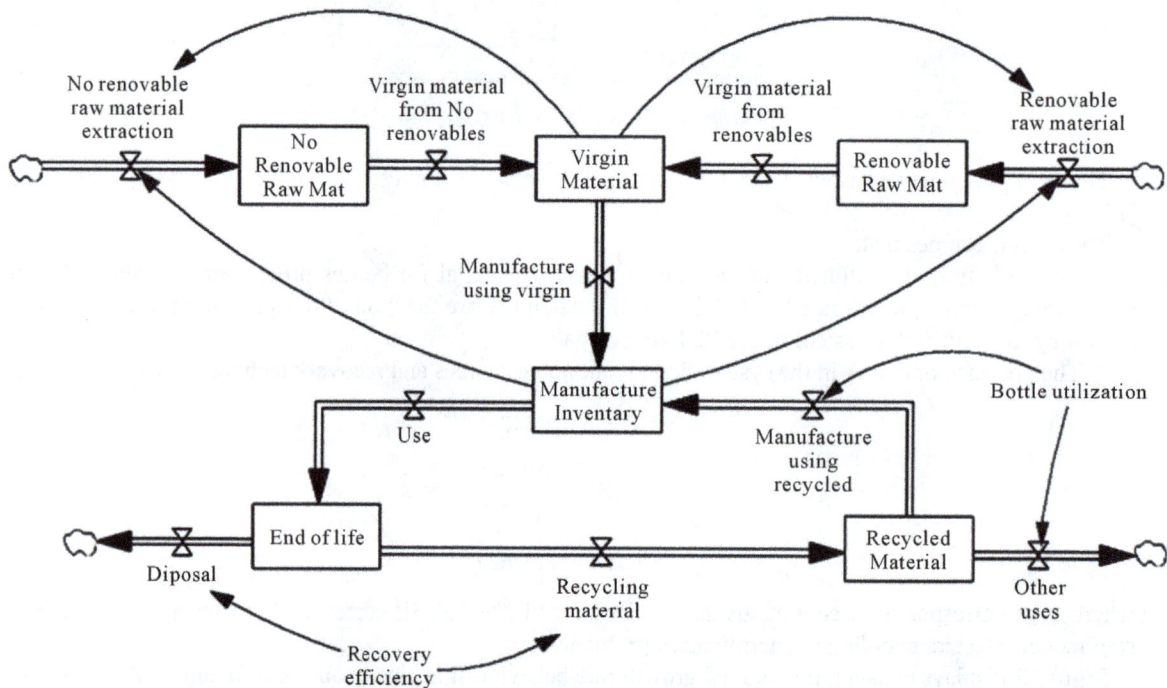

Figure 4. Simplified hypothetical stock-and-flow model showing feedback, delays structures and associated variables.

4. Results

A prospective study *i.e.* System Dynamics for Polyethylene Terephthalate as bottles has been performed, focused on petrochemical origin polymers and biopolymers with a horizon of 50 years.

The raw materials of petrochemical origin will go continue because oil and natural gas undergo major changes as depletion or replacement of its uses. Therefore, in this study we will always use petrochemical PET in some proportion according to model.

Biopolymers manufacturing technology and marketing will grow according with three scenarios, representing policies and/or action application strategies visualizing a petrochemical resources decline at year 30 of the planning horizon. These scenarios are described in **Figure 7**, **Figure 8** and **Figure 9**. To establish those three scenarios, it was necessary to perform a sensitivity analysis of the following variables required by the model in the mass balance:

1) Production of petrochemical inputs for PET: ethylene glycol and p-xylene.

2) PET production from renewable inputs, e.g. biomass.

3) The added output by recycling bottles whose production can be any of the previous two sources.

4) The production of recycled PET, whose consumption is dedicated to other uses different of packaging.

5) The PET not recovered and therefore moved to deposition assuming that these quantities will serve to generate electricity.

Figure 5 show the information fed into the model to perform sensitivity analysis, into the modified parameters were:

- Ef reco.—Presents the greatest impact on material flows as it depends on the availability of the material collection for recycling and therefore the replacement of virgin material.

- O renov inic.—Strong influence on the proportions of materials of petrochemical origin vs. the renewable origin.

- T assimilation technology.—A minor delay in implementing technological contribution is more significant; however, the full impact remains relatively low due to the limitations of input sources.

- T recol.—Shorter time for recollection of material for recycling, more deposition is mitigated, promoting the amount of recycled material then impacting the production of virgin material.

- Ret bene.—To obtain different benefits as economic, political, social or other in less time (delay reduction), promotes the implementation of policies for recycling in the previous five years.

Figure 6 describes the results of the behavior in time of different PET flows accordingly to multi-parameter evaluation, where "Virgin Materials From No Renewables" and "Virgin Materials From Renewables" are new PET production flows respectively via petrochemical and bio-resources; "Other Uses" corresponds to recycled PET that is used in low performance applications like textile and scrub brushes; and, "Disposal" is the PET flow that ends in landfills. Blue, green, gray and yellow stripes are probability boundaries of PET flow in the proposed range of 50 years, where random uniform distributions are applied to model parameters evaluation. For example, the area between green stripes, which includes yellow and green stripes, represents 75% of possible PET flow variable profiles in the planning horizon. "Current" corresponds to the results associated to the initial values of the parameters above mentioned.

In **Figure 7**, the growth rate at which renewable technology is available to integrate the supply of the polymer in the medium term is shown as a red line. This technology would be ready to use 10 years before the possible decline of petrochemical-based polymers. In such a manner that when petrochemical input is declining, the total supply would be 50% for each raw material source.

The supply intersection among petrochemicals and biopolymers will have as a consequence, policies and/or actions that will break the intersection in favor of biopolymers, will require 30 years to complete the planning horizon of 50 years, achieving then, a contribution of 80% of virgin biomaterial to total PET production. The remaining 20% will be petrochemical material.

In the case of PET bottles in this first scenario, the total offer consists of virgin and recycled material. For example, in the year 50 the total supply will be 290,000 tons of PET bottles; 259,000 will correspond to virgin

Figure 5. Sensitivity simulation parameters setup.

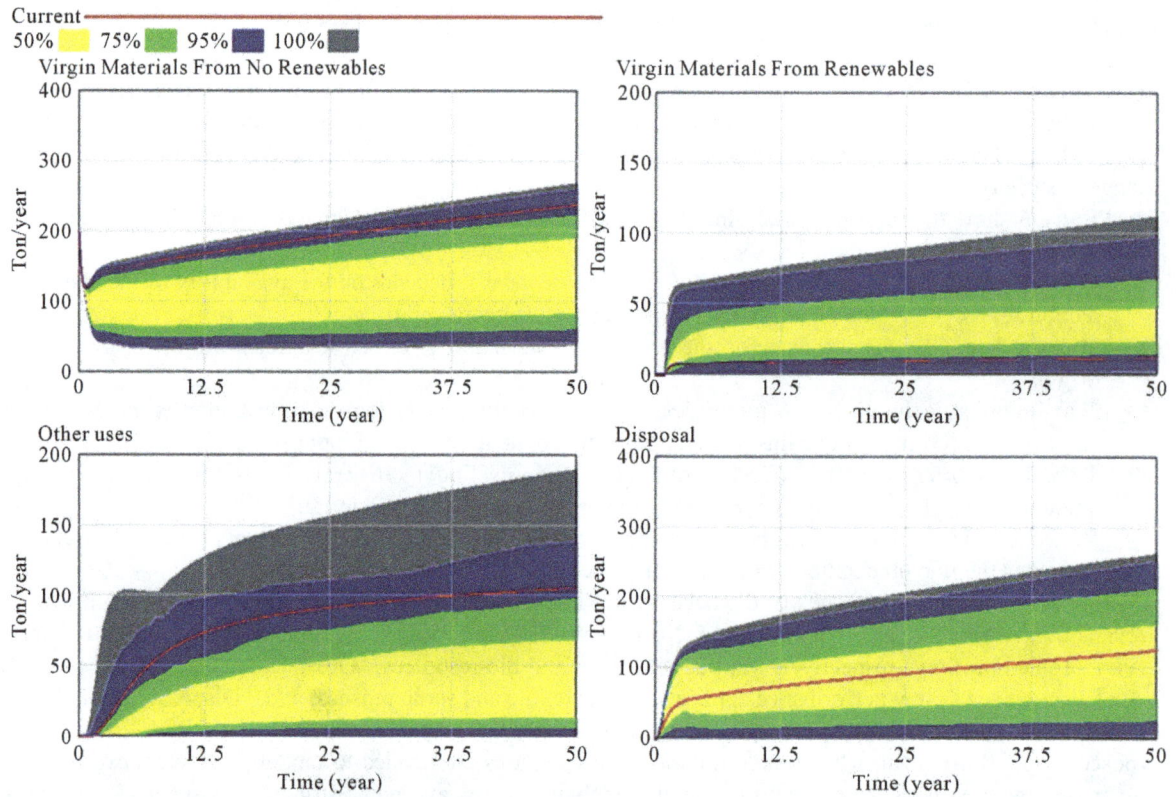

Figure 6. Sensitivity graphs for different model flows.

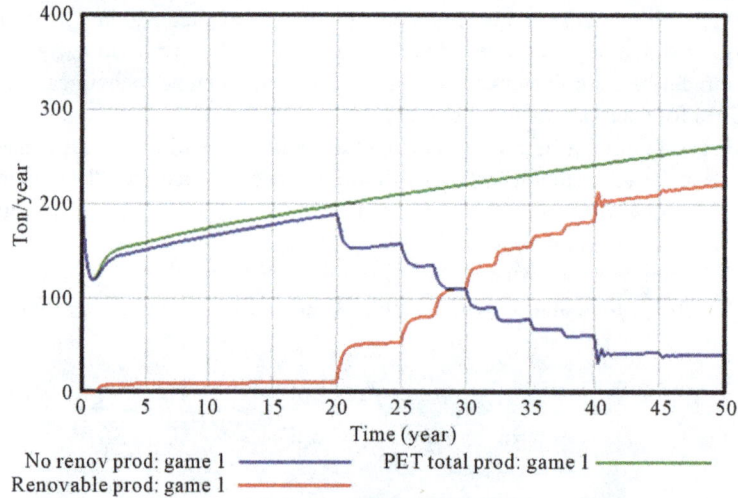

Figure 7. Scenario 1, technology available for implementation 10 years before petrochemicals decline.

material (89.3%), while recycled material will be 10.7%, (33,900 tons).

It should be emphasized that these calculations started from the assumptions that were made at the beginning of the study.

In **Figure 8**, the rate at which technology will be available to integrate the PET supply in a medium-term will be possible at the time of the declining of the petrochemical-based polymer, (30 years) so that when declining arrives, the ratio of the integration of the supply of virgin material will be 94.9% petrochemical source, and thus only 5.1% bio material.

The gap between virgin petrochemical and biomaterials, do not intersect at the end of the Planning Horizon. Nevertheless, the implementation of policies and/or actions that support the implementation of biopolymers will be required. In year 50, that is, 20 years after the technology adjustments, a contribution of these materials will be achieved to 34.9%. So the petrochemical source will be 65.1%.

Accordingly to second scenario, the total offer consists of virgin and recycled material; for the 50th year total supply would be 290,000 tons of PET bottles, of which 259,000 are virgin material. In the third scenario, **Figure 9**, the rate at which the technology for biopolymers, will be available 5 years after PH starts. The possible decline (year 30) of petrochemical origin is maintained so that the ratio of the integration of petrochemical virgin material supply will be 70% and therefore, 30% of bio origin.

The gap between petrochemical and bio virgin production intersects at PH 50th year. Therefore, implementation of policy actions that support bio polymer technologies will be needed. In the year 50, 20 years after the start of encouraging policies, a contribution of almost 50% virgin material production, will be achieved.

PET bottles, for the scenario number three, the total offer consists of virgin and recycled material; for the 50th year the total supply would be 290,000 tons of PET bottles, of which the proportions of virgin and recycled material remain the same.

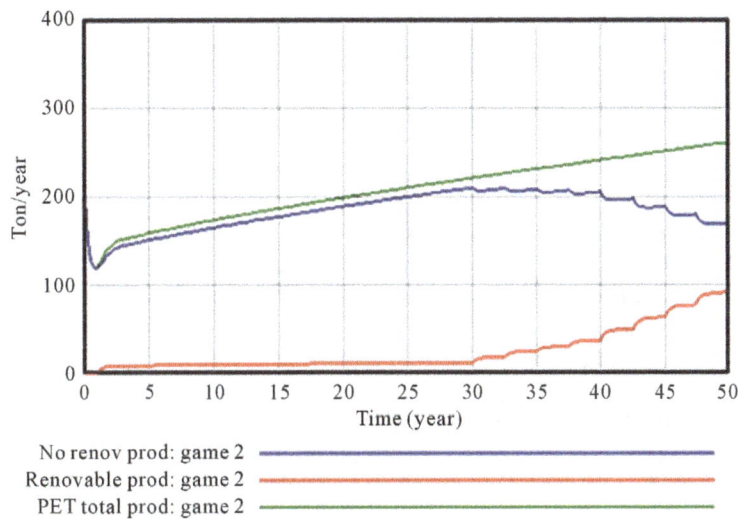

No renov prod: game 2
Renovable prod: game 2
PET total prod: game 2

Figure 8. Scenario 2, technology available for implementation considering petrochemicals decline.

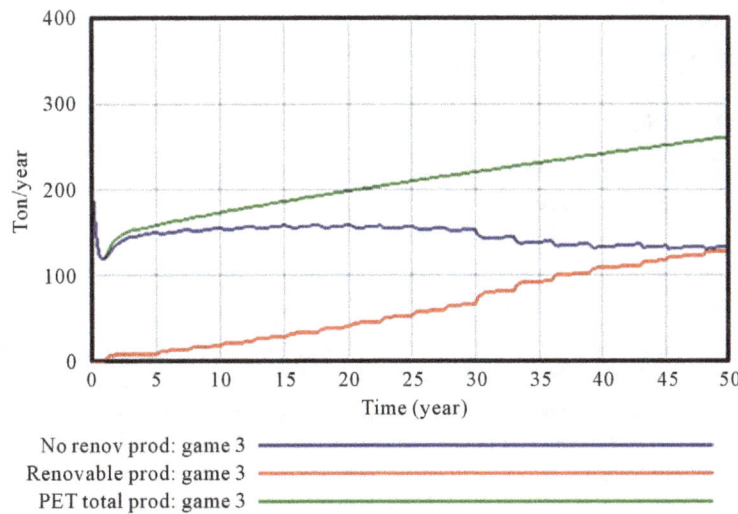

No renov prod: game 3
Renovable prod: game 3
PET total prod: game 3

Figure 9. Scenario 3, gradual technology implementation.

5. Conclusions

This paper has presented a life-cycle analysis for Polyethylene Terephthalate as beverage packaging under system dynamics approach, taking into account production via renewable resources, as a contribution to decision making aid methodology.

Results show the usefulness of the system dynamics to display the behavior of the life cycle in time for plastic materials, in particular PET bottle. Including the life cycle management of the production of PET from renewable sources, significantly impacts of the total production of material, attributed to technology assimilation delay and therefore, the contribution to the total production.

The replacement with renewable materials has a limit. This limit is set by process technological efficiency and by required time to obtain a unit of virgin material. Then, the share of non-renewable sources to obtain PET will be maintained at low level, while development and optimization of technologies to use of renewable sources will continue.

Software tools, "sensitivity testing" and "gamming", prove to be useful to generate scenarios under different premises of policies and actions, joining to the model parameters needed, so that prospective scenario analysis is affordable.

Bottle grade Polyethylene Terephthalate yield, via gas and oil technologies, can be reduced but not excluded from chain production, through generation and implementation of appropriate policies encouraging research and development of new PET synthesis technologies, via renewable and non-renewable resources, as well as those for chemical recycling of waste PET bottle resin.

The policies above mentioned, should take into account beverage industry growth prospective, so that production goal via renewable technology can be considered at the first stage; lately this goal can be reviewed and adjusted, according to new or evolving preferences, policies and resources availability.

Petrochemical polymers declining should be a major contributor to encourage biopolymers and recycling policies to supply PET bottle consumption in the planning horizon.

Improvement of system efficiency is strongly correlated to different fields of knowledge as biotechnology, genetics, process engineering and bio refining.

PET recycling, bottle to make bottles, has been constantly increasing, especially by environmental policies adopted by different corporations leading to a list of successes. An example is the 50,000 ton/year *PETStar* recycling plant start up in 2014, as it can be observed on the NAPCOR reports.

Increasing PET waste recovery efficiency and optimization of processes to improve added value to recycled PET resin, have an important effect in total production of bottles. As recycling sector continues to consolidate, maximizing recovery efficiency and adding value to PET recovered for high performance applications can minimize virgin PET requirements.

Collection activity is primary a logistics problem to guarantee supply for recycle plants.

Some of the conclusions herein obtained were not taken of this paper, but from the whole research. More research is in course about the model behavior under other premises, considerations and preferences, as well as to any other material, looking for the possibility to make the model a generalized one.

Acknowledgements

Financial support from Consejo Nacional de Ciencia y Tecnología (CONACYT, México) (Ph.D. scholarship to P. A. A.-J.) is gratefully acknowledged.

References

[1] Forrester, J.W. (1995) Couterintuitive Behavior of Social Systems. *Technology Review*, **73**, 52-68.

[2] Sandén, B.A. and Karlström, M. (2007) Positive and Negative Feedback in Consequential Life-Cycle Assessment. *Journal of Cleaner Production*, **15**, 1469-1481. http://dx.doi.org/10.1016/j.jclepro.2006.03.005

[3] Changsirivathanathamrong, A., Moore, S. and Linard, K. (2007) Integrating Systems Dynamics with Life Cycle Assessment: A Framework for Improved Policy Formulation and Analysis. Modelling and Simulation Society of Australia and New Zealand (MSSANZ).

[4] Stasinopoulos, P., Compston, P., Newel, B. and Jones, H.M. (2012) A System Dynamics Approach in LCA to Account for Temporal Effects a Consequential Energy LCI of Car Body-in-Whites. *International Journal of Life Cycle Assessment*, **17**, 199-207. http://dx.doi.org/10.1007/s11367-011-0344-0

[5]   Dewulf, J. and Van Langenhove, H. (2004) Thermodynamic Optimization of the Life Cycle of Plastics by Exergy Analysis. *International Journal of Energy Research*, **28**, 969-976. http://dx.doi.org/10.1002/er.1007

[6]   Patel, M., Bastioli, C., Marini, L. and Würdinger, E. (2005) Life-Cycle Assessment of Bio-Based Polymers and Natural Fiber Composites. *Biopolymers Online*. http://dx.doi.org/10.1002/3527600035.bpola014

[7]   Pilz, H., Brandt, B. and Fehringer, R. (2010) The Impact of Plastics on Life Cycle Energy Consumption and Greenhouse Gas Emissions in Europe. Denkstatt Gmbh.

[8]   Harmsen, P. and Hackmann, M. (2013) Green Building Blocks for Biobased Plastics: Biobased Processes and Market Development. Wageningen UR Food & Biobased Research.

[9]   Escobar, C., García, C. and Mareschal, B. (2010) Petrochemical Industry: Assessment and Planning Using Multicriteria Decision Aid Methods. *Technology and Investment*, **1**, 118-134. http://dx.doi.org/10.4236/ti.2010.12015

[10] Brans, J.P., Macharis, C., Kunsch, P.L., Chevalier, A. and Schwaninger, M. (1998) Combining Multicriteria Decision Aid and System Dynamics for the Control of Socio-Economic Processes. An Iterative Real-Time Procedure. *European Journal of Operational Research*, **109**, 428-441. http://dx.doi.org/10.1016/S0377-2217(98)00068-X

[11] Mejía, L., Toledo, C. and Rayle, B. (2012) Decision Making in Sustainable Development: Some Methods to Evaluate Energy and Nonrenewable Resources Waste When Using Some Plastics. *American Journal of Operations Research*, **2**, 399-407. http://dx.doi.org/10.4236/ajor.2012.23048

[12] De Jong, E., Higson, A., Walsh, P. and Wellisch, M. (2012) Bio-Based Chemicals: Value Added Products from Biorefineries. IEA Bioenergy—Task 42 Biorefinery.

[13] National Association for PET Container Resources, NAPCOR (1999-2013) Reports on Postconsumer PET Container Recycling Activity. Florence, KY, USA.

[14] Udo de Hæs, H.A., Heijungs, R., Suh, S. and Huppes, G. (2004) Three Strategies to Overcome the Limitations of Life-Cycle Assessment. *Journal of Industrial Ecology*, **8**, 19-32. http://dx.doi.org/10.1162/1088198042442351

[15] Quintero, D. and López, S. (2010) Análisis estructural: Un apoyo para el modelado con dinámica de sistemas. *Revista Avances en Sistemas e Informática*, **7**, 153-161.

[16] Karayaniddis, G. and Achilias, D. (2007) Chemical Recycling of Poly(Ethylene Terephthalate). *Macromolecular Materials and Engineering*, **292**, 128-146. http://dx.doi.org/10.1002/mame.200600341

[17] Dimonie, D., Socoteanu, R., Pop, S., Fierascu, I., Fierascu, R., Petrea, C., Zaharia, C. and Petrache, M. (2012) Overview on Mechanical Recycling by Chain Extension of POSTC-PET Bottles, Material Recycling—Trends and Perspectives. http://www.intechopen.com/books/material-recycling-trends-and-perspectives/overview-on-mechanical-recycling-by-chain-extension-of-postc-pet-bottles

12

A Novel Statistical Analysis for Residual Stress in Injection Molding

Faisal Alkaabneh, Mahmoud Barghash, Yousef Abdullat

University of Jordan, Amman, Jordan
Email: faisalkaabneh@yahoo.com

Abstract

Residual stresses can reduce the reliability of plastic injection molding parts. This work is an attempt to model the residual stresses as a function of injection molding parameters. More stress is placed on reducing the number of input factors and to include all possible interactions. For this purpose, two-stage experimentation is suggested: a factor screening stage and Response Surface optimization stage. In screening stage Taguchi 3 level experimental design is used to classify the input parameters as significant and non-significant factors. Eight input variables were classified into 3 non-significant and 5 significant factors using this screening stage. Thus for the Response Surface optimization stage: instead of doing 160 experiments in Central Composite, 56 are only needed after the screening stage in half Central Composite Design. The best subset and regression model fitting tools in addition to model verification using randomly selected input setting were used to select a model for predicting residual stresses with a verified Root Mean Square Error (RSME) of nearly 0.93 MPa.

Keywords

Injection Molding, Multi-Stage Experimental Design, Taguchi Experimental Design, Response Surface Methodology, Regression Analysis

1. Introduction

The plastic injection molding process is a widely used polymer processing operation [1]. Because injection molding has many advantages, such as short production cycles, excellent surfaces of the products, and facile molding of complicated shapes, so it is the most popular molding process for making thermoplastic parts [2]. The plastic injection molding process is a cyclic process which consists of three stages. These stages are filling and packing stage, cooling stage and ejection stage [3]. Plastic injection molding is one of the most important methods applied for forming thin-shell plastic products in industry [4]. This is especially suitable for the house hold electronic equipment manufacturing processes. The application of computer, communication and consumer electronics (3C) products such as portable computer and cell phone, etc. was widely used in the whole world. The vigorous development of 3C products has had the trend for products to be light, thin, short and small [5].

As the wall thickness of plastic parts becomes thinner, the injection molding operation becomes more difficult. Hence, the industry has the demand for techniques of plastic injection molding to produce plastic parts with thin wall features [6] [7]. For example, today cell phone dimensions are getting smaller and their appearances are getting more esthetic. Besides these properties, users care about phones' durability. Users do not want that falling to the ground, taking impact, exposing to force cause damage to the cell phone. The part which is produced by using injection molding method does not have the desired strength and resistance to impact, because of the residual stress problem encountered those days in the industry. Residual stresses are internal stresses of the molded part in the absence of external forces. They are caused mainly by non-uniform temperature profile in the cavity during filling, packing, and cooling stages [8]. It is known that the residual stress is produced due to high pressure, temperature change, and relaxation of polymer chains, resulting in shrinkage and warpage of the part [9]. Internal stresses in injection molding components are a principal cause of shrinkage and warpage, and are considered to be responsible for environmental stress cracking in plastic parts [10] [11]. The best way to prevent high residual stress problem is to choose the plastic material and injection parameters right. The decrease in the thickness of the part also weakens the strength of the part; however, increasing thickness of plastic parts is not a feasible solution as the demand of the cell phone design requires parts to be thin, light and small. This problem can be solved by choosing the appropriate material for the durability. But a material may be appropriate for the strength may not be appropriate for the application of that part. Thus, one important solution is to minimize residual stresses by optimizing the process factors of the injection molding process.

Manufacturing process optimization is of three main parts, experimentation, modeling and searching. Experimentation may be actual running of the experimentation on the injection molding process or using Finite element software's prior to finalizing the design. The modern computational and simulation tools for injection-molding process have matured to a stage where they can provide substantial insight into the process metrics and can be profitably utilized to help improve the design of injection-molded components [12]. Computer aided engineering (CAE) has made a major impact on the design and manufacturing process in the injection molding industry in terms of both quality improvement and cost reduction based on applications of various computer simulation techniques [13]. In the past, the optimization of the process parameters was considered to be a "black art", which relied heavily on the experience and knowledge of experts and required a trial-and-error process [14]. Currently, Design of experiments (DOE) is grabbing much attention within the perspective of manufacturing process optimization. DOE has been a very useful tool to design and analyze complicated industrial design problems. It helps us to understand process characteristics and to investigate how inputs affect responses based on statistical backgrounds. In addition, it has been used to systematically determine the optimal process parameters with fewer testing trials [15]. A widely DOE methodology is Taguchi Experimental Design (DOE), which various industries have used over the years to improve products or manufacturing processes. It is a powerful and effective method to solve challenging quality problems [16]. Taguchi analysis is a useful methodology for optimizing process that does not include interaction factors. One main reason for using Taguchi experimentation is the low number of experiments which excludes interactions, and this scatters some shadows on the accuracy of the regression model built using this experimentation type. Thus, it can improve some processes but may not give the optimal. A more comprehensive approach is Response Surface which includes all interaction effects but requires heavy experimentation and data collection. The regression models that can be further used for optimization are more accurate in the Response Surface experimental design. But, Response Surface has extensive experimentation and cost which places some restrictions on the use of this important technique.

A recent published article [17], where the authors developed, for the first time, a new framework to analyze and optimize different manufacturing process setting of plastic injection modeling to minimize different defects via a multi-objective multi-criteria process through using integrated Analytical Hierarchy Process (AHP) wish Taguchi DOE.

This work suggests an intermediate approach that uses both advantages of two main experimentation types whereby, Taguchi Experimental Design (TED) is used for significant factor determination and Response Surface is used for modeling the process residual stresses. This study introduces an innovative method to model the residual stresses based on injection molding process settings; namely filling time, melt temperature, mold temperature, switch to pack (%), pressure holding time, holding pressure magnitude, cooling time, and cooling inlet temperature. This method firstly, uses experiment runs based on Taguchi Orthogonal Array (OA) with only main effects of the input variables included. Secondly, the input parameters are classified using Analysis Of Variance (ANOVA) into significant and non-significant factors. Thirdly, Response Surface design of experiment is conducted using only significant factors identified in the previous step. Fourthly, regression models is developed based on data generated from the Response Surface design of experiment and then validation tests are conducted to choose the best regression model based on the minimum Root Mean Square Error (RSME).

This paper successfully applies multi-stage experimentation, namely, Taguchi and central composite design to model and optimize the residual stress. The residual stress is extremely important as it reduces the reliability of parts in case of tension or improves it in case of compression. The number of experiments and thus time to design and time to customer can be reduced appreciably by this technique.

2. Research Methods

2.1. Taguchi Method and Signal to Noise S/N Ratio

Taguchi methods have been used widely in engineering analysis to optimize the performance characteristics through the setting of design parameters. Taguchi method is also strong tool for the design of high quality systems. To optimize designs for quality, performance, and cost, Taguchi method presents a systematic approach that is simple and effective [18]. The Taguchi method consists of three stages which are system, parameters, and tolerance designs, respectively. The system design involves the application of scientific and engineering knowledge required in manufacturing of a product. The parameter design is employed to find optimal process values for improving of the quality characteristics.

The tolerance design is used for determining and analyzing of the tolerances in optimal settings recommended by the parameter design [19]. Taguchi recommends the use of the S/N ratio for determination of the quality characteristics implemented in engineering design problems. The S/N ratio characteristics with signed-target type can be divided into three stages: the smaller is the better, the nominal is the best, and the larger is the better [20]. The quality characteristics are evaluated through the S/N ratio obtained in the Taguchi experimental plan. ANOVA then can be used to evaluate the experimental errors and test of significance to understand the effect of various factors [21].

2.2. Response Surface Methodology

Response surface methodology is an integration of mathematical and statistical techniques for modeling and optimizing the response variable models involving several quantitative independent variables. It is well adapted to making an analytical model for complicated problems. In industry, RSM is a very useful tool for quality and productivity improvement, in which often we wish to discover functional relationships between the response and independent variables. Upon determining the relationship, we can easily resolve practical quality and productivity problems by using appropriate statistical techniques [22]. In general, for predicting the optimal point, a second-order polynomial function was popularly used and fitted to correlate the relationship between independent variables (X_i) and response (Y). The quadratic response surface is always described as follows.

$$Y = b_o + \sum_{i=1}^{n} b_i X_i + \sum_{i=1}^{n} b_{ii} X_i^2 + \sum \sum_{i<j} b_{ij} X_i X_j \tag{1}$$

where n is the number of design variables, and b_o, b_i, b_{ii}, and b_{ij} represent the coefficients of constant, linear, quadratic, and cross product terms, respectively. To build the empirical response models, the necessary data are generally collected by the design of experiments, followed by the statistical single or multiple regression technique. The more popular statistical approach such as analysis of variance (ANOVA) is adopted to justify the significance of the empirical model [23].

2.3. Regression Model Development

The purpose of multidimensional analysis of regression is to determine the quantitative relations between the investigated values and the variables, which directly influence them to assess the results of their activity and to predict the behavior of the investigated variables [24]. Regression analysis is also one of the most widely used statistical tools because it provides simple methods for establishing functional relationship among variables. It can be employed to develop a suitable model for predicting dependent variables from a set of independent variables [25].

3. Methodology

Figure 1 shows the proposed scheme in our study. In phase 1 (planning) Taguchi experimental plan was first conducted using reduced number of experiments, time and cost based on orthogonal arrays to determine the most significant factors. In phase 2, a Response Surface Methodology (RSM) Central Composite design is conducted based on the identified significant factors in phase 1. In Phase 3 (modeling quality characteristic) data obtained from Response Surface experiments are used to develop 3 regression models then validation tests are conducted to choose the best model yielding minimum Root Mean Square Error (RSME).

4. Case Study

4.1. Part Geometry and Finite Element Model

Finite Element (FE) analysis of the cell phone cover part is performed using SimpoeMesh [26]. Geometry of the part employed in this current study having width, length, height and thickness of 83, 145, 17 and 2 mm respectively was shown in **Figures 2(a)-(c)**. FE model of the part is created by discrediting the geometry into smaller simple elements. The FE model shown in **Figure 2(d)** includes 6884 tetrahedron elements having average aspect ratio of 3.2068. The FE analyses were performed using Simpoe-Mold software.

4.2. Mold and Material Description

The study analysis is conducted for ABS + PC (P) SABIC/Cycoloy CY6110 material; its properties are given in **Table 1**. Water is used as the cooling fluid, steel 420SS as mold material. The gate location is shown **Figure 3(a)** having 2 mm diameter. The cooling channels are shown in **Figure 3(b)** and cooling channels properties are given in **Table 2**.

4.3. Phase 1: Screening Stage Design and Analysis

4.3.1. Taguchi Screening Experiments Experimental Design
Taguchi Method was developed for saving time, cost and effort. Minitab 15 software was used for statistical calculations [27]. The factor levels for the eight included variables are given in **Table 3**. The levels for each factor are set based on literature, materials data sheets, and experience. Tests were organized using Taguchi's L27 (3^8) orthogonal array (**Table 4** columns 2 - 9). This orthogonal array is efficient for screening purpose. It can give the significance of the different input variables with minimum number of experiments. **Table 4**-column

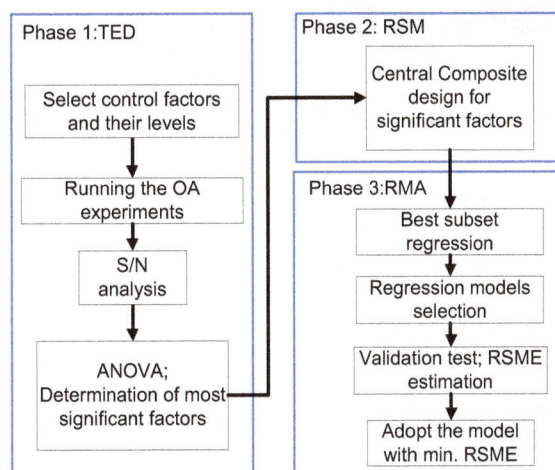

Figure 1. The proposed methodology.

(a)

(b)

(c)

(d)

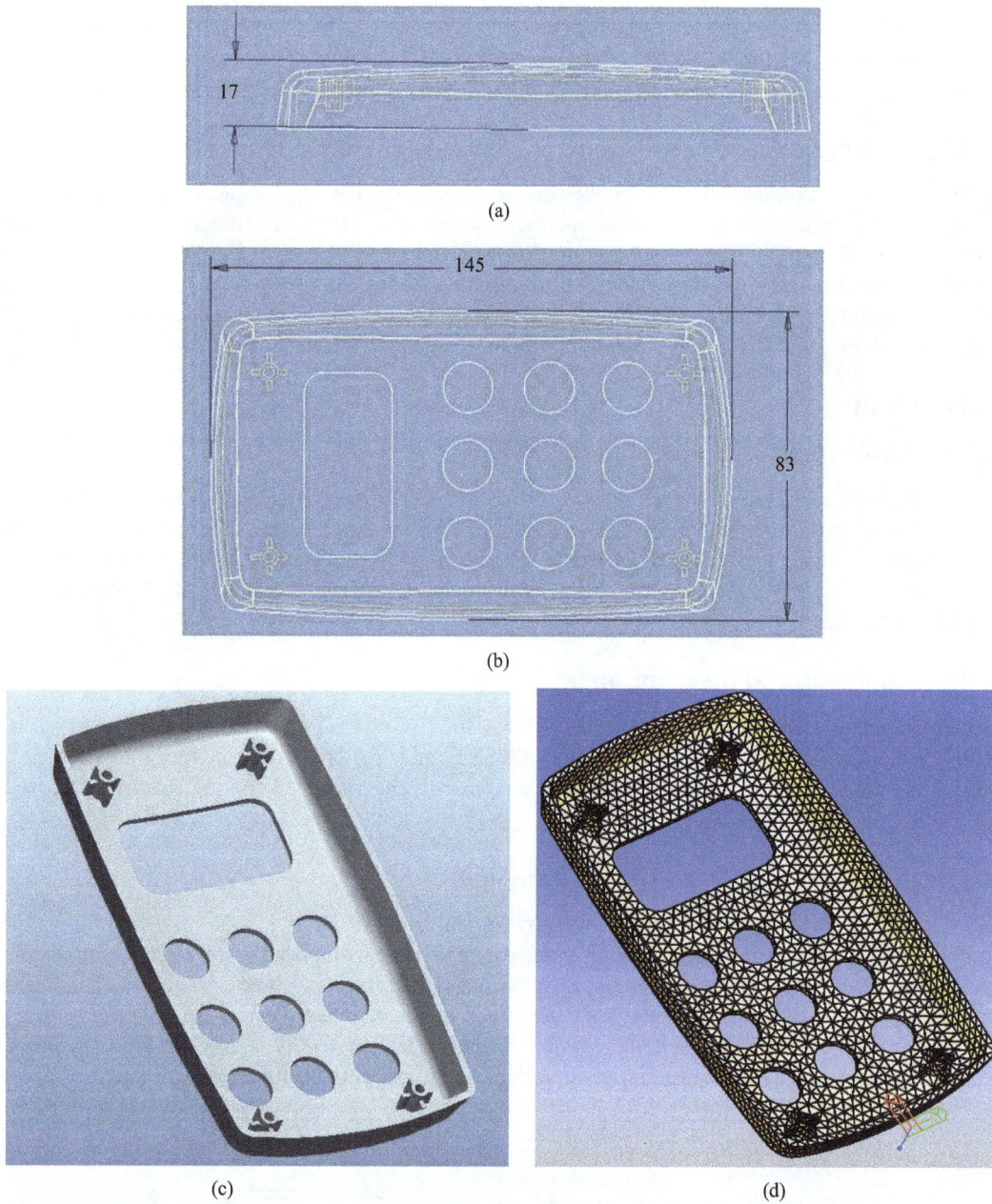

Figure 2. (a) and (b) CAD drawings of the part, (c) general view and (d) finite element model.

Table 1. Material properties of ABS + PC.

Material property	Performance
Material structure	Amorphous
Elastic module (MPa)	2822
Poisson ratio	0.40
Max shear stress (MPa)	0.50
Max shear rate (s)	50000
Ejection temperature (C)	95

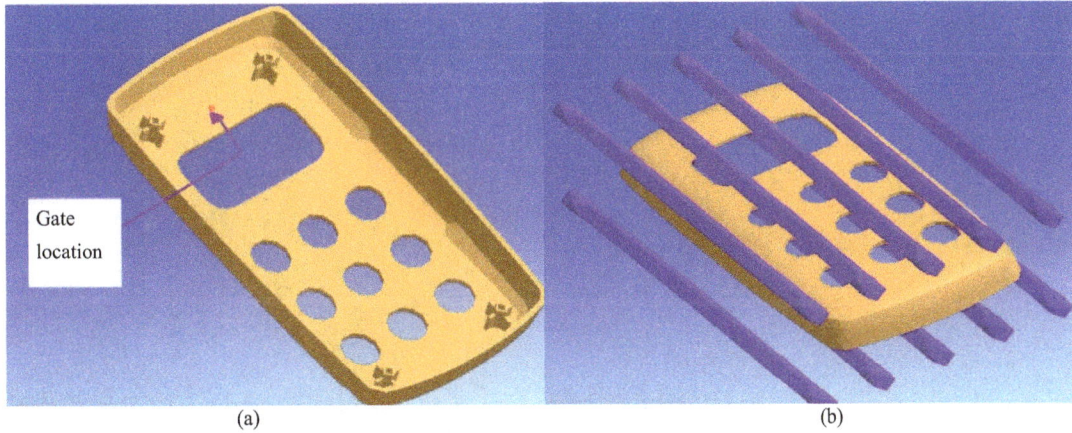

Figure 3. (a) Gate location, (b) cooling channels configuration.

Table 2. Selected cooling channel parameters.

Property	Unit	Value, name
Number of channels	-	10
Distance between cooling channels and the part	mm	12
Distance between cooling channels center	mm	24
Type of channels	-	Longitude.

Table 3. The process parameters and their levels.

Process factor	Level 1	Level 2	Level 3	Unit
Filling time (A)	1	2.5	4.0	s
Melting temperature (B)	250	275	300	°C
Mold temperature (C)	60	72.5	85	°C
Switch to pack % (D)	50	75	100	%
Pressure holding time (E)	2	6	10	s
Cooling time (F)	6	9	12	s
Holding pressure magnitude (G)	60	80	100	MPa
Cooling inlet temperature (H)	10	15	20	°C

10 shows the experimental results obtained from Simpoe-Mold [28]. The test results were evaluated in terms of S/N ratio (**Table 4** column 11). The S/N ratio was calculated by smaller the better quality characteristic for determining the effect of each factor on the selected injection molding quality characteristic according to the equation [29]:

$$S / N = -10 \log(\sum_{i=1}^{n} y_i^2) \qquad (2)$$

where y_i is the measured property and n corresponds to the number of samples in each test trial.

4.3.2. Screening Stage Factor Analysis

Figure 4 shows the factors plot curves using MINITAB. It is clear that factors A, D, E, F, and G are of the highest effect on the residual stresses and the other factors have slight effect within the specified range of the experimentation.

According to **Figure 4** the optimal value for the residual stresses can be obtained using the combination of A

Table 4. The Taguchi L_{27} orthogonal array, experimental results, and S/N ratio.

Exp. No.	A	B	C	D	E	F	G	H	Residual stresses (MPa)	S/N ratio
1	1	1	1	1	1	1	1	1	20.4127	−26.198
2	1	1	1	1	2	2	2	2	17.6756	−24.9475
3	1	1	1	1	3	3	3	3	21.566	−26.6754
4	1	2	2	2	1	1	1	2	23.0203	−27.2422
5	1	2	2	2	2	2	2	3	20.4502	−26.214
6	1	2	2	2	3	3	3	1	24.4753	−27.7746
7	1	3	3	3	1	1	1	3	24.2603	−27.6979
8	1	3	3	3	2	2	2	1	23.3738	−27.3746
9	1	3	3	3	3	3	3	2	25.9399	−28.2794
10	2	1	2	3	1	2	3	1	23.5955	−27.4566
11	2	1	2	3	2	3	1	2	16.4261	−24.3107
12	2	1	2	3	3	1	2	3	27.5606	−28.8058
13	2	2	3	1	1	2	3	2	23.3139	−27.3523
14	2	2	3	1	2	3	1	3	15.937	−24.0481
15	2	2	3	1	3	1	2	1	27.4976	−28.7859
16	2	3	1	2	1	2	3	3	21.8192	−26.7768
17	2	3	1	2	2	3	1	1	17.4536	−24.8377
18	2	3	1	2	3	1	2	2	30.9188	−29.8045
19	3	1	3	2	1	3	2	1	25.0033	−27.9599
20	3	1	3	2	2	1	3	2	29.1737	−29.2998
21	3	1	3	2	3	2	1	3	27.4683	−28.7766
22	3	2	1	3	1	3	2	2	27.1661	−28.6805
23	3	2	1	3	2	1	3	3	31.5451	−29.9786
24	3	2	1	3	3	2	1	1	22.5222	−27.0522
25	3	3	2	1	1	3	2	3	21.6857	−26.7235
26	3	3	2	1	2	1	3	1	27.6303	−28.8277
27	3	3	2	1	3	2	1	2	24.8103	−27.8926

Main Effects Plot (data means) for SN ratios

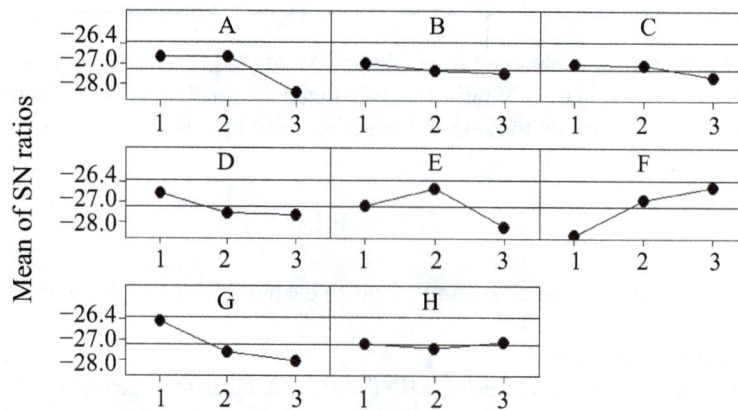

Signal-to-noise: Smaller is better

Figure 4. Plot of process effects for residual stresses.

(Filling time) 1s, B (Melt temperature) 250°C, C (Mold temperature) 60°C, D (Switch to pack %) 50%, E (Pressure holding time) 6s, F (Cooling time) 12s, G (Holding pressure magnitude) 60 MPa, and H (Cooling inlet temperature) 20°C, is 13.3111 MPa.

Table 5 shows the ANOVA significance analysis for the factors. Considering the 5% P-significance level, factors A, D, E, F, and G are considered significant and this result coincides perfectly with Figure 4.

Factors screening requires less experiments and is effective in this sense to identify if the input machine setting is significant enough to be included in further analysis. As, is seen in this work, half of input factors are not significant and should not be included in the model building experimental design part.

4.4. Response Surface Methodology

As discussed above the RSM is high costly and time consuming. The cost increases exponentially as the number of factors increase. Thus, in this work we have included a factor screening phase whereby, we exclude any non-significant factors prior to modeling.

In this study, the approximation of the mathematical model will be proposed using the fitted third-order polynomial regression model, which is called the cubic model. The necessary data for building the response model are generally collected by the experimental design [30].

The significant factors regarding residual stresses for Response Surface experiments are screened from the injection process parameters through the Taguchi experiments based on P-value of 0.05 that is 95% confidence interval. Filling time, switch to pack percentage, pressure holding time, holding pressure magnitude, and cooling time are found to be the significant factors in these screening tests. In this study, the experimental design adopts the centered central composite design (CCD) in order to fit the cubic model of the RSM. The factorial portion of CCD is a full factorial design with all combinations of the factors at two levels and composed of the eight star (axial) points, and six central points in cubes (coded level 0) which is the midpoint between the high and low levels. The star points are at the face of the cube portion on the design which corresponds to an alpha (α) value of 2.366. This type of design is commonly called the face-centered CCD.

Table 6 shows the five process factors and their levels. The experimental plans is generated using stipulated conditions based on the face centered CCD and involves 56 total runs as shown in Table 7. Table 7 shows the experiments layout and results.

Table 5. ANOVA results.

	DF	Seq SS	Adj SS	Adj MS	F	P
A	2	12.3309	12.3309	6.1654	16.18	0.001
B	2	0.8437	0.8437	0.4219	1.11	0.368
C	2	1.4884	1.4884	0.7442	1.95	0.192
D	2	4.4536	4.4536	2.2268	5.85	0.021
E	2	10.9439	10.9439	5.4719	14.36	0.001
F	2	17.9833	17.9833	8.9916	14.36	0.000
G	2	12.6835	12.6835	6.3418	23.6	0.001
H	2	0.2655	0.2655	0.1327	16.65	0.714
Residual Error	10	3.8097	3.8097	0.381	0.35	
Total	26	64.8026				

Table 6. Process factors and their levels for the full factorial experimental design.

Process factor	Low level	High level	Unit
Filling time (A)	1	4.0	s
Switch to pack % (B)	25	75	%
Pressure holding time (C)	2	10	s
Cooling time (D)	9	15	s
Holding pressure magnitude (E)	40	80	MPa

Table 7. Experiments layout and their results.

Exp. No	A	B	C	D	E	Residual Stresses
1	1.87	39.43	4.31	10.73	51.55	15.3154
2	3.13	39.43	4.31	10.73	51.55	17.7162
3	1.87	60.57	4.31	10.73	51.55	15.1058
4	3.13	60.57	4.31	10.73	51.55	17.9076
5	1.87	39.43	7.69	10.73	51.55	17.0077
6	3.13	39.43	7.69	10.73	51.55	18.5774
7	1.87	60.57	7.69	10.73	51.55	16.6253
8	3.13	60.57	7.69	10.73	51.55	19.4672
9	1.87	39.43	4.31	13.27	51.55	13.7617
10	3.13	39.43	4.31	13.27	51.55	14.487
11	1.87	60.57	4.31	13.27	51.55	12.2111
12	3.13	60.57	4.31	13.27	51.55	14.6087
13	1.87	39.43	7.69	13.27	51.55	14.299
14	3.13	39.43	7.69	13.27	51.55	15.5651
15	1.87	60.57	7.69	13.27	51.55	14.4523
16	3.13	60.57	7.69	13.27	51.55	16.348
17	1.87	39.43	4.31	10.73	68.45	17.3996
18	3.13	39.43	4.31	10.73	68.45	17.3468
19	1.87	60.57	4.31	10.73	68.45	17.4586
20	3.13	60.57	4.31	10.73	68.45	16.7691
21	1.87	39.43	7.69	10.73	68.45	19.8274
22	3.13	39.43	7.69	10.73	68.45	19.5139
23	1.87	60.57	7.69	10.73	68.45	19.407
24	3.13	60.57	7.69	10.73	68.45	20.097
25	1.87	39.43	4.31	13.27	68.45	17.1021
26	3.13	39.43	4.31	13.27	68.45	17.3039
27	1.87	60.57	4.31	13.27	68.45	16.8999
28	3.13	60.57	4.31	13.27	68.45	15.7888
29	1.87	39.43	7.69	13.27	68.45	16.5648
30	3.13	39.43	7.69	13.27	68.45	16.4264
31	1.87	60.57	7.69	13.27	68.45	16.6773
32	3.13	60.57	7.69	13.27	68.45	17.0083
33	2.5	50	6	12	60	16.2736
34	2.5	50	6	12	60	16.2736
35	2.5	50	6	12	60	16.2736
36	2.5	50	6	12	60	16.2736
37	2.5	50	6	12	60	16.2736
38	2.5	50	6	12	60	16.2736
39	1	50	6	12	60	13.8783
40	4	50	6	12	60	17.4748
41	2.5	25	6	12	60	14.2905
42	2.5	75	6	12	60	16.497

Continued

43	2.5	50	2	12	60	15.8673
44	2.5	50	10	12	60	16.3209
45	2.5	50	6	9	60	19.8856
46	2.5	50	6	15	60	12.5724
47	2.5	50	6	12	40	14.8686
48	2.5	50	6	12	80	14.9246
49	2.5	50	6	12	60	16.3101
50	2.5	50	6	12	60	16.3101
51	2.5	50	6	12	60	16.3101
52	2.5	50	6	12	60	16.3101
53	2.5	50	6	12	60	16.3101
54	2.5	50	6	12	60	16.3101
55	2.5	50	6	12	60	16.3101
56	2.5	50	6	12	60	16.3101

Phase 3: Regression Modeling and Model Selection of the Residual Stresses

The best subset tool is used in modeling the residual stresses. **Table 8** shows the results for the best subset. Column 1 shows the number of variables included in the test, columns 2 and 3 are the non-adjusted and adjusted regression correlation factor, respectively, and column 4 is called the Mallows CP. In statistics, Mallows suggested for model selection, among input variables, and the goal is to find the best model involving a subset of these predictors. The lower, the CP factor the better the model is [31].

Although many of the mentioned models are good in terms of the adjusted R square, the Mallows CP suggests that the best model is the twenty third model. And because of the close value of the CP for the last three models, they are included in further processing for the selection of the best model. These models for the recommended variables are fitted using Minitab and the fitness results are given in equations 3 - 5.

$$Model \ 1 = 5.71 + 6.793 * A + 8.03 * 10^{-2} * C - 1.772 * D + 0.473 * E - 9.902 * 10^{-2} * A * E +$$
$$6.416 * 10^{-3} * B * C - 3.88 * 10^{-3} * C * E - 3.7 * 10^{-6} * B^3 + 1.135 * 10^{-3} * C^3 + \tag{3}$$
$$1.833 * 10^{-3} * D^3 - 1.139 * 10^{-5} * E^3$$

$$Model \ 2 = 76.36 + 6.793 * A + 1.6535 * C - 2.262 * D + 3.072 * E + 5.451 * 10^{-2} * E^2 - 9.902 * 10^{-2} A * E * +$$
$$6.396 * 10^{-3} * B * C - 0.15053 * C * D + 2.315 * 10^{-2} * D * E - 3.68 * 10^{-6} * B^3 + 1.146 * 10^{-3} * C^3 + \tag{4}$$
$$1.842 * 10^{-3} * D^3 - 3.1233 * 10^{-5} * E^3$$

$$Model \ 3 = 12.47 + 6.793 * A + 1.9707 * C - 2.271 * D + 0.1704 * E - 9.902 * 10^{-2} * A * E +$$
$$0.15053 * C * D - 2.315 * 10^{-2} * D * E - 9.9 * 10^{-7} * B^3 + 1.169 * 10^{-3} * C^3 + \tag{5}$$
$$1.863 * 10^{-3} * D^3 - 1.125 * 10^{-5} * E^3$$

1) Verification results

Table 9 shows the verification results for some randomly selected experiments. Columns 2 through 6 show the selected values for the Filling time (A), Switch to pack % (B), Pressure holding time (C), Holding pressure magnitude (D), and Cooling time (E). The predictions for the Models 1 - 3 in **Table 9** are shown in columns 7 - 9. The actual results are in column 10, while **Table 10** shows the absolute errors associated with each model, row 7 shows the RSME for the three models. Form **Table 10**, the best model is model 3 with The RSME value of 0.930751.

Table 8. Best subset results for residual stress modeling.

Vars	R-Sq	R-Sq (adj)	Mallows CP	S	A	B	C	D	E	AB	AC	AD	AE	BC	BD	BE	CD	CE	DE	A3	B3	C3	D3	E3
1	42.4	41.4	59.7	1				X																
1	40.7	39.6	63.1	1																			X	
2	57.6	56	32.3	1				X					X											
2	57.4	55.8	32.6	1				X									X							
3	69.9	68.2	10.4	1				X			X								X					
3	69.3	67.5	11.6	1				X	X		X													
4	71.8	69.6	8.7	1	X		X										X		X					
4	71.7	69.5	8.9	1	X		X	X									X							
5	77.5	75.3	-0.3	1	X		X	X					X				X							
5	76.6	74.3	1.4	1	X		X	X					X						X					
6	78.8	76.2	-0.8	1	X		X	X					X				X		X					
6	78.4	75.7	0	1	X		X	X					X				X							X
7	79.3	76.3	0.1	1	X		X	X					X				X		X					X
7	79.2	76.2	0.3	1	X		X	X					X				X		X			X		
8	79.8	76.3	1.2	1	X		X	X					X				X		X				X	X
8	79.8	76.3	1.3	1	X		X	X				X	X				X		X					X
9	80.3	76.5	2.2	1	X		X	X			X	X	X				X		X					X
9	80.3	76.4	2.3	1	X	X	X	X				X	X				X		X					X
10	80.8	76.5	3.3	1	X		X	X				X	X	X			X		X				X	X
10	80.7	76.4	3.4	1	X	X	X	X				X	X				X		X				X	X
11	81.1	76.4	4.7	1	X		X	X	X			X	X	X			X		X				X	X
11	81.1	76.3	4.7	1	X		X	X	X			X	X			X	X		X				X	X
12	81.4	76.2	6.1	1	X		X	X	X		X	X	X	X			X		X				X	X
12	81.4	76.1	6.2	1	X		X	X	X			X	X		X	X	X		X				X	X
13	81.6	75.9	7.7	1	X		X	X	X		X	X	X	X		X	X		X				X	X
13	81.6	75.9	7.8	1	X	X	X	X	X		X	X	X	X			X		X				X	X
14	81.7	75.5	9.5	1	X		X	X	X		X	X	X	X		X	X		X	X	X			X
14	81.7	75.5	9.5	1	X	X	X	X	X		X	X	X	X			X		X	X	X			X
15	81.8	75	11.3	1	X		X	X	X		X	X	X	X	X	X	X		X	X	X			X
15	81.8	75	11.3	1	X		X	X	X		X	X	X	X			X	X	X	X	X	X		X
16	81.9	74.4	13.2	1	X		X	X	X		X	X	X	X	X	X	X	X	X			X	X	X
16	81.9	74.4	13.2	1	X		X	X	X	X	X	X	X	X	X	X	X		X			X	X	X
17	81.9	73.8	15.1	1	X		X	X	X	X	X	X	X	X	X	X	X	X	X			X	X	X
17	81.9	73.8	15.1	1	X		X	X	X		X	X	X	X	X	X	X	X	X	X		X	X	X
18	82	73.2	17	1	X		X	X	X	X	X	X	X	X	X	X	X	X	X	X		X	X	X
18	81.9	73.1	17.1	1	X	X	X	X	X	X	X	X	X	X	X	X		X	X	X		X	X	X
19	82	72.4	19	1	X		X	X	X	X	X	X	X	X	X	X	X	X	X	X	X	X	X	X
19	82	72.4	19	1	X	X	X	X	X	X	X	X	X	X	X	X	X	X	X		X	X	X	X
20	82	71.7	21	1	X	X	X	X	X	X	X	X	X	X	X	X	X	X	X	X	X	X	X	X

2) Model graphical illustration and optimization

Figures 5(a)-(c) shows a 3D fitted Response Surface of the residual stresses for the significant variables.

Table 9. Random points verification of the suggested models.

Exp. No.	A	B	C	D	E	Model 1	Model 2	Model 3	Results
1	1.5	60	7	10	50	16.30118	16.86167	16.94736	16.5994
2	2	50	9	12	60	17.1164	17.08319	17.05788	15.5094
3	3	40	3	14	70	14.70965	16.14324	15.81389	16.2522
4	3.5	70	5	13	75	15.21857	15.40207	15.98266	16.97746
5	4	35	2	11	65	17.65731	17.05032	16.61571	17.4100

Table 10. Error calculation for each suggested model.

Exp. No.	Error 1	Error 2	Error 3
1	0.298225	0.262268	0.347957
2	1.606999	1.57379	1.548475
3	1.542553	0.10896	0.438315
4	1.75889	1.575395	0.994798
5	0.247307	0.359676	0.79429
RSME	1.281071	1.016731	0.930751

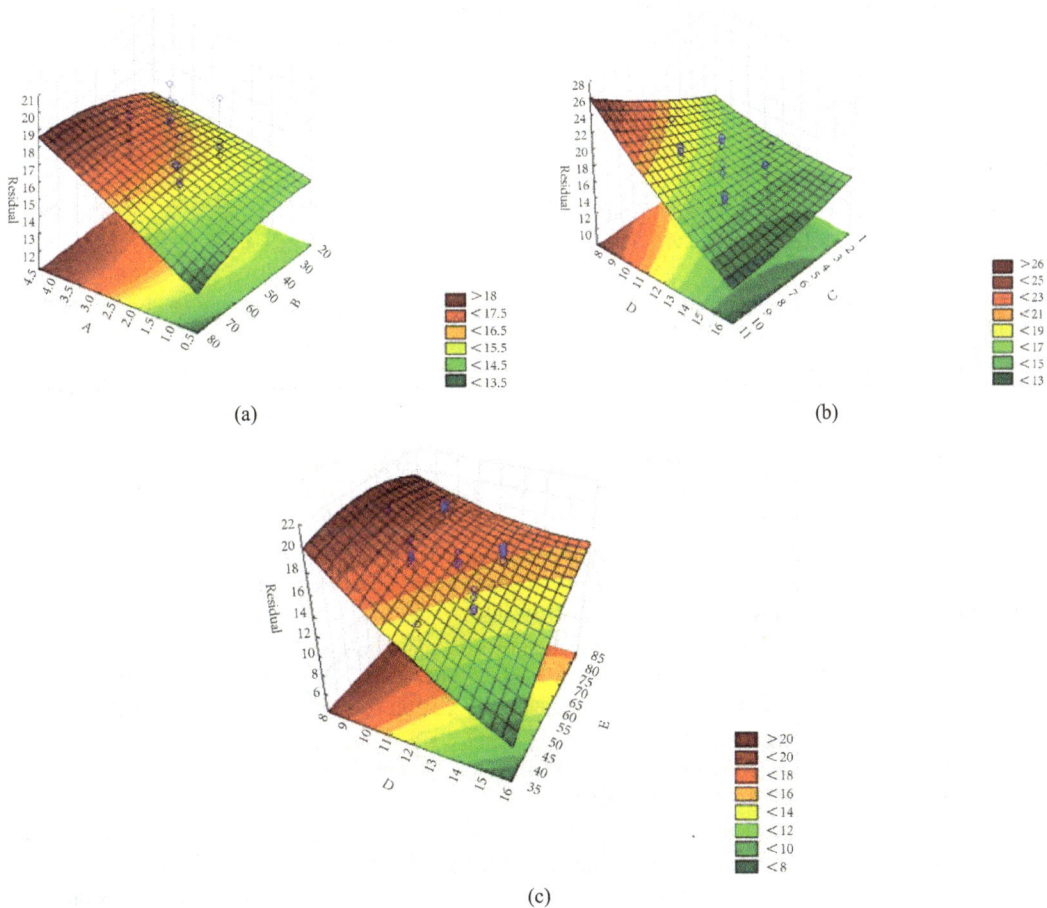

Figure 5. 3D illustration plots for the residual stress versus other input variables. The optimal combination discussed above (Filling time, 1s; Switch to pack percentage, 80%; Pressure holding time, 11s; Cooling time, 16s; and Holding pressure magnitude, 35 MPa) was verified and it gave residual stresses of 7.0125 MPa.

According to **Figure 5** to minimize the residual stresses, variable A (Filling time) must be as low as possible, 0 s., however, since the filling time cannot have a zero value it is more realistic to have 1s. as filling time, variable

B (Switch to pack percentage) must be as high as possible, thus the selected value for variable B is 80%, variable C (Pressure holding time) must be as high as possible, thus 11s, as pressure holding time will be selected, variable D (Cooling time) must be as high as possible, thus 16s, as cooling time will be selected, and variable E (Pressure holding magnitude) must be as low as possible, thus 35 MPa will be selected.

The optimal combination discussed above (Filling time, 1s; Switch to pack percentage, 80%; Pressure holding time, 11s; Cooling time, 16s; and Holding pressure magnitude, 35 MPa) was verified and it gave residual stresses of 7.0125 MPa.

The optimal value for the residual stresses obtained using TED was 13.3111 MPa; however, the optimal value for the residual stresses obtained using RSM was 7.0125 MPa. It is known that TED is not capable for global optimization and it excludes the interactions among variables; on the contrary the RSM is capable for global optimization as well as revealing the interactions among variables.

5. Conclusions

The suggested model building stage for the injection molding included a screening stage to reduce the number of variables included in the modeling stage. This has direct effects on the number of experiments and the modeling cost. It actively reduced the number of input variables from 8 to 5 and the number of Response surface experiments from 160 to 56. Filling time, switch to pack percentage, pressure holding time, holding pressure magnitude, and cooling time are found to be the significant factors. The best subset and verification based model selection was successful in building a model with a verification result of nearly 0.93.

There was a controversial difference in optimal behavior between the Taguchi stage (Phase 1) and Response Surface stage (Phase 2). With the expectation of the Response Surface stage, the outperforming Taguchi expectations definitely justifies the phase 2 suggested in this work.

References

[1]   Chen, X., Lam, Y.C. and Li, D.Q. (2000) Analysis of Thermal Residual Stress in Plastic Injection Molding. *Journal of Materials Processing Technology*, **101**, 275-280. http://dx.doi.org/10.1016/S0924-0136(00)00472-6

[2]   Kurt, M., Kamber, S., Kaynak, Y., Atakok, G. and Girit, O. (2009) Experimental Investigation of Plastic Injection Molding: Assessment of the Effects of Cavity Pressure and Mold Temperature on the Quality of the Final Products. *Materials and Design*, **30**, 3217-3124. http://dx.doi.org/10.1016/j.matdes.2009.01.004

[3]   Hassan, H., Regnier, N., Pujos, C., Arquis, E. and Defaye, G. (2010) Modeling the Effect of Cooling System on the Shrinkage and Temperature of the Polymer by Injection Molding. *Applied Thermal Engineering*, **30**, 1547-1557. http://dx.doi.org/10.1016/j.applthermaleng.2010.02.025

[4]   Oktem, H., Erzurumlu, T. and Uzman, I. (2007) Application of Taguchi Optimization Technique in Determining Plastic Injection Molding Process Parameters for a Thin Shell Part. *Materials and Design*, **28**, 1271-1278. http://dx.doi.org/10.1016/j.matdes.2005.12.013

[5]   Chiang, K.-T. and Chang, F.-P. (2006) Application of Grey-Fuzzy Logic on the Optimal Process Design of an Injection- Molded Part with a Thin Shell Feature. *International Communications in Heat and Mass Transfer*, **33**, 94-101. http://dx.doi.org/10.1016/j.icheatmasstransfer.2005.08.006

[6]   Ozcelik, B. and Erzurumlu, T. (2005) Determination of Effecting Dimensional Parameters on Warpage of Thin Shell Plastic Parts Using Integrated Response Surface Method and Genetic Algorithm. *International Communications in Heat and Mass Transfer*, **32**, 1085-1094. http://dx.doi.org/10.1016/j.icheatmasstransfer.2004.10.032

[7]   Barghash, M.A. and Alkaabneh, F.A. (2014) Shrinkage and Warpage Detailed Analysis and Optimization for the Injection Molding Process Using Multistage Experimental Design. *Quality Engineering*, **26**, 319-334. http://dx.doi.org/10.1080/08982112.2013.852679

[8]   Kim, C-H. and Youn, J-R. (2007) Determination of Residual Stresses in Injection-Moulded Flat Plate: Simulation and Experiments. *Polymer Testing*, **26**, 862-868. http://dx.doi.org/10.1016/j.polymertesting.2007.05.006

[9]   Choi, D.-S. and Im, Y.-T. (1999) Prediction of Shrinkage and Warpage in Consideration of Residual Stress in Integrated Simulation of Injection Molding. *Composite Structures*, **47**, 655-665. http://dx.doi.org/10.1016/S0263-8223(00)00045-3

[10] Wang, T.-H. and Young, W.-B. (2005) Study on Residual Stresses of Thin-Walled Injection Molding. *European Polymer Journal*, **41**, 2511-2517. http://dx.doi.org/10.1016/j.eurpolymj.2005.04.019

[11] Kamal, M., Lai-Fook, R. and Hernandez-Aguilar, J. (2002) Residual Thermal Stresses in Injection Moldings of Thermoplastics: A Theoretical and Experimental Study. *Polymer Engineering & Science*, **42**, 1098-1114.

http://dx.doi.org/10.1002/pen.11015

[12] Mathivanan, D. and Parthasarathy, N.-S. (2009) Sink-Mark Minimization in Injection Molding through Response Surface Regression Modeling and Genetic Algorithm. *International Journal of Advanced Manufacturing Technology*, **45**, 867-874. http://dx.doi.org/10.1007/s00170-009-2021-z

[13] Changyu, S., Lixia, W. and Qian, L. (2007) Optimization of Injection Molding Process Parameters Using Combination of Artificial Neural Network and Genetic Algorithm Method. *Journal of Materials Processing Technology*, **183**, 412-418. http://dx.doi.org/10.1016/j.jmatprotec.2006.10.036

[14] Zhao, P., Zhou, H., Li, Y. and Li, D. (2010) Process Parameters Optimization of Injection Molding Using a Fast Strip Analysis as a Surrogate Model. *International Journal of Advanced Manufacturing Technology*, **49**, 949-959. http://dx.doi.org/10.1007/s00170-009-2435-7

[15] Park, K. and Ahn, J. (2004) Design of Experiment Considering Two-Way Interactions and Its Application to Injection Molding Processes with Numerical Analysis. *Journal of Materials Processing Technology*, **146**, 221-227. http://dx.doi.org/10.1016/j.jmatprotec.2003.10.020

[16] Chen, C.-P., Chuang, M.-T., Hsiao, Y.-H., Yang, Y.-K. and Tsai, C.-H. (2009) Simulation and Experimental Study in Determining Injection Molding. *Expert Systems with Applications*, **36**, 10752-10759. http://dx.doi.org/10.1016/j.eswa.2009.02.017

[17] AlKaabneh, F.A., Barghash, M. and Mishael, I. (2013) A Combined Analytical Hierarchical Process (AHP) and Taguchi Experimental Design (TED) for Plastic Injection Molding Process Settings. *The International Journal of Advanced Manufacturing Technology*, **66**, 679-694.

[18] Erzurumlu, T. and Ozcelik, B. (2006) Minimization of Warpage and Sink Index in Injection Molded Thermoplastic Parts Using Taguchi Optimization Method. *Materials and Design*, **27**, 853-861. http://dx.doi.org/10.1016/j.matdes.2005.03.017

[19] Ozcelik, B., Ozbay, A. and Demirbas, E. (2010) Influence of Injection Parameters and Mold Materials on Mechanical Properties of ABS in Plastic Injection Molding. *International Communications in Heat and Mass Transfer*, **37**, 1359-1365. http://dx.doi.org/10.1016/j.icheatmasstransfer.2010.07.001

[20] Taguchi, G. (1990) Introduction to Quality Engineering. McGraw-Hill, New York.

[21] Liao, S., Hsieh, W., Wang, J. and Su, Y. (2004) Shrinkage and Warpage Prediction of Injection-Molded Thin-Wall Parts Using Artificial Neural Networks. *Polymer Engineering and Science*, **44**, 2029-2040. http://dx.doi.org/10.1002/pen.20206

[22] Lin, B.-T., Jean, M.-D. and Chou, J.-H. (2007) Using Response Surface Methodology for Optimizing Deposited Partially Stabilized Zirconia in Plasma Spraying. *Applied Surface Science*, **253**, 3254-3262. http://dx.doi.org/10.1016/j.apsusc.2006.07.021

[23] Chen, C.-C., Su, P.-L. and Lin, Y.-C. (2009) Analysis and Modeling of Effective Parameters for Dimension Shrinkage Variation of Injection Molded Part with Thin Shell Feature Using Response Surface Methodology. *International Journal of Advanced Manufacturing Technology*, **45**, 1087-1095. http://dx.doi.org/10.1007/s00170-009-2045-4

[24] Postawa, P. and Koszkul, J. (2005) Change in Injection Moulded Parts Shrinkage and Weight as a Function of Processing Conditions. *Journal of Materials Processing Technology*, **162**, 109-115. http://dx.doi.org/10.1016/j.jmatprotec.2005.02.241

[25] Palardy, G., Hubert, P., Haider, M. and Lessard, L. (2008) Optimization of RTM Processing Parameters for Class A Surface Finish. *Composites*: *Part B*, **39**, 1280-1286. http://dx.doi.org/10.1016/j.compositesb.2007.12.003

[26] Simpoe Mesh Version 3.0.

[27] Minitab 15, MINITAB Inc., 2007.

[28] Simpoe-Mold 2011 R2.0.

[29] Fung, C.-P. and Kang, P.-C. (2005) Multi-Response Optimization in Friction Properties of PBT Composites Using Taguchi Method and Principle Component Analysis. *Journal of Materials Processing Technology*, **170**, 602-610. http://dx.doi.org/10.1016/j.jmatprotec.2005.06.040

[30] Chiang, K.-T. and Chang, F.-P. (2007) Analysis of Shrinkage and Warpage in an Injection-Molded Part with a Thin Shell Feature Using the Response Surface Methodology. *International Journal of Advanced Manufacturing Technology*, **35**, 468-479. http://dx.doi.org/10.1007/s00170-006-0739-4

[31] Mallows, C.L. (1973) Some Comments on CP. *Technometrics*, **15**, 661-675.

13

An $O(n)$ Time Algorithm for Scheduling UET-UCT of Bipartite Digraphs of Depth One on Two Processors

Ruzayn Quaddoura

Department of Computer Science, Faculty of Information Technology, Zarqa University, Zarqa, Jordan
Email: ruzayn@zu.edu.jo

Abstract

Given n unit execution time (UET) tasks whose precedence constraints form a directed acyclic graph, the arcs are associated with unit communication time (UCT) delays. The problem is to schedule the tasks on two identical processors in order to minimize the makespan. Several polynomial algorithms in the literature are proposed for special classes of digraphs, but the complexity of solving this problem in general case is still a challenging open question. We present in this paper an $O(n)$ time algorithm to compute an optimal schedule for the class of bipartite digraphs of depth one.

Keywords

Scheduling, Makespan, Precedence Constraints, Bipartite Graph, Optimal Algorithm

1. Introduction

The problem of scheduling a set of tasks on a set of identical processors under a precedence relation has been studied for a long time. A general description of the problem is the following. There are n tasks that have to be executed by m identical processors subject to precedence constraints and (may be without) communication delays. The objective is to schedule all the tasks on the processors such that the makespan is the minimum. Generally, this problem can be represented by a directed acyclic graph $G = (V, E)$ called a task graph. The set of vertices V corresponds to the set of tasks and the set of edges E corresponds to the set of precedence constrains. With every vertex i, a weight p_i is associated that represents the execution time of the task i, and with every edge (i, j), a weight c_{ij} is associated that represents the communication time between the tasks i and j. If $(i, j) \in E$ and the task i starts its execution at time t on a processor P, then either j starts its execution on P at time greater than or equal to $t + p_j$, or j starts its execution on some other processor at time greater than or equal to $t + p_j + c_{ij}$.

According to the three field notation scheme introduced in [1] and extended in [2] for scheduling problems with communication delays, this problem is denoted as $P_m \mid prec, p_i, c_{ij} \mid C_{\max}$.

A large amount of work in the literature studies this problem with a restriction on its structure: the time of execution of every task is one unit execution time (UET), the number of processors m is fixed, the communication delays are neglected, constant or one unit (UCT), or special classes of task graph are considered. We find In this context, the problem $P_2 \mid prec, p_i = 1 \mid C_{\max}$, is polynomial [3] [4], *i.e.* when the communication delays are not taken into account. On the contrary, the problem $P_3 \mid prec, p_i = 1 \mid C_{\max}$ remains an open question [5].

The problem of two processors scheduling with communication delays is extensively studied [6] [7]. In particular, it is proven in [8] that the problem $P_2 \mid prec = \text{binary tree}, p_i = 1, c_{ij} = c \mid C_{\max}$ is NP-hard where c is a large integer, whereas this problem is polynomial when the task graph is a complete binary tree.

A challenging open problem is the two processors scheduling with UET-UCT, *i.e.* the problem $P_2 \mid prec, p_i = 1, c_{ij} = 1 \mid C_{\max}$ for which the complexity is unknown. However, several polynomial algorithms have been shown for special classes of task graphs, especially for trees [9] [10], interval orders [11] and a subclass of series parallel digraphs [12]. In this paper we present an $O(n)$ time algorithm to compute an optimal algorithm for the class of bipartite digraphs of depth one, that is the digraphs for which every vertex is either a source (without predecessors) or a sink (without successors).

2. Scheduling UET-UCT for a Bipartite Digraph of Depth One on Two Processors

2.1. Preliminaries

A schedule UET-UCT on two processors for a general directed acyclic digraph $G = (V, E)$ is defined by a function $\sigma : V \to \mathbb{N}^+ \times \{P_1, P_2\}$, $\sigma(v) = (t_v, P_i), i = 1, 2$ where t_v is the time for which the task v is executed and P_i the processor on which the task v is scheduled. A schedule σ is feasible if:

a) $\forall u, v \in V$, if $u \neq v$ then $\sigma(u) \neq \sigma(v)$

b) If $(u, v) \in E$ then $t_u + 1 \leq t_v$ if u and v are scheduled on the same processor, and $t_u + 2 \leq t_v$ if u and v are scheduled on distinct processors.

A time t of a schedule σ is said to be idle if one of the processors is idle during this time. The makespan $C_{\max}$ or the length of a schedule σ is the last non-idle time of σ, that is:

$$C_{\max} = \max \{ t : \exists v \in V \ \sigma(v) = (t, P_i), i = 1 \text{ or } 2 \}$$

A schedule σ is optimal if $C_{\max}$ is the minimum among all feasible schedules.

Let $G = (B \cup W, E)$ be a bipartite digraph of depth one. Since every vertex of G is either a source or a sink, there exists always a feasible schedule such that the sources B are executed before executing the sinks W. Our algorithm for solving the problem under consideration produces an optimal schedule satisfies this condition and that we called a natural schedule defined as follows.

Definition 1 *Let* $G = (B \cup W, E)$ *be a bipartite digraph of depth one. A natural schedule of G is obtained by scheduling first the sources B then the sinks W starting from the processor* P_1 *and alternating between* P_1 *and* P_2 *such that the resulting schedule is optimal.*

The definition of a natural schedule σ of a bipartite digraph of depth one $G = (B \cup W, E)$ implies the following properties:

1) The number of sources executed on P_1 is $\lceil |B|/2 \rceil$ and the number of sources executed on P_2 is $\lfloor |B|/2 \rfloor$.

2) If $|B|$ is even then:

    a) σ contains at most 2 idle times, the first is at time $|B|/2 + 1$, and the second is at time $C_{\max}$.

    b) If $|B|/2 + 1$ is an idle time then P_2 is idle at this time (may be P_1 also).

3) If $|B|$ is odd then:

    a) σ contains at most 3 idle times, the first is at time $\lceil |B|/2 \rceil$, the second is at time $\lceil |B|/2 \rceil + 1$, and the third is at time $C_{\max}$.

    b) If $\lceil |B|/2 \rceil$ or $\lceil |B|/2 \rceil + 1$ is an idle time then P_2 is idle at this time and P_1 is non.

4) $\lceil |B \cup W|/2 \rceil \leq C_{\max} \leq \lceil |B \cup W|/2 \rceil + 1$.

Without loss of generality, we can suppose that both idle times $C_{\max}$ and $\lceil |B \cup W|/2 \rceil + 1$, if exist, are distinct. In this supposition, $C_{\max} = \lceil |B \cup W|/2 \rceil$ if and only if σ has at most the idle time $C_{\max}$ otherwise.

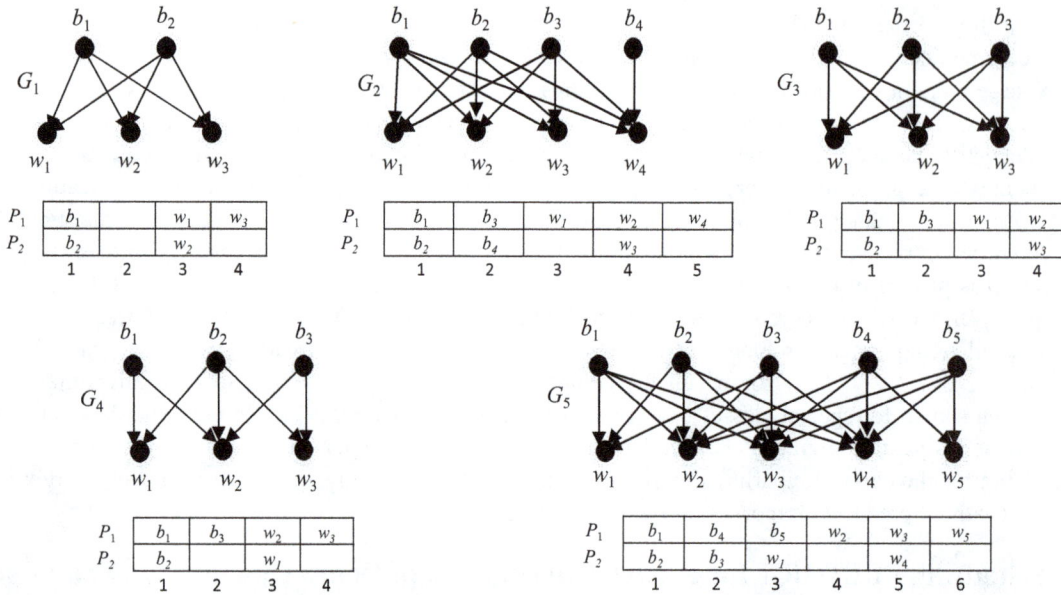

Figure 1. Bipartite digraphs of depth one and their corresponding natural schedules.

$C_{\max} = \lceil |B \cup W|/2 \rceil + 1$. **Figure 1** illustrates some bipartite digraphs of depth one and their corresponding natural schedules.

2.2. Scheduling Algorithm

The idea of solving the problem $P_2 \mid prec = bipartite\ of\ depth\ one, p_j = 1, c_{ij} = 1 \mid C_{\max}$ is to determine the necessary and sufficient conditions to exist idle times in a natural schedule of the task graph. In the following, we consider $G = (B \cup W, E)$ is a bipartite digraph of depth one where B is the set of sources and W is the set of sinks, and σ is a natural schedule for G. A vertex $b \in B$ ($w \in W$) is called universal if $d(b) = |W|$ $(d(w) = |B|)$. We distinguish two cases, $|B|$ is even and $|B|$ is odd.

Lemma 2 Assume that $|B|$ is even.

1) The two processors P_1 and P_2 are idle at time $|B|/2 + 1$ if and only if G is a bipartite complete.

2) The processor P_2 only is idle at time $|B|/2 + 1$ if and only if one of the following holds:

a) Every vertex of B is universal except exactly one.

b) Every vertex of W is universal except exactly one.

Proof. 1) If G is a bipartite complete then obviously P_1 and P_2 must be idle at time $|B|/2 + 1$. The inverse, let b and w be a source and a sink of G. If b is not adjacent to w, then we can schedule b on P_2 at time $|B|/2$ and w on P_1 at time $|B|/2 + 1$, a contradiction. So b must be adjacent to w, therefore G is a bipartite complete.

2) Assume that P_2 only is idle at time $|B|/2 + 1$ and the conditions a and b are not hold. Then, there exist $b_1, b_2 \in B$ and $w_1, w_2 \in W$ such that b_1, b_2, w_1 and w_2 are all not universal. If $\{b_1, b_2, w_1, w_2\}$ is a stable set, i.e. no two vertices are adjacent, we can schedule b_1, b_2 on P_1, P_2 at time $|B|/2$ and w_1, w_2 on P_1, P_2 at time $|B|/2 + 1$, a contradiction. If $\{b_1, b_2, w_1, w_2\}$ is not a stable set then we can suppose that $b_1 w_1 \in E$. Since b_1 and w_1 are not universal, there exist $w \in W$ and $b \in B$ such that $b_1 w \notin E$ and $b w_1 \notin E$. But now, we can schedule b_1, b on P_1, P_2 respectively at time $|B|/2$, and w_1, w on P_1, P_2 respectively at time $|B|/2 + 1$, a contradiction.

The inverse, suppose that every vertex of B is universal except exactly one. Since $|B|$ is even, there exist $b_1, b_2 \in B$ scheduled on P_1, P_2 at time $|B|/2$. By 1, the two processors can't be idle at time $|B|/2 + 1$. If P_2 is not idle at time $|B|/2 + 1$, then both b_1 and b_2 are not universal, a contradiction. In a similar way we prove the case b. □

Notice that if B (or W) contains exactly one non-universal vertex b then the vertex of W which is independent of b is also non-universal but it is not necessary unique (see **Figure 1**). Algorithm Schedule_$|B|$_is_even (G) constructs a natural schedule for $G = (B \cup W, E)$ if $|B|$ is even, Lemma 2 proves its correctness.

Algorithm Schedule_|B|_is_even (G)

$B_1 = \{b \in B : d(b) = |W|\}$

$B_2 = \{b \in B : d(b) < |W|\}$

$W_1 = \{w \in W : d(w) = |B|\}$

$W_2 = \{w \in W : d(w) < |B|\}$

If $|B_1| = |B|$ and $|W_1| = W$ then

    Schedule B_1 alternately on P_1 and P_2 at times $1, 2, \cdots, |B|/2$

    Schedule W_1 alternately on P_1 and P_2 at times $|B|/2 + 1, \cdots, \lceil |B \cup W|/2 \rceil + 1$

Else if $|B_2| = 1$ or $|W_2| = 1$ then

    Let $b \in B_2$ and $w \in W_2$ such that $bw \notin E$

    Schedule b on P_2 at time $|B|/2$ and w on P_1 at time $|B|/2 + 1$

    Schedule $B - \{b\}$ alternately on P_1 and P_2 at times $1, 2, \cdots, |B|/2$

    Schedule $W - \{w\}$ alternately on P_1 and P_2 at times $|B|/2 + 1, \cdots, \lceil |B \cup W|/2 \rceil + 1$

Else let $b_1, b_2 \in B_2$ and $w_1, w_2 \in W_2$ such that $b_1 w_1 \notin E, b_2 w_2 \notin E$

    Schedule b_1 on P_1 and b_2 on P_2 at time $|B|/2$

    Schedule w_2 on P_1 and w_1 on P_2 at time $|B|/2 + 1$

    Schedule $B - \{b_1, b_2\}$ alternately on P_1 and P_2 at times $1, 2, \cdots, |B|/2 - 1$

    Schedule $W - \{w_1, w_2\}$ alternately on P_1 and P_2 at times $|B|/2 + 2, \cdots, \lceil |B \cup W|/2 \rceil$

    Figure 1 shows the construction of natural schedules of the graphs G_1 and G_2 resulting from the algorithm Schedule_|B|_is_even (G).

    Lemma 3 *Assume that $|B|$ is odd.*

1) *The processor P_2 is idle at times $\lceil |B|/2 \rceil$ and $\lceil |B|/2 \rceil + 1$ if and only if G is a bipartite complete.*

2) *The processor P_2 is idle at time $\lceil |B|/2 \rceil$ and not idle at time $\lceil |B|/2 \rceil + 1$ if and only if*

    a) *There is $w \in W$ such that $d(w) = |B| - 1$.*

    b) *For every $w \in W$, $d(w) = |B|$ or $|B| - 1$.*

3) *The processor P_2 is idle at time $\lceil |B|/2 \rceil + 1$ and not idle at time $\lceil |B|/2 \rceil$ if and only if*

    a) *There is $w \in W$ such that $d(w) \leq |B| - 2$*

    b) *For every $w \in W$ for which $d(w) \leq |B| - 2$ and for every $b \in \bar{N}(w)$, $d(b) = |W| - 1$ where $\bar{N}(w)$ is the set of non-neighbors of w.*

Proof. 1) If G is a bipartite complete then obviously P_2 is idle at times $\lceil |B|/2 \rceil$ and $\lceil |B|/2 \rceil + 1$. The inverse, let b and w be a source and a sink of G. If b is not adjacent to w, then we can schedule b on P_1 at time $\lceil |B|/2 \rceil$ and w on P_2 at time $\lceil |B|/2 \rceil$ or at time $\lceil |B|/2 \rceil + 1$ according to the adjacency relation between the source scheduled on P_1 at time $\lceil |B|/2 \rceil - 1$ and w, a contradiction. So b must be adjacent to w, therefore G is a bipartite complete.

2) Assume that P_2 is idle at time $\lceil |B|/2 \rceil$ and not idle at time $\lceil |B|/2 \rceil + 1$. The vertex $w \in W$ scheduled on P_2 at time $\lceil |B|/2 \rceil + 1$ is of degree less than or equal to $|B| - 1$, since it must be independent of the vertex $b \in B$ scheduled on P_1 at time $\lceil |B|/2 \rceil$. If there is a vertex $w \in W$ such that $d(w) \leq |B| - 2$, then we can schedule w on P_2 at time $\lceil |B|/2 \rceil$ and two vertices from B independent of w on P_1 at times $\lceil |B|/2 \rceil - 1$ and $\lceil |B|/2 \rceil$, a contradiction.

The inverse, by 1, the processor P_2 is not idle at time $\lceil |B|/2 \rceil$ or at time $\lceil |B|/2 \rceil + 1$. If P_2 is not idle at time $\lceil |B|/2 \rceil$ then the vertex scheduled at this time would be of degree less than or equal to $|B| - 2$, a contradiction.

3) Assume that P_2 is idle at time $\lceil |B|/2 \rceil + 1$ and not idle at time $\lceil |B|/2 \rceil$. The vertex $w \in W$ scheduled on P_2 at time $\lceil |B|/2 \rceil$ is of degree less than or equal to $|B| - 2$, since it must be independent of the two vertices $b_1, b_2 \in B$ scheduled on P_1 at times $\lceil |B|/2 \rceil - 1$ and $\lceil |B|/2 \rceil$. Let $w \in W$ such that $d(w) \leq |B| - 2$ and let $b \in \bar{N}(w)$ such that $d(b) \leq |W| - 2$. Now, we can schedule w on P_2 at time $\lceil |B|/2 \rceil$, b and another vertex b' independent of w on P_1 at time $\lceil |B|/2 \rceil$ and $\lceil |B|/2 \rceil - 1$ respectively, and schedule a vertex w' independent of b on P_2 at time $\lceil |B|/2 \rceil + 1$, a contradiction. The inverse, by 1 and 2, the processor P_2 can't be idle at time $\lceil |B|/2 \rceil$ and any vertex scheduled on P_2 at time $\lceil |B|/2 \rceil$ must be of degree less than or equal to $|B| - 2$. The processor P_2 is idle at time $\lceil |B|/2 \rceil + 1$, otherwise the vertex scheduled on P_1 at time $\lceil |B|/2 \rceil$ is of degree less than or equal to $|W| - 2$, a contradiction. $\square$

    To construct a natural schedule for G when $|B|$ is odd, we need to the Procedure Two_Vertices (G). This procedure return 1 if the condition 3.b of Lemma 3 holds, and return two vertices w, b such that $d(w) \leq |B| - 2$, $b \in \bar{N}(w)$ and $d(b) \leq |W| - 2$ if this condition is not hold.

Procedure Two_Vertices (G)

$W_3 = \{w \in W : d(w) \le |B| - 2\} = \{w_1, \cdots, w_{|W_3|}\}$

For $i = 1$ to $|W_3|$

$\quad \bar{N}(w_i) = \{b_1^i, b_2^i, \cdots, b_k^i\}$

$\quad$ For $j = 1$ to k

$\quad\quad$ If $d(b_j^i) \le |W| - 2$ then Return (w_i, b_j^i)

Return 1.

Algorithm Schedule_|B|_is_odd (G) constructs a natural schedule for $G = (B \cup W, E)$ if $|B|$ is odd. Lemma 3 proves its correctness.

Algorithm Schedule_|B|_is_odd (G)

$B_1 = \{b \in B : d(b) = |W|\}$

$B_2 = \{b \in B : d(b) < |W|\}$

$W_1 = \{w \in W : d(w) = |B|\}$

$W_2 = \{w \in W : d(w) = |B| - 1\}$

$W_3 = \{w \in W : d(w) \le |B| - 2\}$

If $|B_1| = |B|$ and $|W_1| = |W|$ then

$\quad$ Schedule a vertex $b \in B$ on P_1 at time $\lceil |B|/2 \rceil$

$\quad$ Schedule a vertex $w \in W$ on P_1 at time $\lceil |B|/2 \rceil + 1$

$\quad$ Schedule $B - \{b\}$ alternately on P_1 and P_2 at times $1, \cdots, \lceil |B|/2 \rceil - 1$

$\quad$ Schedule $W - \{w\}$ alternately on P_1 and P_2 at times $\lceil |B|/2 \rceil + 2, \cdots, \lceil |B \cup W|/2 \rceil + 1$

Else if $|W_3| = 0$ then

$\quad$ Let $w \in W_2$ and $b \in B_2$ such that $bw \notin E$

$\quad$ Schedule b on P_1 at time $\lceil |B|/2 \rceil$

$\quad$ Schedule w on P_2 at time $\lceil |B|/2 \rceil + 1$

$\quad$ Schedule $B - \{b\}$ alternately on P_1 and P_2 at times $1, \cdots, \lceil |B|/2 \rceil - 1$

$\quad$ Schedule $W - \{w\}$ alternately on P_1 and P_2 at times $\lceil |B|/2 \rceil + 1, \cdots, \lceil |B \cup W|/2 \rceil + 1$

Else If Two_Vertices (G) = 1 then

$\quad$ Let $w \in W_3$ and $b_1, b_2 \in B_2$ such that $b_1 w, b_2 w \notin E$

$\quad$ Schedule b_1, b_2 on P_1 at times $\lceil |B|/2 \rceil - 1$ and $\lceil |B|/2 \rceil$

$\quad$ Schedule w on P_2 at time $\lceil |B|/2 \rceil$

$\quad$ Schedule $B - \{b_1, b_2\}$ alternately on P_1 and P_2 at times $1, \cdots, \lceil |B|/2 \rceil - 1$

$\quad$ Schedule a vertex $w' \in W - \{w\}$ on P_1 at time $\lceil |B|/2 \rceil + 1$

$\quad$ Schedule $W - \{w, w'\}$ alternately on P_1 and P_2 at times $\lceil |B|/2 \rceil + 2, \cdots, \lceil |B \cup W|/2 \rceil + 1$

Else let (w, b) = Two_Vertices (G)

$\quad$ Let $b_1 \in B_2 - \{b\}$ and $w_1 \in W_2 \cup W_3 - \{w\}$ such that $b_1 w, bw_1 \notin E$

$\quad$ Schedule b_1, b on P_1 at times $\lceil |B|/2 \rceil - 1$ and $\lceil |B|/2 \rceil$ respectively

$\quad$ Schedule w, w_1 on P_2 at times $\lceil |B|/2 \rceil$ and $\lceil |B|/2 \rceil + 1$ respectively

$\quad$ Schedule $B - \{b_1, b\}$ alternately on P_1 and P_2 at times $1, \cdots, \lceil |B|/2 \rceil - 1$

$\quad$ Schedule $W - \{w, w_1\}$ alternately on P_1 and P_2 at times $\lceil |B|/2 \rceil + 1, \cdots, \lceil |B \cup W|/2 \rceil$

Figure 1 shows the construction of natural schedules of the graphs G_3, G_4 and G_5 resulting from the algorithm Schedule_|B|_is_odd (G).

2.3. Complexity

We assume that $G = (B \cup W, E)$ is represented by its adjacency lists, so the set of neighbors and the set of non neighbors of every vertex of G are known already. In this supposition, we can check easily that any step (except the step Two_Vertices (G)) of the two algorithms Schedule_|B|_is_even (G) and Schedule_|B|_is_ odd (G) can be executed either within a constant time or within an $O(n)$ time where $|B \cup W| = n$. Let's prove that the Procedure Two_Vertices (G) runs within $O(|B|)$ time.

The worst case of this Procedure occurs when its result is 1. In this case, for any $w_1, w_2 \in W_3$, $\bar{N}(w_1) \cap \bar{N}(w_2) = \varnothing$, otherwise, a vertex b independent of w_1 and w_2 would be of degree less than or equal to $|W| - 2$. So the number of comparisons of if statement in this procedure is equal to at most $|\bar{N}(w_1)| + \cdots + |\bar{N}(w_{|W_3|})| \le |B|$. Therefore, the procedure Two_Vertices (G) runs within $O(|B|)$ time and the total time of our scheduling algorithm is $O(n)$.

3. Conclusion

We have presented an $O(n)$ time algorithm for the optimal schedule of bipartite digraphs of depth one with UET-UCT on two processors. The complexity of this problem for general directed acyclic graphs is still an open question. We believe that our algorithm can be used to solve this problem in general as follow: Consider a topological sort of a directed acyclic graph G. The linear ordering defined by this topological sort decomposes G into consecutives bipartite digraphs of depth one. The schedule obtained by the concatenation of the schedules of these bipartite digraphs is a feasible schedule or may be modified to a feasible schedule of G. Now, if we can determine the necessary and sufficient conditions to exist idle times in this feasible schedule then we can determine the complexity of this problem. This is a useful guide and foundation for future research.

Acknowledgements

This research is funded by the Deanship of Research and Graduate Studies in Zarqa University/Jordan. The author is grateful to anonymous referee's suggestion and improvement of the presentation of this paper.

References

[1] Graham, R.L., Lawler, E.L., Lenstra, J.K. and Rinnooy Kan, A.H.G. (1979) Optimization and Approximation in Deterministic Scheduling: A Survey. *Annals of Discrete Mathematics*, **5**, 287-326. http://dx.doi.org/10.1016/S0167-5060(08)70356-X

[2] Veltman, B., Lageweg, B.J. and Lenstra, L.K. (1990) Multiprocessor Scheduling with Communication Delays. *Parallel Computing*, **16**, 173-182. http://dx.doi.org/10.1016/0167-8191(90)90056-F

[3] Coffman Jr., E.G. and Graham, R.L. (1972) Optimal Scheduling for Two-Processor Systems. *Acta Informatica*, **1**, 200-213. http://dx.doi.org/10.1007/BF00288685

[4] Fujii, M., Kasami, T. and Ninomiya, K. (1969) Optimal Sequencing of Two Equivalent Processors. *SIAM Journal on Applied Mathematics*, **17**, 784-789. http://dx.doi.org/10.1137/0117070

[5] Garey, M.R. and Johnson, D.S. (1979) Computers and Intractability: A Guide to the Theory of NP-Completeness. Freeman.

[6] Chrétienne, P. and Picouleau, C. (1995) Scheduling with Communication Delays: A Survey. In: *Scheduling Theory and Its Applications*, John Wiley & Sons.

[7] Norman, M.G., Pelagatti, S. and Thanisch, P. (1995) On the Complexity of Scheduling with Communication Delay and Contention. *Parallel Processing Letters*, **5**, 331-341. http://dx.doi.org/10.1142/S012962649500031X

[8] Afrati, F., Bampis, E., Finta, L. and Mili, I. (2005) Scheduling Trees with Large Communication Delays on Two Identical Processors. *Journal of Scheduling*, **8**, 179-190. http://dx.doi.org/10.1007/s10951-005-6366-3

[9] Varvarigou, T., Roychowdhury, V.P., Kailath, T. and Lawler, E. (1996) Scheduling in and out Forests in the Presence of Communication Delays. *IEEE Transactions on Parallel and Distributed Systems*, **7**, 1065-1074. http://dx.doi.org/10.1109/71.539738

[10] Veldhorst, M. (1993) A Linear Time Algorithm to Schedule Trees with Communication Delays Optimally on Two Machines. Technical Report COSOR 93-07, Department of Math, and Computer Science, Eindhoven University of Technology, Eindhoven.

[11] Ali, H. and El-Rewini, H. (1993) The Time Complexity of Scheduling Interval Orders with Communication Is Polynomial. *Parallel Processing Letters*, **3**, 53-58. http://dx.doi.org/10.1142/S0129626493000083

[12] Finta, L., Liu, Z., Mills, I. and Bampis, E. (1996) Scheduling UET-UCT Series Parallel Graphs on Two Processors. *Theoretical Computer Science*, **162**, 323-340. http://dx.doi.org/10.1016/0304-3975(96)00035-7

14

A Note on Standard Goal Programming with Fuzzy Hierarchies: A Sequential Approach

Maged George Iskander

Faculty of Business Administration, Economics and Political Science, The British University in Egypt, El-Sherouk City, Egypt

Email: maged.iskander@bue.edu.eg

Abstract

In the paper [Standard goal programming with fuzzy hierarchies: a sequential approach, Soft Computing, First online: 22 March 2015], it has been assumed that the normalized deviations should lie between zero and one. In some cases, this assumption may not be valid. Therefore, additional constraints must be incorporated into the model to ensure that the normalized deviations should not exceed one. This modification is illustrated by the given numerical example.

Keywords

Fuzzy Goal Programming, Imprecise Hierarchy, Normalized Deviations

1. Introduction

The problem of fuzzy goal programming when the importance relation between the fuzzy goals is vague has initially been investigated by Aköz and Petrovic [1] and followed by Li and Hu [2] and Cheng [3]. A suggested sequential approach in fuzzy goal programming, when the importance hierarchy of the goals is imprecise, has been presented by Arenas-Parra *et al.* [4]. In their article, the model of goal programming with fuzzy hierarchy (GPFH) is given as:

$$\text{Maximize } \lambda \sum_{i=1}^{k}\left(1 - \frac{n_i}{m_i - \underline{f_i}}\right) + \left(1 - \lambda\right)\sum_{\substack{(i,j)=1 \\ i \neq j}}^{k}\sum_{r=1}^{3} b_{\tilde{R}_r(i,j)}\mu_{\tilde{R}_r(i,j)}$$

subject to:

$$f_i(x) + n_i - p_i = m_i, \quad i = 1, \cdots, k,$$

$$1 - \left(\frac{n_i}{m_i - \underline{f_i}} - \frac{n_j}{m_j - \underline{f_j}} \right) \geq \mu_{\tilde{R}_1(i,j)}, \quad \text{if } b_{\tilde{R}_1(i,j)} = 1,$$

$$\frac{1 - \left(\dfrac{n_i}{m_i - \underline{f_i}} - \dfrac{n_j}{m_j - \underline{f_j}} \right)}{2} \geq \mu_{\tilde{R}_2(i,j)}, \quad \text{if } b_{\tilde{R}_2(i,j)} = 1, \tag{1}$$

$$\frac{n_j}{m_j - \underline{f_j}} - \frac{n_i}{m_i - \underline{f_i}} \geq \mu_{\tilde{R}_3(i,j)}, \quad \text{if } b_{\tilde{R}_3(i,j)} = 1,$$

$$0 \leq \mu_{\tilde{R}_r(i,j)} \leq 1, \quad r = 1, 2, 3,$$

$$n_i, p_i \geq 0, \quad n_i \times p_i = 0, \quad i = 1, \cdots, k,$$

$$x \in X,$$

where $0 \leq \lambda \leq 1$, and $f_i(x)$ is an ith linear function of an x vector of decision variables, $i = 1, \cdots, k$. n_i and p_i are the negative and positive deviations, respectively, where m_i is the aspiration level and $\underline{f_i}$ is the anti-ideal value for the ith fuzzy goal constraint. $b_{\tilde{R}_r(i,j)}$ ($r = 1, 2, 3$) is a binary variable associated with the membership function of the rth importance relation (slightly, moderately, significantly) of the ith goal more than the jth goal; while $\mu_{\tilde{R}_r(i,j)}$ is the membership function of the rth imprecise relation between the ith and the jth fuzzy goals. X is the set of system constraints which define the feasible set of the problem.

This model is implemented for each class of *Phase I*. Hence, it is assumed that the normalized deviation for the ith fuzzy goal constraint must lie between zero and one, *i.e.*

$$0 \leq n_i / (m_i - \underline{f_i}) \leq 1. \tag{2}$$

This assumption may be violated, especially when the anti-ideal value is close to the aspiration level. In this case, $n_i / (m_i - \underline{f_i})$ may exceed one, due to a small denominator value, which means that the value of the achieved goal is worse than the anti-ideal value of that goal. Accordingly, for each class, the following constraints should be incorporated in the GPFH model

$$n_i \leq m_i - \underline{f_i}, \tag{3}$$

if the negative deviation is required to be minimized for the ith fuzzy goal constraint, *i.e.*, if $f_i(x) \geq m_i$; or

$$p_i \leq \underline{f_i} - m_i, \tag{4}$$

if the positive deviation is required to be minimized for the ith fuzzy goal constraint, *i.e.*, if $f_i(x) \leq m_i$.

Notably, constraints (3) and (4) correspond to the non-negativity of the membership functions of the fuzzy goal constraints given by Aköz and Petrovic [1].

Proposition: The normalized deviations constraints might limit the feasible set of the problem. This may worsen the value of the achievement function of each class, and therefore affect the results of the suggested sequential approach.

In the next section, this note is verified by the given illustrative example.

2. Illustrative Example

The GPFH model (*Phase I*) is solved using the following example that is given by Arenas-Parra *et al.* [4]

Goal 1: $4x_1 + 2x_2 + 8x_3 + x_4 \leq 35$
Goal 2: $4x_1 + 7x_2 + 6x_3 + 2x_4 \geq 100$

Goal 3: $x_1 - 6x_2 + 5x_3 + 10x_4 \geq 120$
Goal 4: $5x_1 + 3x_2 + 2x_4 \geq 70$
Goal 5: $4x_1 + 4x_2 + 4x_3 \geq 40$

subject to:

$$\left. \begin{array}{l} 7x_1 + 5x_2 + 3x_3 + 2x_4 \leq 98, \\ 7x_1 + x_2 + 2x_3 + 6x_4 \leq 117, \\ x_1 + x_2 + 2x_3 + 6x_4 \leq 130, \\ 9x_1 + x_2 + 6x_4 \leq 105, \\ x_i \geq 0,\, i = 1, \cdots, 4, \end{array} \right\} X$$

where *Class I* contains goals (1, 2, and 4). Accordingly, the assumed anti-ideal values for those goals are $\underline{f_1} = 261.33$, $\underline{f_2} = 0$, $\underline{f_4} = 0$. Also, the GPFH model for *Class I* assumes that Goal 1 is *moderately more important than* Goal 2; and Goal 2 is *moderately more important than* Goal 4. Finally, the parameter λ_I is set equal to 0.8.

Thus, the information of the model for *Class I* is as follows:

$$\text{Maximize } AF_I = \lambda_I \left(1 - \frac{P_1}{226.33} + 1 - \frac{n_2}{100} + 1 - \frac{n_4}{70} \right) + (1 - \lambda_I) \left[\mu_{\tilde{R}_2(1,2)} + \mu_{\tilde{R}_2(2,4)} \right]$$

subject to:

$$4x_1 + 2x_2 + 8x_3 + x_4 + n_1 - p_1 = 35,$$

$$4x_1 + 7x_2 + 6x_3 + 2x_4 + n_2 - p_2 = 100,$$

$$5x_1 + 3x_2 + 2x_4 + n_4 - p_4 = 70,$$

$$\frac{1 - \left(\dfrac{p_1}{226.33} - \dfrac{n_2}{100} \right)}{2} \geq \mu_{\tilde{R}_2(1,2)},$$

$$\frac{1 - \left(\dfrac{n_2}{100} - \dfrac{n_4}{70} \right)}{2} \geq \mu_{\tilde{R}_2(2,4)},$$

$$0 \leq \mu_{\tilde{R}_2(1,2)} \leq 1, \quad 0 \leq \mu_{\tilde{R}_2(2,4)} \leq 1,$$

$$n_k, p_k \geq 0, \quad n_k \times p_k = 0, \quad k = 1, 2, 4,$$

$$x \in X.$$

Our comment is verified by just resolving the GPFH model for *Class I* in *Phase I*. Assume that the anti-ideal values of the first and the fourth fuzzy goal constraints $\underline{f_1}$ and $\underline{f_4}$ are 40 and 63 instead of 261.33 and 0, respectively. In this case, the normalized p_1 is $p_1/5$, while the normalized n_4 becomes $n_4/7$.

Then, the solution obtained is: $\mu_{\tilde{R}_2(1,2)} = 0.463$, $\mu_{\tilde{R}_2(2,4)} = 1$, $p_1 = 0.375$, $n_2 = 0$, $n_4 = 9$, $G_1 = 35.375$, $G_2 = 100$, $G_4 = 61$, $AF_I^* = 1.604$. Hence, $n_4/7 = 1.286$, which is greater than 1.

Accordingly, by incorporating the following three constraints

$$p_1 \leq 5,$$

$$n_2 \leq 100,$$

$$n_4 \leq 7,$$

and by solving the model, the solution becomes: $\mu_{\tilde{R}_2(1,2)} = 0.325$, $\mu_{\tilde{R}_2(2,4)} = 1$, $p_1 = 1.750$, $n_2 = 0$, $n_4 = 7$, $G_1 = 36.750$, $G_2 = 105$, $G_4 = 63$, $AF_I^* = 1.585$.

It is realized that incorporating the normalized deviations constraints leads to a worse value of AF_I^*, which verifies the proposition.

3. Conclusion

The normalized deviations constraints must be included in the GPFH model in all classes of *Phase I* as well as in *Phase II* to ensure that the achieved value of each goal should never become worse than the anti-ideal value of that goal.

References

[1]   Aköz, O. and Petrovic, D. (2007) A Fuzzy Goal Programming Method with Imprecise Goal Hierarchy. *European Journal of Operational Research*, **181**, 1427-1433. http://dx.doi.org/10.1016/j.ejor.2005.11.049

[2]   Li, S. and Hu, C. (2009) Satisfying Optimization Method Based on Goal Programming for Fuzzy Multiple Objective Optimization Problem. *European Journal of Operational Research*, **197**, 675-684. http://dx.doi.org/10.1016/j.ejor.2008.07.007

[3]   Cheng, H.-W. (2013) A Satisficing Method for Fuzzy Goal Programming Problems with Different Importance and Priorities. *Quality and Quantity*, **47**, 485-498. http://dx.doi.org/10.1007/s11135-011-9531-0

[4]   Arenas-Parra, M., Bilbao-Terol, A. and Jiménez, M. (2015) Standard Goal Programming with Fuzzy Hierarchies: A Sequential Approach. *Soft Computing*. http://dx.doi.org/10.1007/s00500-015-1644-2

15

A New Approach of Solving Single Objective Unbalanced Assignment Problem

Ventepaka Yadaiah[1], V. V. Haragopal[2]

[1]Department of Mathematics, Osmania University, Hyderabad, India
[2]Department of Statistics, Osmania University, Hyderabad, India
Email: v.yadaiah@gmail.com, haragopalvajjha@gmail.com

Abstract

In this paper, we discuss a new approach for solving an unbalanced assignment problem. A Lexi-search algorithm is used to assign all the jobs to machines optimally. The results of new approach are compared with existing approaches, and this approach outperforms other methods. Finally, numerical example (Table 1) has been given to show the efficiency of the proposed methodology.

Keywords

Assignment Problem, Lexi-Search Algorithm, Jobs Clubbing Method

1. Introduction

Consider a problem which consists of a set of "n" machines $M = \{M_1, M_2, M_3, \cdots, M_n\}$. A set of "m" jobs $J = \{J_1, J_2, J_3, \cdots, J_m\}$ which are to be considered assign for execution on "n" available machines. The execution cost of each job on all the machines are known and mentioned in the matrix, namely assigned cost matrix (ACM) of order, where $m > n$. The unbalanced assignment problem is a special type of linear programming problem in which our objective is to assign number of salesmen to number of areas at a minimum cost (time). The mathematical formulation of the problem suggests that this is a 0 - 1 programming problem. It is highly degenerate all the algorithms developed to find optimal solution of transportation problem, applicable to unbalanced assignment problem. However, due to its highly degeneracy nature a specially designed algorithm, widely known as Hungarian method proposed by Kuhn [1], is used for its solution, and Kadhirvel and Balamurugan [2] solved the unbalanced assignment problems using triangular fuzzy Numbers. Different methods have been presented for Assignment Problem and various articles have been published on the subject [3]-[7].

The objectives are to determine the optimal assignment cost, in such a way that all the jobs are to be allotted on the available machines in an optimum way. The mathematical formulation of the assignment problem [8] [9] is as follows.

Table 1. Assigned cost matrix (ACM).

	M_1	M_2	M_3	M_4	M_5
J_1	300	290	280	290	210
J_2	250	310	290	300	200
J_3	180	190	300	190	180
J_4	320	180	190	240	170
J_5	270	210	190	250	160
J_6	190	200	220	190	140
J_7	220	300	230	180	160
J_8	260	190	260	210	180

2. Model Construction of Simple Assignment Problem

Minimize (Maximize):

$$Z = \sum_{i=1}^{m}\sum_{i=1}^{n} C_{ij} X_{ij}$$

Subject to

$$\sum_{j=1}^{n} X_{ij} = 1; \text{ for } i = 1, 2, 3, \cdots, m$$

$$\sum_{i=1}^{m} X_{ij} = 1; \text{ for } j = 1, 2, 3, \cdots, n$$

where $X_{ij} = \begin{cases} 1, & \text{if the } i^{th} \text{ job is assigned to the } j^{th} \text{ machine.} \\ 0, & \text{if the } i^{th} \text{ job is not assigned to the } j^{th} \text{ machine.} \end{cases}$

Problem definition:

Also, if the numbers of jobs are not equal to number of machines, then it is known as an unbalanced assignment problem. Now consider the assumptions of choosing an unbalanced assignment problem as:

• The completion of a program from computational point of view means that the all jobs are assigned to various machines and final optimal assignment cost has been obtained.

• The number of jobs are more than number of machines.

The variants of assignment problem are considered by various researchers like Kagade & Bajaj [10] and Avanish Kumar [11]. From the work of these authors, they found that the approach of clubbing the costs of the jobs was implemented for multi objective problems and single objective problems, where as this paper considers the clubbing of jobs for an assignment problem by the exact solution problem with Lexi-search approach [12] [13].

3. Methodology

To determine the assignment cost as well as combination of job (s) Vs machine (s) of an unbalanced assignment problem for a set of "n" machines $M = \{M_1, M_2, M_3, \cdots, M_n\}$. A set of "m" jobs $J = \{J_1, J_2, J_3, \cdots, J_m\}$ which are to be considered as assigned for execution on "n" available machines with an execution cost C_{ij}, where $i = 1, 2, \cdots, m$ and $j = 1, 2, \cdots, n$ are mentioned in the ACM of order, where $m > n$. First of all, we obtain the sum of each row and each column of the ACM store and the results should be arranged in the array, namely, Sum_Row and Sum_Column. Then we select the first m rows (jobs) on the basis of Sum_Row that is, starting with the most minimum to next minimum to the array Sum_Row and deleting rows (jobs) correspo- nding to the remaining (m-n) jobs. Store results in the new array that will be the array for the first sub problem (**Table 2**). Repeating this process until the remaining jobs become less than "n" machines, when remaining jobs are less than n then deleting (n-m) columns (machines) on the basis of Sum_Column. That is, corresponding to value (s) most maximum to next maximum to form the last sub problem (**Table 3**). Store the results in the new array that shall be the array for the last sub problem. which are now balanced assignment problems, in this way for the

Table 2. First sub-problem $N_1ACM(,)$.

	M_1	M_2	M_3	M_4	M_5
J_3	180	190	300	190	180
J_4	320	180	190	240	170
J_5	270	210	190	250	160
J_6	190	200	220	190	140
J_7	220	300	230	180	160

Table 3. Second sub problem $N_2ACM(,)$.

	M_2	M_4	M_5
J_1	290	290	210
J_2	310	300	200
J_8	190	210	180

defined assignment problem. Now we apply the Lexi-search approach to obtain the exact optimum solution of each sub problem (**Tables 4-7**). Finally, add the total assignment cost of each sub problem to obtain the optimal assignment cost along with assignment sets. And also we check the assignment cost for jobs clubbing problem (**Table 8**) through Lexi-search approach (**Tables 9-11**), getting the same value. To solve this problem we follow the following algorithm.

Algorithm

Step-1: Consider "m" jobs on "n" machines costs given as a matrix (ACM), which is an unbalanced assignment problem where $m > n$.

Step-2:

Step-2.1: Obtain the sum of each row and column of the ACM and the store the results in the arrays namely *Sum_Row* and *Sum_Column*.

Step-2.2: Select the first m rows (jobs) on the basis of *Sum_Row*. That is, starting with the most minimum to next minimum to the array *Sum_Row* and deleting rows (jobs) corresponding to the remaining (m-n) jobs. Store the results in the new array that shall be the array for the first sub problem.

Step-2.2.1: If there is no remaining jobs, *i.e.*, (m-n = 0), then go to step-3.

Step-2.2.2: If the remaining (m-n) jobs are still more than n, then repeat step-2.2 for the remaining jobs to form next sub-problem (s), else, step-2.3.

Step-2.3: If remaining jobs are less than n then deleting (n-m) columns (machines) on the basis of *Sum_Column*. That is corresponding to value (s) most maximum to next maximum to form the last sub problem. Store the results in the new array that shall be the array for the last sub problem.

Step-3: If the total effectiveness of ACM is to be maximized, change the sign of each cost element in the effectiveness matrix and go to step-4, otherwise go directly to step-5 if ACM has the total value as minimum.

Step-4: Arrange all the jobs $J_1, J_2, J_3, \cdots, J_n$ according to their cost (*i.e.* available jobs). This arrangement consists of n columns and m rows. Each column represents a machine, and the elements in that column are the costs arranged in increasing order according to their jobs.

Step-5: Include the job from the first machine in the partial solution value (psv) "w". If the cost itself is greater than or equal to trial value (TRV) then stop. Otherwise go to next step.

Step-6: Calculate the bound.

Step-7: If the sum of bound and psv is greater than or equal to TRV then drop the job added in step 5, and go to step 5. Otherwise go to next step, *i.e.* go to Sub block (GS).

Step-8: Include the next available job (from the last job included in the partial solution "w") into the partial solution.

Table 4. Alphabet table $N_1 ACM(,)$.

M_1	M_2	M_3	M_4	M_5
$J_3 - 180$	$J_4 - 180$	$J_4 - 190$	$J_7 - 180$	$J_6 - 140$
$J_6 - 190$	$J_3 - 190$	$J_5 - 190$	$J_3 - 190$	$J_5 - 160$
$J_7 - 220$	$J_6 - 200$	$J_6 - 220$	$J_6 - 190$	$J_7 - 160$
$J_5 - 270$	$J_5 - 210$	$J_7 - 230$	$J_4 - 240$	$J_4 - 170$
$J_4 - 320$	$J_7 - 300$	$J_3 - 300$	$J_5 - 250$	$J_3 - 180$

Table 5. Search table $N_1 ACM(,)$.

M_1	M_2	M_3	M_4	M_5	A.B	Remark
$J_3 - 180$	$J_4 - 180$	$J_4 - 190RP$	$J_7 - 180$	$J_6 - 140$		
		$J_5 - 190$				
Bound = 870	Bound = 870	Bound = 870:	Bound = 870	Bound = 870:	870	
(180 + 180 + 190 + 180 + 140)	(180 + 180 + 190 + 180 + 140)	(180 + 180 + 190 + 180 + 140)	(180 + 180 + 190 + 180 + 140)	(180 + 180 + 190 + 180 + 140)		
			$J_3 - 190RP$			
			$J_6 - 190$			GNSB
			Bound: 900 > 870			
			(180 + 180 + 190 + 190 + 160)			
		$J_6 - 220$				GNSB
		Bound: 920 > 870				
		(180 + 180 + 220 + 180 + 160)				
	$J_3 - 190RP$					GNSB
	$J_6 - 200$					
	Bound: 910 > 870					
	(180 + 200 + 190 + 180 + 160)					
$J_6 - 190$						GNSB
Bound: 920 > 870						
(190 + 180 + 190 + 180 + 180)						
$J_7 - 220$						
Bound: 920 > 870						
(220 + 180 + 190 + 190 + 140)						
$J_5 - 270$						
Bound: 1030 > 870						
(270 + 180 + 220 + 180 + 180)						
$J_4 - 320$						
Bound: 1020 > 870						
(320 + 190 + 190 + 180 + 140)						

Table 6. Alphabet table $N_2ACM(,)$.

M_2	M_4	M_5
$J_8 - 190$	$J_8 - 210$	$J_8 - 180$
$J_1 - 290$	$J_1 - 290$	$J_2 - 200$
$J_2 - 310$	$J_2 - 300$	$J_1 - 190$

Table 7. Search table $N_2ACM(,)$.

M_2	M_4	M_5	A.B	Remark
$J_8 - 190$	$J_8 - 210RP$	$J_8 - 180RP$		
Bound: 680	$J_1 - 290$	$J_2 - 200$	680	
$(190 + 290 + 200)$	Bound: 680	Bound: 680		
	$(190 + 290 + 200)$	$(190 + 290 + 200)$		
	$J_2 - 300$			
	Bound: $700 > 680$			GNSB
	$(190 + 300 + 210)$			
$J_1 - 290$				
Bound: $700 > 680$				
$(290 + 210 + 200)$				
$J_2 - 310$				
Bound: $730 > 680$				
$(310 + 210 + 210)$				

Table 8. Jobs clubbing modified problem is $N_3ACM(,)$.

	M_1	M_2	M_3	M_4	M_5
$J_1 * J_7$	520	590	510	470	370
$J_2 * J_6$	440	510	510	490	340
J_2	180	190	300	190	180
$J_4 * J_8$	580	370	450	450	350
J_5	270	210	190	250	160

Table 9. Alphabet table $N_3ACM(,)$.

M_1	M_2	M_3	M_4	M_5
$J_3 - 180$	$J_3 - 190$	$J_5 - 190$	$J_3 - 190$	$J_5 - 160$
$J_5 - 270$	$J_5 - 210$	$J_3 - 300$	$J_5 - 250$	$J_3 - 180$
$J_2 * J_6 - 440$	$J_4 * J_8 - 370$	$J_4 * J_8 - 450$	$J_4 * J_8 - 450$	$J_2 * J_6 - 340$
$J_1 * J_7 - 520$	$J_2 * J_6 - 510$	$J_1 * J_7 - 510$	$J_1 * J_7 - 470$	$J_4 * J_8 - 350$
$J_4 * J_8 - 580$	$J_1 * J_7 - 590$	$J_2 * J_6 - 510$	$J_2 * J_6 - 510$	$J_1 * J_7 - 370$

Table 10. Search table $N_3ACM(,)$.

M_1	M_2	M_3	M_4	M_5	A.B	Remark
J_3-180	$J_3-190RP$	$J_5-210RP$	$J_3-190RP$	$J_5-210RP$	1650	JB
Bound: 1650	J_5-210	$J_3-190RP$	$J_5-250RP$	$J_3-190RP$		
$(180+210+450$ $+470+340)$	Bound: 1650	J_4*J_8-450	$J_4*J_8-450RP$	J_2*J_6-340		
	$(180+210+450$ $+470+340)$	Bound: 1650	J_1*J_7-470	Bound: 1650		
		$(180+210+450+$ $470+340)$	Bound: 1650	$(180+210+450$ $+470+340)$		
			$180+210+450+$ $470+340)$			
				$J_4*J_8-350RP$		GNSB
				$J_1*J_7-370RP$		
			J_2*J_6-490			GNSB
			Bound: 1700 > 1650			
			$(180+210+450+$ $490+370)$			
		J_1*J_7-510				JB
		Bound: 1690 > 1650				
		$(180+210+510+$ $450+340)$				
		J_2*J_6-510				GNSB
		Bound: 1720 > 1650				
		$(180+210+510+$ $450+370)$				
	J_4*J_8-370	J_5-190	$J_5-190RP$	$J_5-160RP$	1550	JB
	Bound: 1550 < 1650	Bound: 1550 < 1650	$J_3-300RP$	$J_3-180RP$		
	$(180+370+190$ $+470+340)$	$(180+370+190+$ $470+340)$	$J_4*J_8-450RP$	J_2*J_6-340		
			J_1*J_7-510	Bound: 1550 < 1650		
			Bound: 1550 < 1650	$(180+370+190$ $+470+340)$		
			$(180+370+190+$ $470+340)$			
				$J_4*J_8-350RP$		GNSB
				$J_1*J_7-370RP$		
			J_2*J_6-490			GNSB
			Bound: 1600 > 1550			
			$(180+370+190+$ $490+370)$			

Step-9: If partial solution value is greater than or equal to the TRV then drop the job added in step-8, and go to step-7. Otherwise go to step-10.

Step-10: If the sum of bound and psv is greater than or equal to TRV then drop the newly added job in step-8, and go to step -7.otherwise go to step 11.

Step-11: If the partial solution contains n − 1 jobs add the dummy job to the partial solution if it is greater than or equal to TRV then drop the dummy job and last two jobs from the partial solution. That is Jump out to the next higher order blocks (JO). If "w" contains only one job, go to step-5, otherwise go to step-8. Otherwise go to the next step.

Step-12: Now calculate the bound.

Step-13: If the sum of bound and psv is greater than or equal to TRV then drop the dummy job and also last job from "w", and go to step-8. Otherwise go to step-14.

Table 11. Search table $N_3 ACM(,)$.

M_1	M_2	M_3	M_4	M_5	A.B	Remark
		$J_3 - 300 RP$				JB
		$J_4 * J_8 - 450 RP$				
		$J_1 * J_7 - 510$				
		Bound: $1650 > 1550$				
		$(180 + 370 + 510 + 250 + 340)$				
		$J_3 - 300 RP$				GNSB
		$J_4 * J_8 - 450 RP$				
		$J_2 * J_6 - 510$				
		Bound: $1680 > 1550$				
		$(180 + 370 + 510 + 250 + 370)$				
	$J_2 * J_6 - 510$					
	Bound: $1700 > 1550$					
	$(180 + 510 + 190 + 450 + 370)$					
	$J_1 * J_7 - 590$					
	Bound: $1750 > 1550$					
	$(180 + 590 + 190 + 450 + 340)$					
$J_5 - 270$						
Bound: $1720 > 1550$						
$(270 + 190 + 450 + 470 + 340)$						
$J_2 * J_6 - 440$						
Bound: $1640 > 1550$						
$(440 + 190 + 190 + 450 + 370)$						
$J_1 * J_7 - 520$						
Bound: $1690 > 1550$						
$(520 + 190 + 190 + 450 + 340)$						
$J_4 * J_8 - 580$						
Bound: $1770 > 1550$						
$(580 + 190 + 190 + 470 + 340)$						

Step-14: Include the latest possible job from the dummy job in "w"

Step-15: If psv is greater than or equal to TRV then drop the last dummy job and also the job from which the ith dummy job was assigned, and go to step-8. Otherwise go to next step.

Step-16: Now calculate the bound.

Step-17: If sum of bound and psv is greater than or equal to TRV then drop the recently added job in "w" and go to step-14. Otherwise go to next step.

Step-18: Include the latest available job from the last job in "w"

Step-19: Now calculate the bound.

Step-20: If the sum of bound and psv is greater than or equal to TRV then drop the latest job, and go to step-18. Otherwise go to next step.

Step-21: If the number of elements in "w" is less than "n" go to step-18. Otherwise go to next step.

Step-22: Replace TRV by partial solution value and trial solution by w. Now go to step-18.

4. Illustration

A company is faced with the problem of assigning five different machines to eight different jobs (**Table 1**). The costs are estimated as follows (in hundreds of rupees):

Solve the problem assuming that the objective is to minimize the total cost. Now obtain the sum of each row and column of $ACM(,)$, *i.e.*, the sum of each row and each column is as follows:

$$Sum\_Row = \begin{array}{cccccccc} J_1 & J_2 & J_3 & J_4 & J_5 & J_6 & J_7 & J_8 \\ 1370 & 1350 & 1040 & 1100 & 1080 & 0940 & 1090 & 1100 \end{array}$$

$$Sum\_Column = \begin{array}{ccccc} M_1 & M_2 & M_3 & M_4 & M_5 \\ 1990 & 1870 & 1960 & 1850 & 1400 \end{array}$$

We partition the matrix $ACM(,)$ to define the first sub-problem $N_1 ACM(,)$ by selecting rows corresponding J_3, J_4, J_5, J_6, J_7 and second sub problem $N_2 ACM(,)$ by selecting rows corresponding to the jobs J_1, J_2, J_8 and by deleting columns corresponding to M_1, M_3, then the modified matrices are as follows:

Sub-Problem-I: $N_1 ACM(,)$

Sub-Problem-II: $N_2 ACM(,)$

4.1. Now apply the Lexi-search method for Sub-Problem-I: $N_1 ACM(,)$

RP: Repitition, GNSB or JB: Go to next super block (JB), A.B or TRV = Absolute bound or Trail bound.

The final optimal assignments of $N_1 ACM(,)$ as follows:

$J_3 \to M_1$, $J_4 \to M_2$, $J_5 \to M_3$, $J_6 \to M_5$, $J_7 \to M_4$

4.2. Now Apply the Lexi-Search Method for Sub-Problem-II: $N_2 ACM(,)$

The final optimal assignments $N_2 ACM(,)$ is: $J_1 \to M_4$, $J_2 \to M_5$, $J_8 \to M_2$

The final optimal assignments assigned cost matrix (ACM) is : $J_1 \to M_4$, $J_2 \to M_5$, $J_3 \to M_1$, $J_4 \to M_2$, $J_5 \to M_3$, $J_6 \to M_5$, $J_7 \to M_4$, $J_8 \to M_2$. The Hungarian method gives us total assignment cost as 890 along with the other one job assigned to dummy machine, in other words the job that is assigned to dummy machine under the Hungarian method was ignored for further processing. While, the original problem was divided into two sub problems, which are balanced assignment problem in nature. Now for the two sub problems with the use of Lexi-search approach, the total cost 870 is recorded for the sub problem-I along with none of the jobs assigned to dummy machine, and the total cost 680 was recorded for the second sub problem-II along with none of the jobs assigned to dummy machine. Now the total cost of the assigned cost matrix (ACM) is 870 + 670 = 1550.

5. Job Clubbing Method

Jobs Clubbing Modified Problem Is $N_3 ACM(,)$: Lexi-Search Approach

The final optimal assignments $N_3 ACM(,)$ as follows: $J_3 \to M_1$, $J_4 * J_8 \to M_2$, $J_5 \to M_3$, $J_1 * J_7 \to M_4$, $J_2 * J_6 \to M_5$

Total assignment cost = 1550.

Problem	Hungarian method	Lexi-search method
Unbalanced assignment problem	Uses the dummy assignment	Never uses the dummy assignment
Jobclubbing	Gives optimum	Gives exact optimum

6. Conclusion

The above illustration was taken by the defined algorithm and implemented on several sizes of the problems to test the effectiveness of the algorithm. This approach was implemented on different sizes of unbalanced assignment problems. From the above, we notice that the standard Hungarian method uses the dummy assignment which may not be possible in some applications, whereas this new approach never assigns the dummy machine in getting the optimum value. The time complexity with the Lexi-search method is verified and found that they are the same in getting optimum. Here, the optimum value of the original unbalanced assignment problems varies from that of balanced assignment problems either in Hungarian method or Lexi-search approach. The only advantage is that the Lexi-search method gives an exact optimum value with the same time complexity. Therefore the present paper suggests a new approach of clubbing the jobs for solving the unbalanced assignment problem with Lexi-search methodology.

References

[1]   Kuhn, H.W. (1955) The Hungarian Method for the Assignment Problem. *Naval Research Logistic Quarterly*, **2**, 83-97. http://dx.doi.org/10.1002/nav.3800020109

[2]   Kadhirvel, K. and Balamurugan, K. (2013) Method for Solving Unbalanced Assignment Problems Using Triangular Fuzzy Numbers. *Journal of Engineering Research and Applications*, **3**, 359-363.

[3]   Turkensteen, M., Ghosh, D., Goldengorin, B. and Sierksma, G. (2008) Tolerance-Based Branch and Bound Algorithm for the ATSP. *European Journal of Operational Research*, **189**, 775-788. http://dx.doi.org/10.1016/j.ejor.2006.10.062

[4]   Basirzadeh, H. (2012) Ones Assignment Method for Solving Assignment Problems. *Applied Mathematical Sciences*, **6**, 2345-2355.

[5]   Frank, A. (2004) On Kuhn's Hungarian Method—A Tribute from Hungary. Wiley Inter Science, Published Online.

[6]   Pentico, D.W. (2007) Assignment Problem: A Golden Anniversary Survey. *European Journal of Operation Research*, **176**, 774-793. http://dx.doi.org/10.1016/j.ejor.2005.09.014

[7]   Singh, S., Dubey, G.C. and Rajesh Shrivastava, R. (2012) A Comparative Analysis of Assignment Problem. *IOSR Journal of Engineering* (IOSRJEN), **2**, 1-15. http://dx.doi.org/10.9790/3021-02810115

[8]   Gillett Billy, E. (2000) Introduction to Operations Research—A Computer Oriented Algorithm Approach. Tata Mc-Graw Hill, New Delhi.

[9]   Taha, H.A. (1971) Operation Research: An Introduction. MacMillan Inc., New York.

[10] Kagade, K.L. and Bajaj, V.H. (2010) A New Approach to Solve Fuzzy Multi-Objective Unbalanced Assignment Problem. *International Journal of Agriculture Statistics*, **6**, 31-40.

[11] Kumar, A. (2006) A Modified Method for Solving the Unbalanced Assignment Problems. *Journal of Applied Mathematics and Computation*, **176**, 76-82. http://dx.doi.org/10.1016/j.amc.2005.09.056

[12] Pandit, S.N.N. (1963) Some Quantitative Combinatorial Search Problems. PhD Thesis, IIT, Khargpur.

[13] Ramesh, M. (1997) Lexi-Search Approach to Some Combinatorial Programming Problem. PhD Thesis, University of Hyderabad, Hyderabad.

16

An Alternative Approach to the Lottery Method in Utility Theory for Game Theory

William P. Fox

Department of Defense Analysis, Naval Postgraduate School, Monterey, USA
Email: wpfox@nps.edu

Abstract

In game theory, in order to properly use mixed strategies, equalizing strategies or the Nash arbitration method, we require cardinal payoffs. We present an alternative method to the possible tedious lottery method of von Neumann and Morgenstern to change ordinal values into cardinal values using the analytical hierarchy process. We suggest using Saaty's pairwise comparison with combined strategies as criteria for players involved in a repetitive game. We present and illustrate a methodology for moving from ordinal payoffs to cardinal payoffs. We summarize the impact on how the solutions are achieved.

Keywords

Analytical Hierarchy Process, Cardinal Utility, Game Theory, Payoff Matrix

1. Introduction

We teach a three-course sequence in mathematical modeling at the Naval Postgraduate School. In our final course, Models of Conflict, we present an introduction to the following topics: decision theory, multi-attribute decision making with the analytical hierarchy process (AHP) and technique of order preference by similarity to ideal solution (TOPSIS), and game theory.

In game theory, we spend about two lessons on utility theory including the lottery method by von Neumann and Morgenstern. Our students find the back & forth lottery method tedious and they usually do not feel they have the true expertise to narrow in on a true lottery preference. We are currently using Straffin's chapter 9 for utility theory [1].

For years, our student's projects and research in two-person non-zero sum games have used ordinal payoffs. They feel comfortable prioritizing the outcomes in an ordinal manner. They can rank first to last place. If no pure strategies solutions existed, the students assumed that the ordinal payoffs were cardinal payoffs to illustrate the methodologies to obtain equilibrium with mixed strategy solutions.

To add more realism to these projects and eventual research, we present a method to obtain cardinal payoffs that is not tedious and follows from material we have already presented in class using multi-attribute decision making, AHP. In this paper, we describe the issue more fully and describe our methodology using AHP. We provide an example illustrating the technique.

2. Ordinal versus Cardinal Utility

Ordinal utility is a method that ranks outcomes. We tell our students it is like knowing the names of how people finish in a race, 1st, 2nd, 3rd, …, last. Cardinal utility uses interval scale values where we would now replace the order of finish with the *times* they ran the race. With the times, we know how much faster each runner is compared to the other runners.

Often real data is not available for analysis in a game theory scenario. Perhaps the best students can initially do is "rank order" the outcomes from 1 to n for each player in the game.

3. Lottery Method Illustrated

Consider an example where we have a choice between going to McDonald's or going to Burger King. Assume that we limit ourselves to the following meal choices:

Burger King: Whopper &French Fries Combo *(x)*, Whopper Jr. &French Fries Combo *(y)*.
McDonalds: Big Mac & French Fries Combo *(w)*, Quarter pounder &French Fries Combo *(z)*.

Step 1. We need an ordinal preference of these choices. Let's assume the row preferences are:
$$z > x > y > w.$$

Step 2. Use the lottery method to assign values: start by assigning z and w arbitrarily keeping in mind that z gets a higher value than w. We could use a scale from [0, 100] and assign 100 to Z and 0 to W, as an example.

Step 3. Next, consider x. Would you prefer x for certain or a lottery which gives you z at 50% of the time and w at 50% of the time. ½ z ½ w? If Rose likes x over the lottery then x ranks higher than the midpoint between z and w. So we use number greater than 50. So you try, would you prefer x for certain or a lottery that gives ¼ w ¾ z? Now, if Rose prefers the lottery then x has value between 50 and 75. We continue until we narrow the value to a point. When Rose is indifferent between the certainty and the lottery we are done. Assume this occurs at 40% w and 60% z. We then would take 60% of 100 for the value of x.

Step 4. We do the same thing for y. Assume, we go through our process and we assign a value of 20 for y.

Step 5. Now, become the column player.

Step 6-Step 9. Repeat step 1 - 4 to obtain values for the column player's preferences.

This could eventually lead to the following payoff matrix assuming the column player's preferences are directly at odds with the row player. The result would be a pure strategy solution where Player 1 gets his 3rd choice and Player 2 gets his 2nd choice, shown in **Table 1**.

4. AHP Method

AHP and AHP-TOPSIS hybrids have been used to rank order alternatives among numerous criteria in many areas of research in business industry, and government including such areas as social networks [2] [3], dark networks [4], terrorist phase planning [5] [6], and terrorist targeting [7].

The following table represents the process to obtain the criteria weights when the Analytic Hierarchy Process is used to determine how to weigh each criterion for the TOPSIS analysis. Using Saaty's 9 point reference scale [8]-[10], displayed in **Table 2**, we obtain subjective judgment to weigh each criterion against all other criterion

Table 1. Payoff Matrix for Lottery example.

			Player 2
		C1	C2
Player 1	R1	(100, 0)	(40, 80)
	R2	(60, 20)	(0, 100)

Table 2. Saaty's 9-point scale.

Intensity of Importance in Pair-Wise Comparisons	Definition
1	Equal importance
3	Moderate importance
5	Strong importance
7	Very strong importance
9	Extreme importance
2, 4, 6, 8	For comparing between the above
Reciprocals of above	In comparison of elements i and j if i is 3 compared to j, then j is 1/3 compared to i.
Rationale	Force consistency; measure values available

lower in importance. We recommend once the list of criteria is obtained that the decision maker ranks these initially in an ordinal fashion to help facilitate an easier pairwise comparison. To insure transitivity hold we use the consistency ratio, CR, from Saaty [8] where a CR < 0.1 is acceptable. **Figure 1** displays the template used.

Let's provide a quick example using this table. Assume we have two criteria that we are comparing: price and color. Price might be much more important to a decision maker than color. If price is compared to color and deemed that it is very strong than we give "price to color" a value of 7 and it's reciprocal, 1/7, is the value of "color to price". Since these are subjective relationships, we should consider sensitivity analysis for the weights. We used Equation (1) the sensitivity analysis for adjusting weights [9]:

$$w_j' = \frac{1 - w_p'}{1 - w_p} w_j \qquad (1)$$

where w_j' is the new weight and w_p is the original weight of the criterion to be adjusted and w_p' is the value after the criterion was adjusted.

Now, assume we have a game where we might know preferences in an ordinal scale only.

Player 2
C1 C2
Player 1 R1 *wx*
R2 *y z*

Also let's assume that this is a zero-sum game.

Player 1's preference ordering is $x > y > w > z$. Now we might just pick values that meet that ordering scheme, such as

10 > 8 > 6 > 4 yielding for following payoff matrix:

Player 2
C1 C2
Player 1 R1 6 10
R2 8 4

The output from the template in **Figure 1** is the important pairwise comparison matrix. All criteria compared to themselves get a value of 1. We obtained the following AHP matrix:

		x	w	y	z
		1	2	3	4
1	x	**1**	3	5	7
2	w	1/3	**1**	2	4
3	y	1/5	1/2	**1**	3
4	z	1/7	1/4	1/3	**1**

AHP Analytic Hierarchy Process n= 4 Criterion

Objective order risk assessment

Only input data in the light green fields!

Please compare the importance of the elements in relation to the above objective and fill in the table: Which element in each pair is more important, A or B, and how much more important is it. (Use

	Criterion	Comment
1	R1C2	
2	R2C1	
3	R1C1	
4	R2C2	0
5		0
6		0
7		0
8		0

Name: Fox
Date: 1/24/2015

	Element A		Element B	More Important	Intensity (1-9)
1	R1C2	compared with	R2C1	A	3
2			R1C1	A	5
3			R2C2	A	7
4				A	8
5				A	9
6				A	9
7					
1	R2C1	compared with	R1C1	A	2
2			R2C2	A	4
3				A	5
4				A	8
5				A	9
6					
1	R1C1	comp. with	R2C2	A	3
2				A	5
3				A	8
4				A	9
5					

Figure 1. AHP template.

From this matrix, we determine the eigenvalues and associated eigenvectors. We get weights (eigenvector) of the following (to 3 decimals)

$x = 0.595$

$w = 0.211$

$y = 0.122$

$z = 0.071$

Player 2

C1 C2

Player 1 R1 0.211 0.595

R2 0.122 0.071

The solution, regardless of the numbers put in for w, x, y, or z is the value in R1C2. The major difference is that the method using AHP is based on real preferences not ordinal preferences. Thus, AHP can help obtain the relative values of the outcomes provided the CR < 0.1. The resulting values are the cardinal utilities values based upon the input preferences. For example, we may conclude here that R1C2 is 4.877 (0.595/0.122) times as important than R2C1.

5. AHP Example in Game Theory

In our game theory course, we initially cover ordinal utility as a method to obtain values for a payoff matrix. Let's apply this to two-person non-zero sum game example from the course.

Example 1. Unites States versus Country X

Consider a game between two players with two strategies each where the best we can initially do is to obtain an ordinal ranking their preferences. The game payoff matrix is listed in **Table 3**.

There are no pure strategies so the players must play equalizing or mixed strategies to find the equilibrium. We find that we are stuck because these are ordinal values. In the past, our students just assume that these values are in fact cardinal values. With that assumption, we find the United States Play ¼ R1 and ¾ R2 while Country X plays ¾ C1 and ¼ C2. The Nash equilibrium is (2.5, 2.5). Further, if we find Prudential strategies, the Securi-

Table 3. Ordinal payoff matrix.

		Country X	
		C1	C2
United States	R1	(2, 4)	(4, 1)
	R2	(3, 2)	(1, 3)

ty Values, to get to Nash Arbitration [11] with these values we find that the United States plays ½ R1, ½ R2 with a security value of 2.5 while Country X plays ½ C1, ½ C2 with a security value of 2.5. Using (2.5, 2.5) we find the Nash Arbitration values are (2.75, 2.875) while playing 3/8 of R1C2 and 5/8 of R1C1, as displayed in **Figure 2** using the AHP method [10] [11].

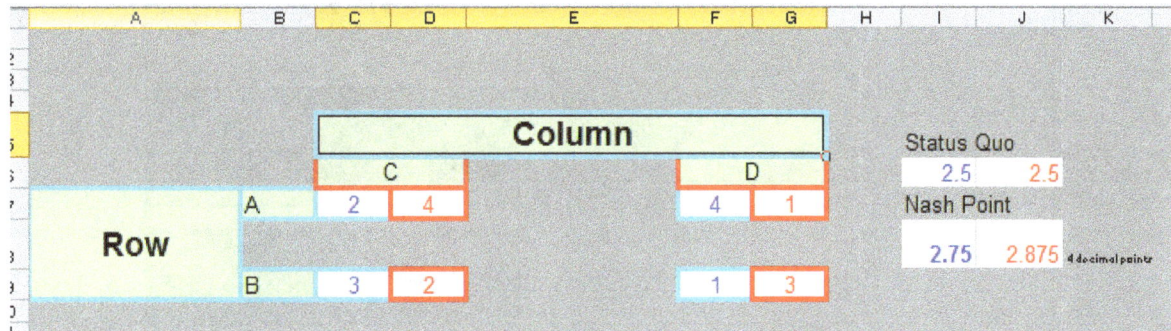

The issue is "what does the Nash arbitration mean" since the initial values were merely ordinal values with no indication how much better a 4 is than a 3, 2, or 1 for each player.

Rather than use the Lottery Method suggested by Morgenstern and von Neumann, we suggest the pairwise comparison method of Saaty for each player's strategies combination. For both the United States and Country X we will need cardinal values for their preferences with these combined strategies: R1C1, R1C2, R2C1, and R2C2.

First, we use Saaty's method [8] for the United States. We utilize a template build for class work [10] [11]. **Figure 2** shows the intensity of the pairwise comparisons for our example with a CR = 0.0899, which is less than 0.1.

The pairwise comparison matrix is

		R1C2	R2C1	R1C1	R2C2
		1	2	3	4
1	R1C2	1	5	6	7
2	R2C1	1/5	1	4	5
3	R1C1	1/6	1/4	1	4
4	R2C2	1/7	1/5	1/4	1

For the U.S., using this matrix, we obtain the following eigenvector as our cardinal values:

R1C2	0.649830851
R2C1	0.17166587
R1C1	0.105026015
R2C2	0.073477264

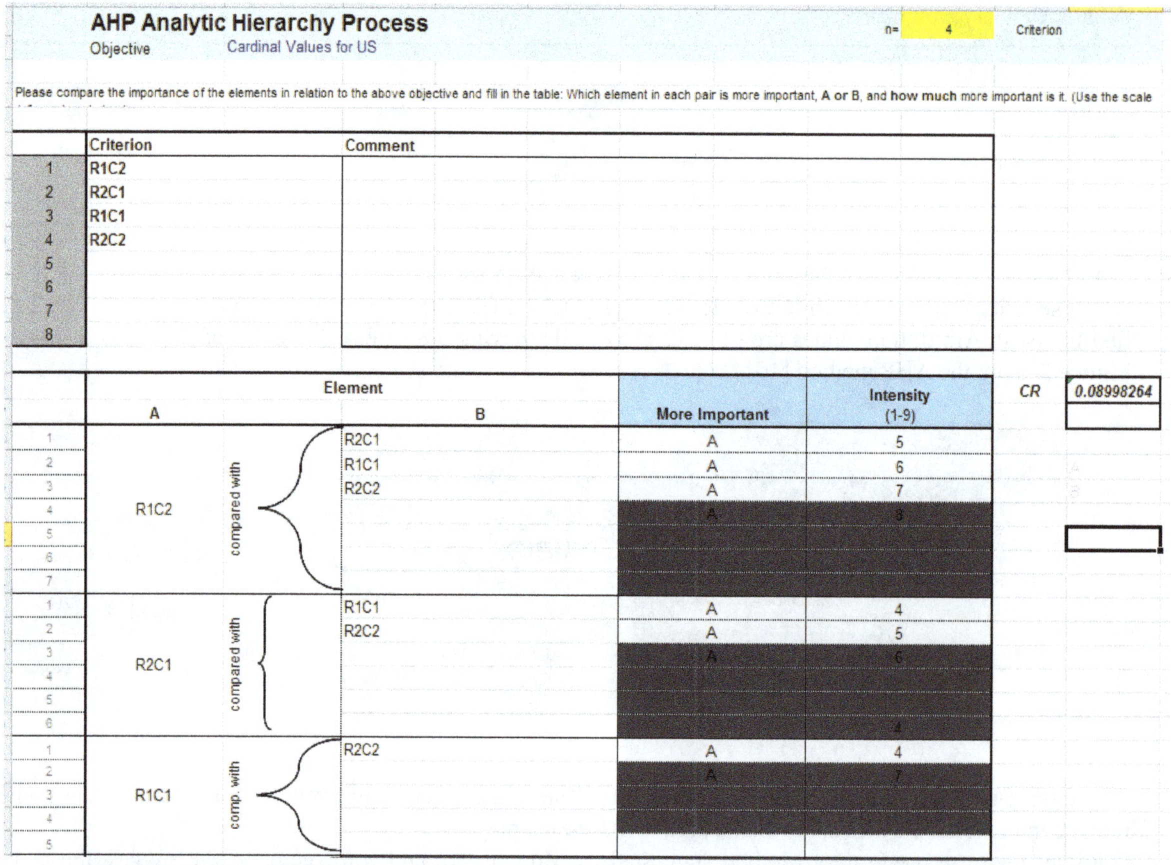

Figure 2. Pairwise comparisons for the United States.

For Country X, we obtain cardinal values as shown by obtaining the intensity of the pairwise comparisons shown in **Figure 3** with a CR = 0.0569, which is less than 0.1. Next, we obtain the eigenvector of the pairwise comparison matrix.

The pairwise comparison matrix is:

		R1C1	R2C2	R2C1	R1C2
		1	2	3	4
1	R1C1	1	3	6	9
2	R2C2	1/3	1	6	8
3	R2C1	1/6	1/6	1	5
4	R1C2	1/9	1/8	1/5	1

The cardinal values, the eigenvector, of the pairwise comparison matrix for Country X are as follows:

R1C1	0.612431976
R2C2	0.243316715
R2C1	0.091240476
R1C2	0.053010833

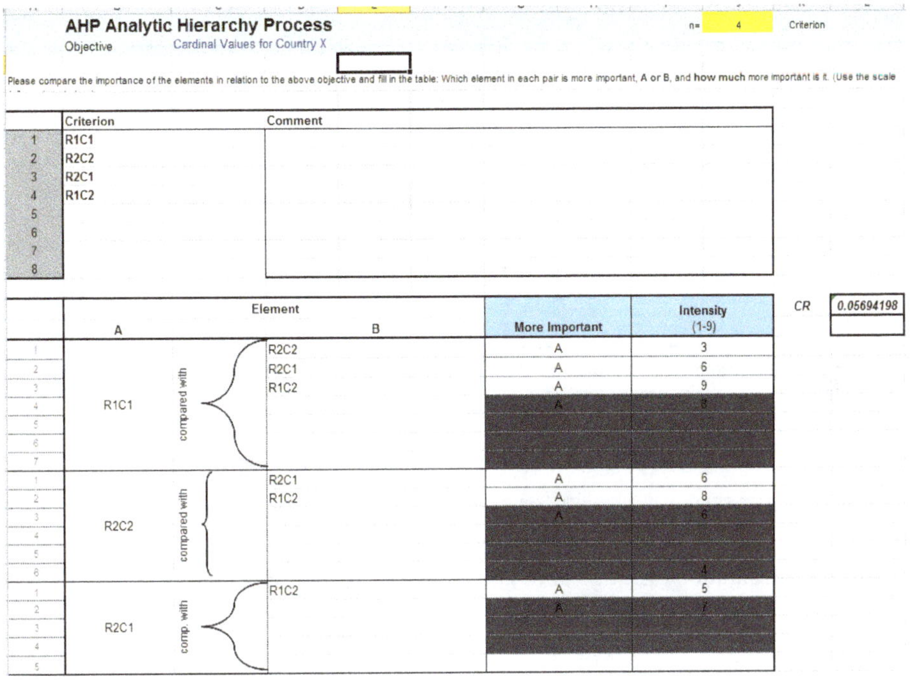

Figure 3. Pairwise comparisons for Country X.

The entire game theory payoff matrix, with cardinal values representing true preferences, is displayed in **Table 4**.

The Nash Equilibrium, Prudential Strategies, and the Nash Arbitration are found using templates built for classroom use [12] and displayed. We find the Nash equilibrium (0.202619, 0.16153).

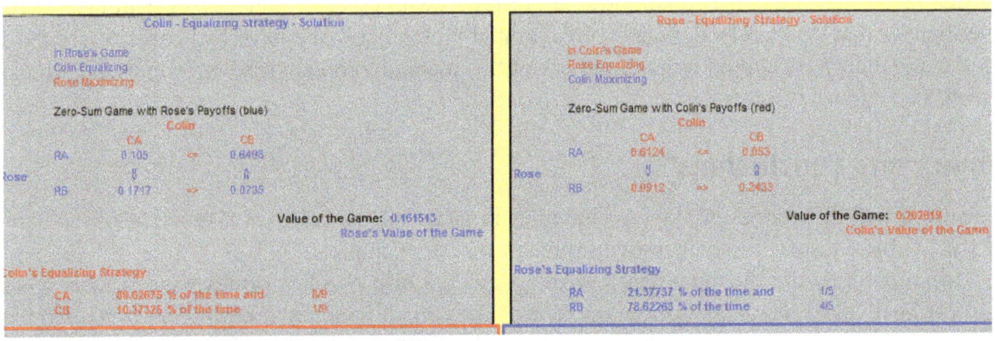

We find the Prudential Strategies or Security Levels are the Nash equilibrium from before.

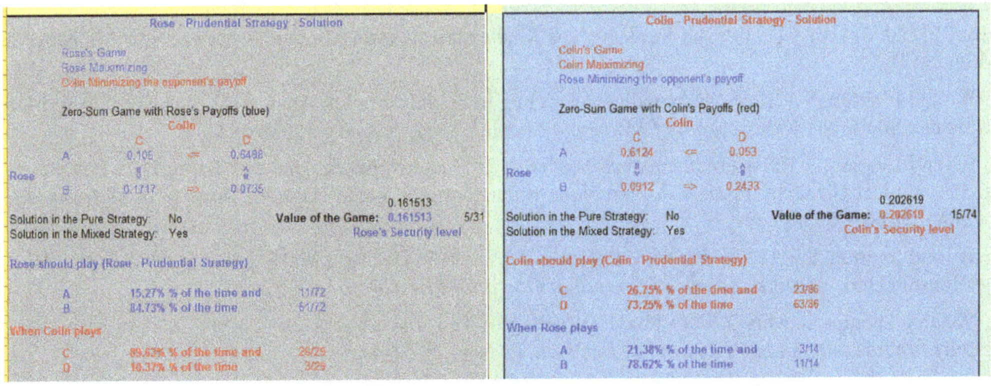

Table 4. Cardinal payoff matrix using AHP results.

		Country X	
		C1	C2
United States	R1	(0.1050, 0.6124)	(0.6498, 0.0530)
	R2	(0.1717, 0.0912)	(0.0735, 0.2433)

Table 5. Summary results.

Results	Ordinal values	Strategies played	Cardinal values	Strategies played
Nash equilibrium	(2.5, 2.5)	¼ R1, ¾ R2, ¾ C1, 1/4 C2	(0.20219, 0.161513)	1/5 R1, 4/5 R2, 8/9 C1, 1/9 C2
Security level	(2.5, 2.5)	½ R1, ½ R2, ½ C1, ½ C2	(0.20219, 0.161513)	11/72 R1, 61/72 R2, 26/29 C1, 3/29 C2
Nash arbitration	(2.75, 2.875)	3/8 R1C2, 5/8 R1C1	(0.373, 0.3368)	0.5075 R1C1, 0.4925 R1C2

We find the Nash Arbitration (0.373, 0.3368) by playing 0.5075 of R1C1 and 0.4925 of R1C2.

We see that our mixed strategies probabilities are different with cardinal preferences than they were with the ordinal preferences that we merely assumed were cardinal preferences. We have had cases where the decisions in AHP and game theory are altered through the use of this method to obtain cardinal values as well as sensitivity analysis of the cardinal weights.

6. Summary and Conclusions

We have showed that differences in playing strategies in game theory occur as a function of the values in the payoff matrix. **Table 5** displays a comparative summary for our example.

In conclusion, not only did the numerical values change but also two key points were seen in this example. First, using cardinal values, the Nash arbitration favored the United States whereas before it favored Country X. Second, how we played our strategies in the game changed substantially.

References

[1] Straffin, P. (2004) Game Theory and Strategy. The Mathematical Association of America: New Mathematics Library, Washington DC.

[2] Fox, W. and Everton, S. (2013) Mathematical Modeling in Social Network Analysis: Using TOPSIS to Find Node Influences in a Social Network. *Journal of Mathematics and Systems Science*, **3**, 531-541.

[3] Fox, W. and Everton, S. (2014) Mathematical Modeling in Social Network Analysis: Using Data Envelopment Analysis and Analytical Hierarchy Process to Find Node Influences in a Social Network. *Journal of Defense Modeling and Simulation*, **2014**, 1-9.

[4] Fox, W. and Everton, S.F. (2014) Using Mathematical Models in Decision Making Methodologies to Find Key Nodes in the Noordin Dark Network. *American Journal of Operations Research*, 1-13 (Online).

[5] Fox, W. and Thompson, M.N. (2014) Phase Targeting of Terrorist Attacks: Simplifying Complexity with Analytical Hierarchy Process. *International Journal of Decision Sciences*, **5**, 57-64.

[6]   Fox, W.P. (2014) Phase Targeting of Terrorist Attacks: Simplifying Complexity with TOPSIS. *Journal of Defense Management*, **4**, 116. http://dx.doi.org/10.4172/2167-0374.1000116

[7]   Fox, W.P. (2014) Using Multi-Attribute Decision Methods in Mathematical Modeling to Produce an Order of Merit List of High Valued Terrorists. *American Journal of Operation Research*, **4**, 365-374. http://dx.doi.org/10.4236/ajor.2014.46035

[8]   Saaty, T. (1980) The Analytical Hierarchy Process. McGraw Hill, New York.

[9]   Alinezhad, A. and Amini, A. (2011) Sensitivity Analysis of TOPSIS Technique: The Results of Change in the Weight of One Attribute on the Final Ranking of Alternatives. *Journal of Optimization in Industrial Engineering*, **7**, 23-28.

[10] Fox, W.P. (2012) Mathematical Modeling of the Analytical Hierarchy Process Using Discrete Dynamical Systems in Decision Analysis. *Computers in Education Journal*, **3**, 27-34.

[11] Fox, W. (2014) Chapter 221, TOPSIS in Business Analytics. *Encyclopedia of Business Analytics and Optimization*, **V**, 281-291.

[12] Feix, M. (2007) Game Theory: Toolkit and Workbook for Defense Analysis Students. MS Thesis, Naval Postgraduate School, Monterey.

Biographical Sketch

Dr. William P. Fox is a professor in the Department of Defense Analysis at the Naval Postgraduate School and teaches a three course sequence in mathematical modeling for decision making. He received his BS degree from the United States Military Academy at West Point, New York, his MS at the Naval Postgraduate School, and his Ph.D. at Clemson University. Previous he has taught at the United States Military Academy and Francis Marion University where he was the chair of mathematics for eight years. He has many publications and scholarly activities including books, chapters of books, journal articles, conference presentations, and workshops. He directs several mathematical modeling contests through COMAP: the HiMCM and the MCM. His interests include applied mathematics, optimization (linear and nonlinear), mathematical modeling, statistical models for medical research, multi-attribute decision making, game theory, and computer simulations. He is president-emeritus of the NPS faculty council. He is currently Past President of the Military Application Society of INFORMS.

17

Applications of Ultrasonic Techniques in Oil and Gas Pipeline Industries: A Review

Wissam M. Alobaidi[1*], **Entidhar A. Alkuam**[2], **Hussain M. Al-Rizzo**[1], **Eric Sandgren**[1]

[1]Systems Engineering Department, Donaghey College of Engineering & Information Technology, University of Arkansas at Little Rock, Little Rock, Arkansas, USA
[2]Department of Physics and Astronomy, College of Arts, Letters, and Sciences, University of Arkansas at Little Rock, Little Rock, Arkansas, USA
Email: *wmalobaidi@ualr.edu

Abstract

The diversity of ultrasound techniques used in oil and gas pipeline plants provides us with a wealth of information on how to exploit this technology when combined with other techniques, in order to improve the quality of analysis. The fundamental theory of ultrasonic nondestructive evaluation (NDE) technology is offered, along with practical limitations as related to two factors (wave types and transducers). The focus is limited to the two main techniques used in pipe plants: First, straight beam evaluation and second, angle beam evaluation. The depth of defect (*DD*) is calculated using straight beam ultrasonic in six different materials according to their relative longitudinal wave (LW) velocities. The materials and respective velocities of LW are: rolled aluminum (6420 m/s), mild steel (5960 m/s), stainless steel-347 (5790 m/s), rolled copper (5010 m/s), annealed copper (4760 m/s), and brass (4700 m/s). In each material eight defects are modeled; the first represents 100% of the material thickness (*D*), 50.8 mm. The other seven cases represent the *DD*, as 87.5% of the material thickness, 75%, 62.5%, 50%, 37.5%, 25%, and 12.5%, respectively. Using angle beam evaluation, several parameters are calculated for six different reflection angles (β_R) (45°, 50°, 55°, 60°, 65° and 70°). The surface distance (SD), ½ skip distance (SKD), full SKD, and 1 ½ SKD, ½ sound path (SP) length, full SP, and 1 ½ SP are calculated for each β_R. The relationship of SKD and SP to the β_R is graphed. A chief limitation is noted that ultrasound testing is heavily dependent on the expertise of the operator, and because the reading of the outcome is subjective, precision may be hard to achieve. This review also clarifies and discusses the options used in solving the industrial engineering problem, with a comprehensive historical summary of the information available in the literature. Merging various NDE inspection techniques into the testing of objects is discussed. Eventually, it is hoped to find a suitable technique combined with ultrasonic inspection to deliver highly effective remote testing.

*Corresponding author.

Keywords

Ultrasonic Testing, Guided Waves, Nondestructive Testing (NDT), Pipe Inspection, Pipe Thickness Measurement

1. Introduction

Ultrasonic testing is one of the important techniques of nondestructive testing (NDT). It uses ultra-high-frequency sonic energy to locate and identify discontinuities in materials that are both on and below the surface of the material (such as metals or plastic, commonly used to make pipes, depending on the application) [1]-[5].

In 2007 D. S. Caravaca *et al.* [6] studied polyethylene pipe joints and detection of improper preparation of the joints using a phased array technique (PAT). The problems that arise with the electrofusion type of bonding used for the pipe are analogous to those that occur in metal welds for steel pipe. This paper aims to offer an ultrasonic method for evaluating polyethylene pipe welds, in pipeline used for gas and water distribution, along with results from both laboratory and field experiments [6].

The sonic energy passes through the substrate. There is a reduction in energy intensity, as well as reflection of the waves by the back wall of the material, and where discontinuities are encountered. The returning signal is captured, mathematically analyzed and presented on a screen, with the resultant waveform showing the location of defects on or within the substrate [7]-[11].

In 2006, Yi-Mei Mao and Pei-Wen Que investigated the possibility of using a then-new sonic signal processing method for inspection of oil pipelines. They compared "ultrasonic signals reflected from defect-free pipelines and from pipelines with defects" [12] and treated the recaptured waveforms with the Hilbert-Huang transform (HHT). The results demonstrated the feasibility of using the technique to successfully locate and determine the extent of discontinuities in oil pipes [12].

Reflected signals are attenuated to different degrees depending on the type of interface they encounter. An interface between a metal and a liquid presents a reduced reflection of the sonic energy, whereas an interface between a gas and a metal causes nearly 100% reflection of the sonic waves [7]. The actual percentage of reflection is dependent on the ratio of the parameters of certain physical properties between the two types of material, for example, the ratio between the metal and the liquid substance at the interface [13]-[16].

Cracks, holes, laminations, slags, cavities, porosities, bursts, lack of fusion, flakes, lack of penetration and other discontinuities that produce sharp boundaries are easily identified by ultrasonic testing. Other types of discontinuities that produce a more diffuse boundary are still identifiable because they will disrupt the sonic waves in a detectable manner [7] [17].

In 2015, Wissam Alobaidi *et al.* surveyed seven types of defects commonly found in pipe joint welds, and five often-used types of welds in manufacturing. The correlation between each defect type and the NDT technologies which best reveal the defects is presented in a table [17]. The ability and shortcomings of four NDT techniques commonly used for testing pipe are compared, one of which is ultrasonic testing, and the table reports whether each technique can detect surface, or subsurface flaws. The paper examines ways in which new quality assurance techniques can be incorporated alongside the standard methods in order to overcome the shortcomings of current methods, with the aim to reduce labor costs and increase line output [17].

Because the sensing mode of ultrasonic evaluation is a mechanical process, the frequency range is limited to avoid permanent damage to the targeted objects. Frequencies used most often range from 0.1 MHz to 25 MHz. Although Ultrasonic testing (UT) is capable of identifying surface defects, it is primarily used to detect and locate discontinuities that are below the surface, especially in metal parts. UT is useful for other types of inspection, including welds, wall thinning, and surface defects, as mentioned above [1] [7].

This review paper presents applications of, and limitations of some ultrasonic techniques; we will demonstrate the fundamental theory of ultrasound and type of waves used; we will thoroughly examine the inspection approaches of the contact and angle beam techniques. Approaches discussed focus on the measurement of defects in oil and gas pipe manufacturing. Primarily we are interested in determining the depth of the defect (*DD*) below the material surface. We will also address the limitations of scanning pipe that depend on the transducers used (contact for pipe body and angle beam for welds). Moreover, the paper presents a literature survey for applications of ultrasonic techniques in the pipe industry. The primary aim for this review is to investigate possible future coupling of one of the ultrasonic techniques with other NDE techniques, to develop a hybrid detection system for discontinuities. This research study is an accumulation of the practical experience of the authors, as well as the practical experience represented by the reviewed papers.

2. Fundamental Theory of Ultrasonic

2.1. Ultrasound

Ultrasonic inspection uses sound as the source for testing the medium under consideration. This is the same kind of sound that creates the motion of our eardrums and allows us to hear. The vibrations used for Ultrasonic Testing (UT) are very much higher frequency than what we can hear. But just like any sound wave that moves through the air, the ultrasonic waves that are sent into metal will propagate through the solid medium. When these vibrations encounter interfaces between discontinuous materials (which represent defects in the materials and welds of pipes, for example) they will be reflected in predictable ways. UT is a commonly used method in industries for quality control purposes. It is useful for testing the integrity of metal parts, both before and after forming into pipes. The roll stock can be tested for invisible defects using straight beam ultrasonic testing, allowing the material to be categorized as acceptable, repairable, or scrap before it is incorporated into pipes [7] [17] [18]. Because air does not transmit ultrasound waves as well as solids or gels the difficulty of introducing the signals into metals is overcome by using water or grease as a conducting medium between the transducer and the material to be tested [13].

Ultrasonic testing is used both during the manufacturing of pipe and for inspection of in-service pipelines. The particulars of pipeline insulation may require ingenious ways of getting the ultrasonic transducers into contact with the pipe to be tested. For example, in 2009, H. Lei *et al.* [19] were interested in developing a device for assessment of the inner walls of underwater oil pipeline. The device, called a pig, uses ultrasonic testing to discover and determine the extent of corrosion within the pipe, while storing the data on a hard disk. Following retrieval of the pig, the recorded data is analyzed in order to determine the reduction in the oil pipe wall, and to identify areas of wall thinning which are then categorized using the American Petroleum Institute (API) standards. The authors claim "perfect performance" of the device [19].

2.2. Waves

Waves commonly used are:
- **Longitudinal Waves (LW):** Another name for these waves is "compression waves". LW is the type of sound wave that we hear, and that is used in manual UT for testing the front end and tail end of the pipe body, and in coil UT for testing the integrity of the plates before they are formed into pipe. The LW pushes the molecules of the tested material in the same direction as the movement of the wave, as shown in **Figure 1**.

The velocity of the ultrasonic LWs is different in different metals; for example, their velocity through copper is roughly 4760 m/s, making them the most rapidly propagating ultrasonic waves used in NDT. Through the analysis of wave velocity the depth of defect (*DD*) can be found [7] [20] [21]. This is what we demonstrated in Section 4.1.
- **Shear waves (SW):** Shear waves, also known as transverse waves, propagate more slowly and at shorter wavelengths than LWs at equal frequencies. The particle motion is at right angles to the movement of the wave, as shown in **Figure 2**. SW is usually used for angle beam UT (for example, to detect discontinuities in both the inner diameter and outer diameter of the weld in pipes). As with LW, the SW velocity varies with the type of metal. Some example velocities and corresponding metal types are: Aluminum, roughly 3040 m/s; Steel, 347 Stainless, roughly 3100 m/s. When SWs reflect from an interface they sometimes become LWs [7] [20] [22] [23]. This is demonstrated in Section 4.2.
- **Rayleigh waves (RW):** These waves, which penetrate the material only to the sub-surface distance of one wavelength (at any given frequency), are also called Surface waves. RW travels along the surface of the tested material at velocities equal to those of SWs. RWs are useful for detecting cracks that break the surface of the tested part. They are also useful for testing pieces with intricate rounded surface features. Any defect in zone α as shown in **Figure 3**, would rest deeper than the wavelength (λ) of the test signal, and would likely not be detectable by RW [7] [20] [22].

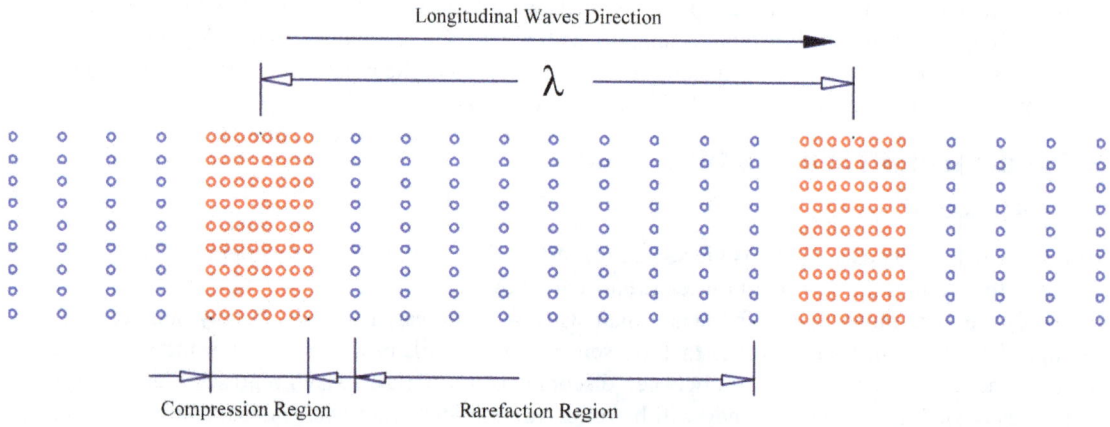

Figure 1. Graphical depiction of parallel motion response of material particles subjected to longitudinal ultrasonic waves, showing compression and rarefaction regions.

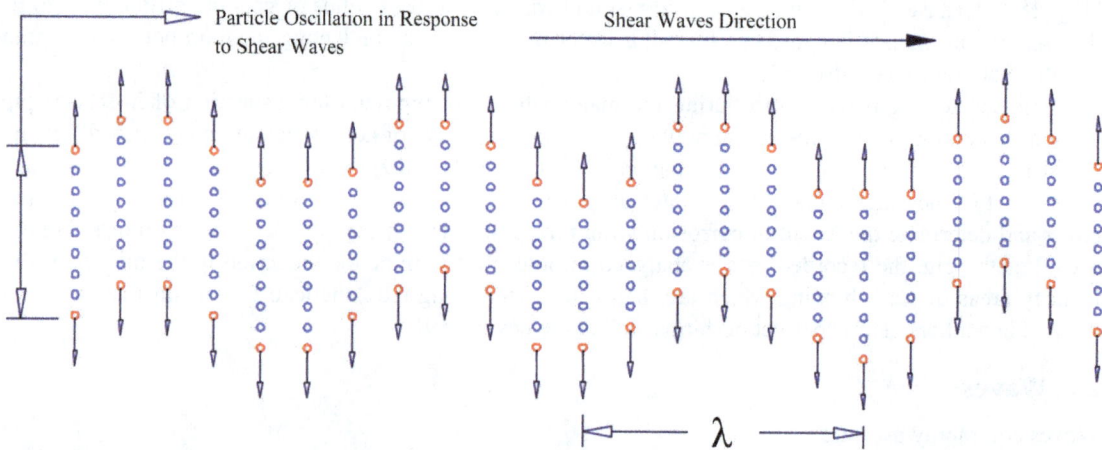

Figure 2. Graphical depiction of perpendicular motion response of material particles subjected to shear ultrasonic waves, showing wavelength.

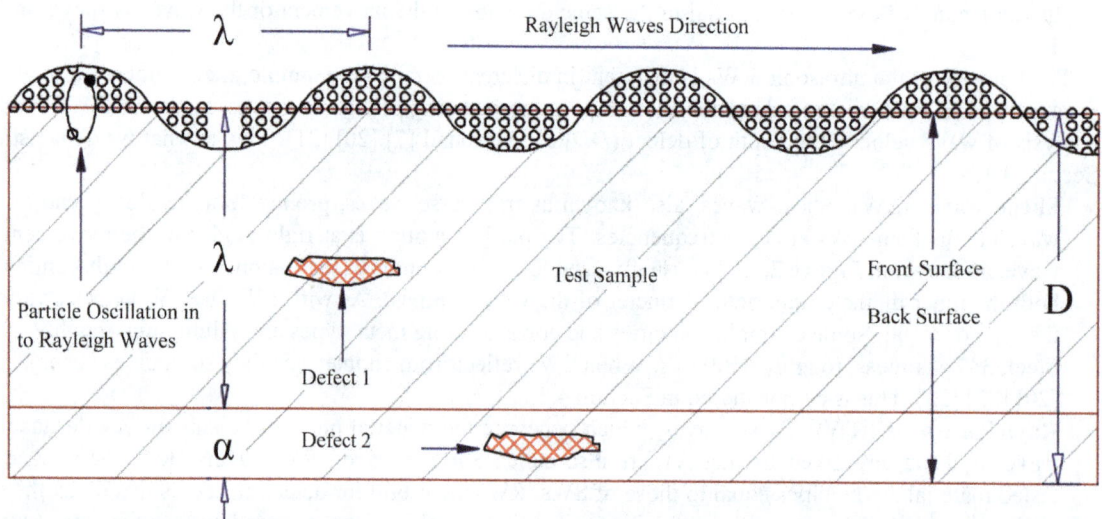

Figure 3. Graphical depiction of limited detection area of Rayleigh waves, showing how they are confined mostly to the surface of a material.

In 2014, N. P. Aleshin *et al.* examined various methods and devices for proficiently introducing ultrasonic signals into pipelines with thicknesses ranging between 6 mm to 20 mm. They covered the choice between Surface and Plate wave modes for conducting these assessments [24].

- **Lamb waves (LMW):** Lamb waves are vibrations that occur from the upper to the lower surface (up to several wavelengths in thickness) of the tested material, usually a plate (composites or metals), so they are also called Plate waves. They propagate not only through the full thickness of the tested material but are capable of propagating from a single point of excitation over significant distances within the material, as shown in **Figure 4**. Because the LMWs travel through the solid in a way that is significantly like the behavior of electromagnetic waves within a waveguide, the characteristics of transmission vary from material to material. The velocity of LMWs is dependent on many factors, including density, plate thickness, and the elastic properties of the material being tested [7] [25].

In 2005, Kevin R. Leonard *et al.* explored use of "helically propagating Lamb waves" transmitted and recaptured with longitudinal transducers. They describe a "meridional-array scheme" [26] used to test the concept of using tomography to detect location and extent of discontinuities in pipes. The research sample used wall-thinning as the defect to be detected. The study also investigated improved reconstruction programs for assessment of helical array signals where "the transmitters and receivers lie along circumferential parallel rings", confirming that frequency compounding reduces noise and artifacts, leading to clearer imagery [26].

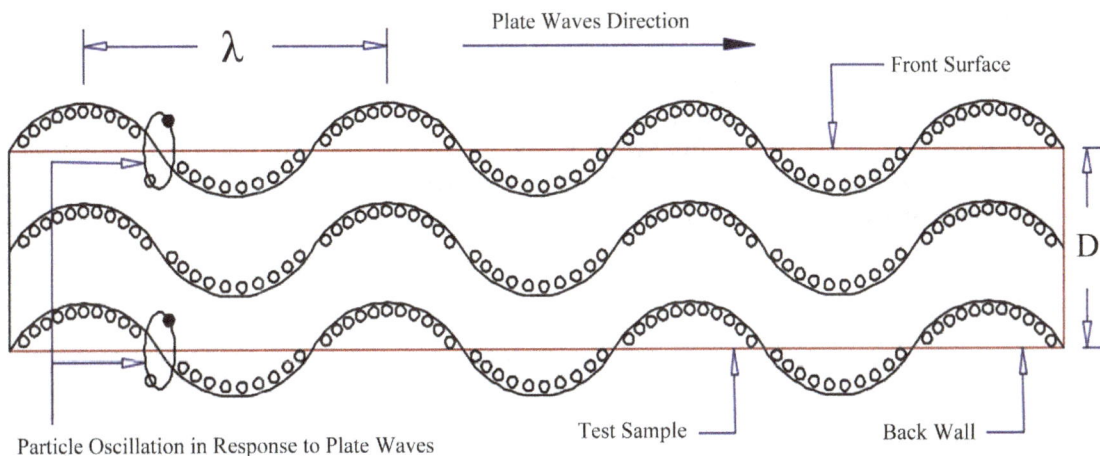

Figure 4. Graphical depiction of ultrasonic lamb waves (plate waves) showing how they move through a test object of a certain thickness which is directly related to the wavelength.

3. Transducers

There are five general categories of ultrasonic transducers used in NDE: straight beam, angle beam, dual element, delay line and immersion transducers. Straight beam and angle beam transducers are used in pipe manufacturing NDE procedures. Usually, the standardized inspection codes will determine the type of transducer the manpower (operator) uses for a particular test. In the case where there is no specification or customer requirements, the operator will select a suitable transducer based on prior experience and knowledge [27]-[29]. Many studies investigated ways to reduce the manpower needed for inspection of pipes by using ultrasonic waves, and also ways to reduce the number of transducers required for an inspection protocol, in order to save capital expense.

In 1998, M. J. S. Lowe *et al.* [30] studied the use of a single transducer for pulse-echo testing of in-service pipeline in order to reveal corrosion in insulated pipes, including oil pipes. They removed only part of the pipe insulation, in order to reduce labor cost. The study focused on selection of the most effective wave modes, and understanding the relationships between the size of a flaw and the signal strength when reflected. The technique was in field trials at the time of publication [30].

In 2004, M. H. S. Siqueira *et al.* investigated the use of a single transducer to replace groups of transducers that are used for rapidly inspecting pipe. The research assessed the use of guided wave pulse-echo conformations with low ultrasonic signal to noise ratios (S/N), together with processing via "frequency bandpass filters and wavelet analysis" [31]. Their results confirmed the practicality of the concept, showing up to 12 dB S/N enhancement of the recaptured signal, allowing analysis even with otherwise unusably noisy signals [31].

In 2006, Younho Cho *et al.* [32] carried out a feasibility study of using guided sonic signals for remote monitoring of stainless steel pipe. They report that their experimental approach, intended to allow them to optimize the guided wave mode resulted in the discovery that "Predicted modes could be successfully generated by controlling frequency, receiver angle and wavelength" [32]. By analyzing scattering patterns mode by mode, they were able to determine that "mode conversion characteristics are distinct depending on dispersive pattern of modes" [32].

4. Approaches

4.1. Straight Beam Evaluation

The ultrasonic signal used for UT is not continuous. A brief pulse of ultrasound is emitted into the test material from a transducer; the signal travels through the test piece thickness and echoes from either the back wall of the piece, or from a discontinuity within the piece. The echoing signal is captured by the transducer only a few microseconds after being emitted. This gives the process the name pulse-echo [33].

The velocity of travel within the tested material must be known in order to calculate both the presence and the depth of the defect. In a flawless test piece, the distance down and back would each equal D, the full thickness of the metal. Thus, the transit time T represents the sonic waves propagating from probes S1 and S2 in both cases in **Figure 5** and the reflection from the back wall of the piece, so for the full thickness the total travel will be $2D$. The factor (V_L) is the velocity of LWs within the type of metal tested, as shown in Equation (1). If the signal finds a defect, the transit time T_D would represent the sonic waves propagating from probe S3 in case 2 and the reflection from the interface of the defect, so the distance traveled will be $2DD$ where DD is the distance from the front surface to the interface with the defect as shown in **Figure 5**, case 2. Thus, the velocity of LWs (V_L) in this case is used to determine DD, and T_D is the time of transit to and from the defect, as shown in Equation (2) [34].

$$V_L = \frac{2D}{T} \tag{1}$$

$$V_L = \frac{2DD}{T_D} \tag{2}$$

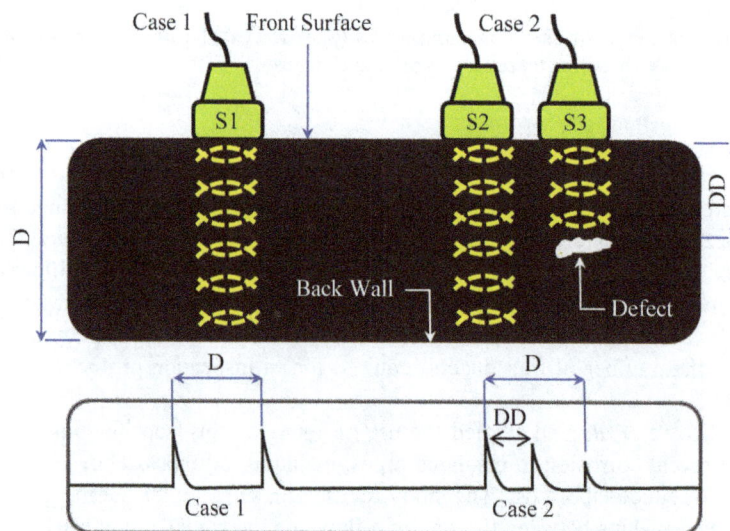

Figure 5. Scheme for the test sample. Two cases are represented. Case 1 has S1 (straight beam #1). Case 2 has S2 (straight beam #2) and S3 (straight beam #3). This figure represents calculation of the value DD, using Equation (2). See Section 6. Discussion.

Straight-beam testing is commonly used in pipe manufacturing to test the roll stock which will be used for building the pipe body. Straight beam testing is effective for detecting cracks that occur parallel to the surfaces of the tested material as well as discontinuities within the body of the material, such as voids or areas of porosity, and inclusions [35].

Figure 6 shows a test sample including eight cases, with the case number shown on the probes from 1 to 8, respectively. This demonstrates the measurement of the various distances calculated as *DD* according to Equation (1) and Equation (2) above, and as graphed in **Figure 7**. The velocities of sonic waves penetrating different materials are compared in **Table 1**. These known constant values are necessary to calculate the distance to defect depending on the material from which the tested object is made. The eight cases are calculated for each of six materials, given different transit times for each, as shown in **Figure 7**, which compares the curves resulting from the eight defects in different materials.

In practical use researchers design mechanisms in order to test their ideas. In 2013, Jin-Sheng Yang *et al.* [36] reported a system to measure variations in the wall thickness of gas pipe using ultrasound based on the standard cleaning pig with some adaptations. To remove the need of a couplant, the system incorporated a wheel made of an elastic substance, which through mechanical tightness of fit conveyed the vibrations to the pipe wall from a standard piezoelectric probe. The paper covers the operational principle and the system design [36].

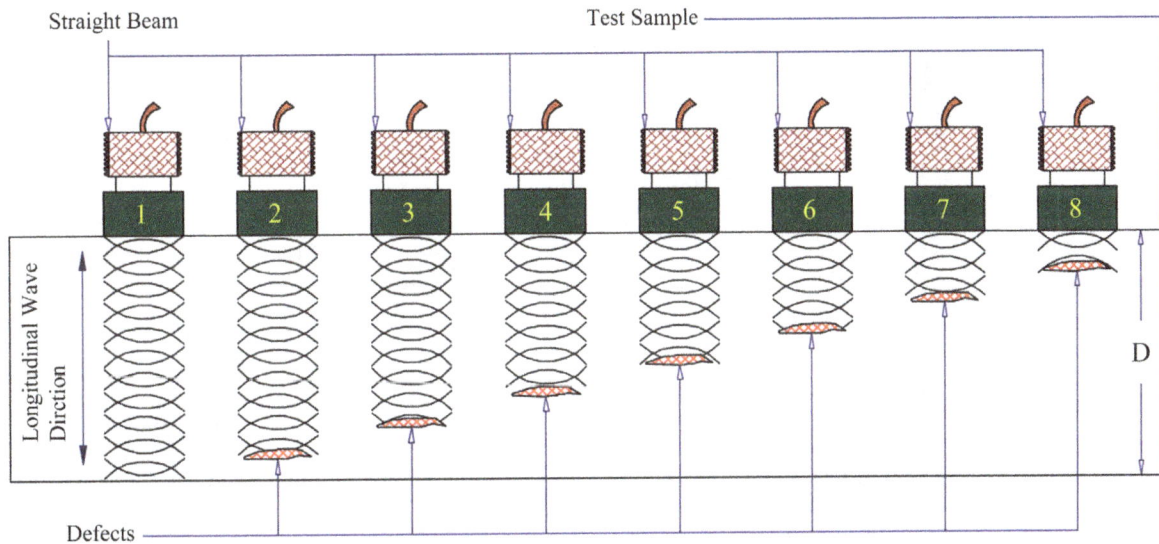

Figure 6. A test sample with 8 cases of defects showing the reflection of test signals from straight beam probes.

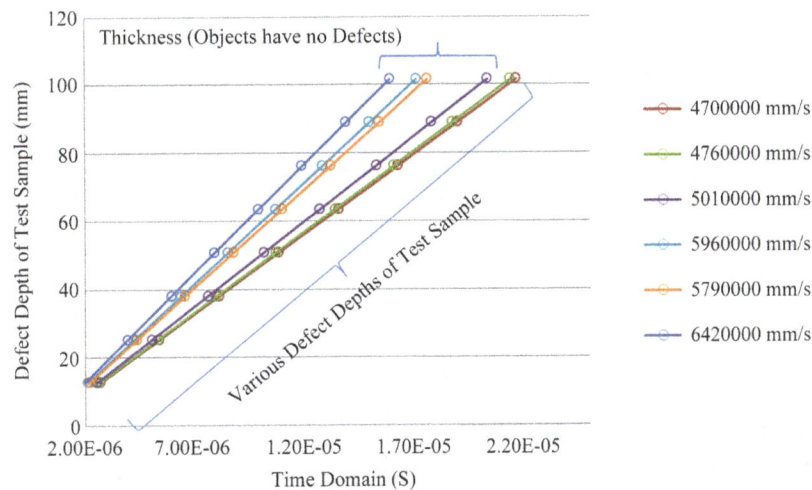

Figure 7. The time domain for the 8 cases from **Figure 6** are graphed for each of the 6 materials in **Table 1**.

Table 1. Longitudinal wave velocities shown for 6 different materials.

Materials	Aluminum, rolled	Steel, mild	Steel, 347 stainless	Copper, rolled	Copper, annealed	Brass
V_L (mm/s)	6,420,000	5,960,000	5,790,000	5,010,000	4,760,000	4,700,000

4.2. Angle Beam Evaluation

Straight beam ultrasonic techniques are best for locating defects in plate-type materials where defects are often parallel to the surface of the object, but they are ineffective for testing welds. The discontinuities in welds are usually at an angle to the surface of the sample under test. Beams approaching the weld interface at an angle are effective at detecting discontinuities within the welding bead. The angle beam transducer is used to generate the test signal in the majority of ultrasonic inspections. To assess discontinuities using angle beam examination, skip distance (SKD) is used to describe the sound path (SP) reflected from the back wall interface (1st leg) and going immediately to reflecting, where it is again reflected from the front wall (2nd leg). SKD is formed by the full SP (1st leg and 2nd leg), and is the distance from the point of excitation to the end of the second leg, as shown in **Figure 8** [17] [37]-[39].

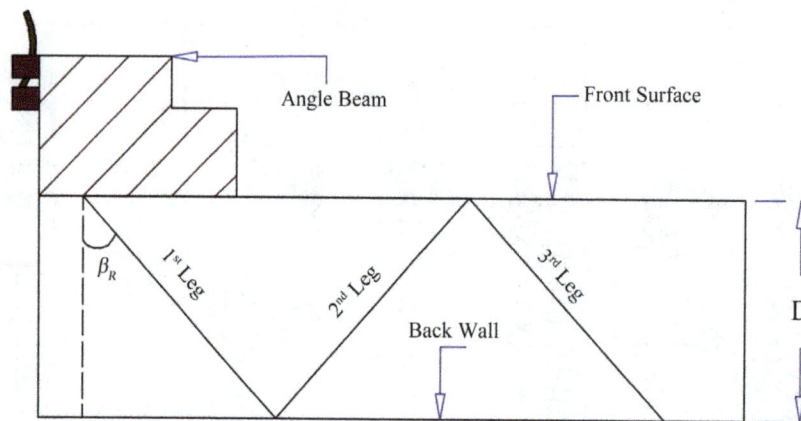

Figure 8. Representation of skip distance through three legs, ½ skip, full skip and 1 ½ skip. D is the material thickness and β_R SP angle.

All the equations below are adapted from [37]-[39].

$$SKD = 2D\text{Tan}(\beta_R) \tag{3}$$

Surface distance (SD) is equal to half SKD.

$$SD = D\text{Tan}(\beta_R) \tag{4}$$

These calculations cannot be completed unless D and β_R to the front surface are known. If the length of the 1st leg is known, the SKD and SD can be computed as.

$$SKD = 2(L1)\text{Sin}(\beta_R) \tag{5}$$

and

$$SD = L1\text{Sin}(\beta_R) \tag{6}$$

The segments of the SP numbered $L1$, $L2$ and $L3$ in **Figure 8** are 1st leg ($L1$), 2nd leg ($L2$) and 3rd leg ($L3$) respectively. And through the trigonometric functions SKD, SD, $L1$, $L2$ and $L3$ can be calculated as shown in **Table 2**, by taking 70, 65, 60, 55, 50 and 45 degrees as the β_R. The first leg and second leg are calculated as follows.

$$L1 = \frac{D}{\text{Cos}(\beta_R)} \tag{7}$$

and

$$L1 + L2 = \frac{2D}{\text{Cos}(\beta_R)} \tag{8}$$

According to the table below, we graphed the SP angle against the SKD. These graphs show the limitations of the scan distance to the pipe weld depending on β_R. Where, SKD and SP length depends on the propagating sonic angle. That means, the larger the angle is, the greater the scanning distance becomes, based on the (½, Full and 1 ½) for SKD and SP as shown in **Figure 9** and **Figure 10**.

To calculate the depth of the discontinuity (DD), we measure vertically from the point of reflection at the interface with the defect, up to the front surface of the test sample [34], as shown in **Figure 11** and **Figure 12**. To calculate the DD's, the values of D, SKD and SD must be known, as shown below. **Figure 11** shows a defect interrupting the path of L1. **Figure 12** shows a defect interrupting the path of L2.

$$DD = L1\text{Cos}(\beta_R) \tag{9}$$

$$DD = 2D - \left[(\text{Cos}\beta_R)(L1 + L2)\right] \tag{10}$$

Table 2. Summary and comparison of the SKD and SP lengths for 6 different beam angles.

β_R	SD (mm)	SKD (mm)			SP (mm)		
		½ SKD	Full SKD	1(½) SKD	½ SP	Full SP	1(½) SP
45	25.4	25.4	50.8	76.2	36	72	108
50	30.27	30.27	60.54	90.81	39.515	79.03	118.545
55	36.275	36.275	72.55	108.825	44.3	88.6	132.9
60	44	44	88	132	50.8	101.6	152.4
65	54.47	54.47	108.94	163.411	60.1	120.2	180.3
70	69.786	69.786	139.572	209.358	74.265	148.53	222.8

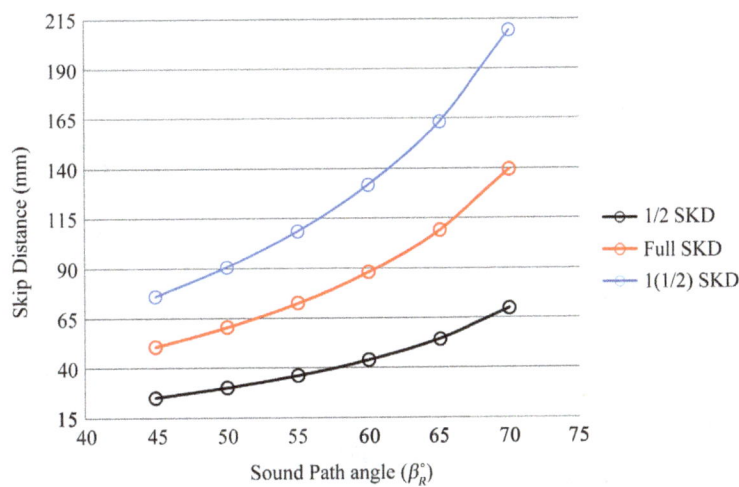

Figure 9. Skip distance plotted against sound path (SP) angle, showing the SP distances achieved with ½, full and 1 ½ SKD, for use by the operator in selecting the SP angle to be used during test depending on the material.

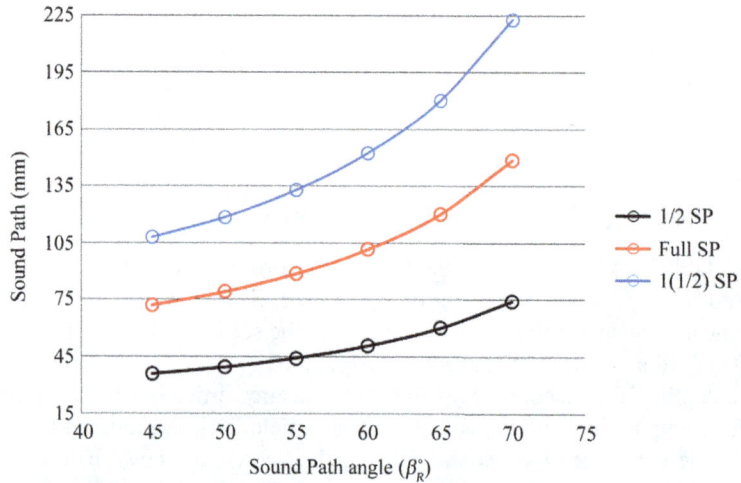

Figure 10. Sound path length plotted against SP angle. This shows the exact length of penetration of sonic waves according to the material for ½, full and 1 ½ SKD.

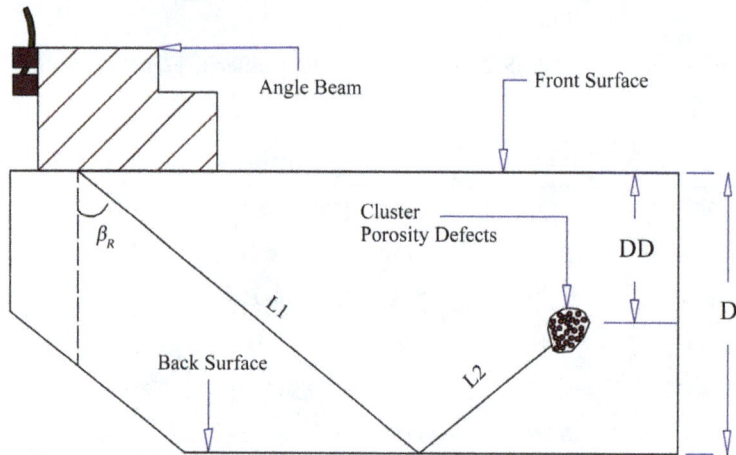

Figure 11. Illustration of the test object and defect depth found by angle beam transducer (defect found by the 2nd leg).

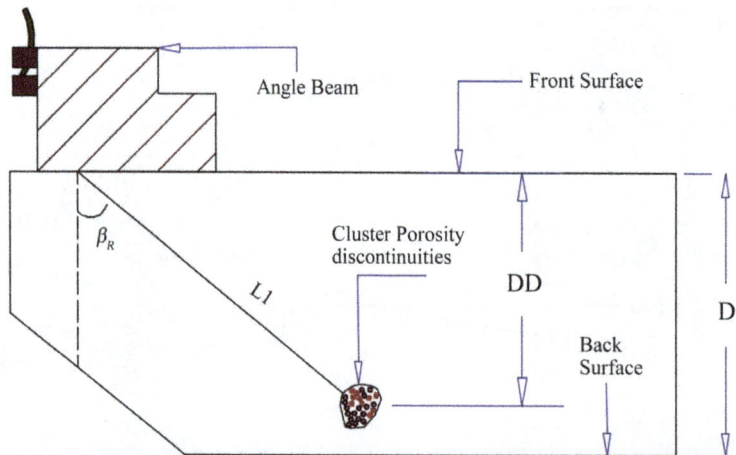

Figure 12. Illustration of the test sample and discontinuity depth as located by angle beam transducer (defect located by the 1st leg).

Figure 13 shows a schematic of a pipe cross section containing a cluster porosity discontinuity in the pipe weld. This is an example of one type of discontinuity that is best discovered by using an Angle Beam Transducer.

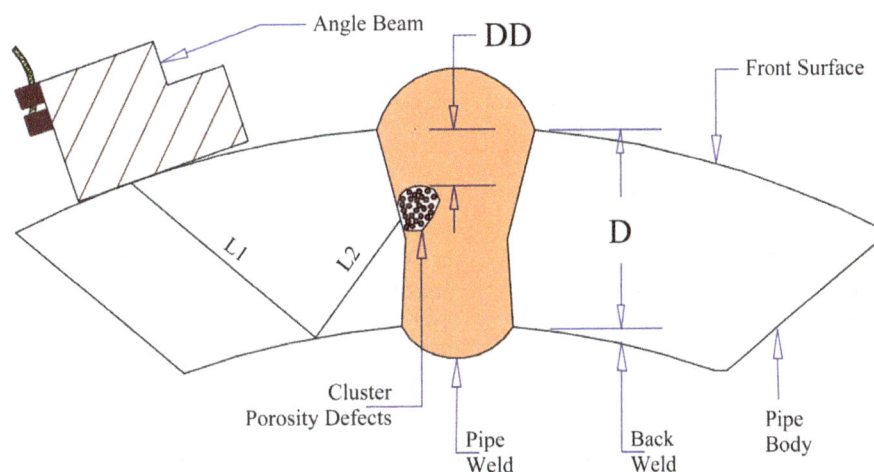

Figure 13. Pipe cross section schematic of discontinuity depth discovered on the 2nd leg by angle beam transducer.

5. Summary

Ten analyses done by ultrasonic NDE specialists are listed in **Table 3** below which summarizes the problems studied by each paper, the name of the first author and the year of publication.

Table 3. Selected prior research into uses of ultrasonic NDT to detect discontinuities in pipeline materials.

No.	Problems	First author name	Year
1	Defect detection in pipes using guided waves	M. J. S. Lowe	1998
2	The use of ultrasonic guided waves and wavelets analysis in pipe inspection	M. H. S. Siqueira	2004
3	Lamb wave tomography of pipe-like structures	Kevin R. Leonard	2005
4	A wall thinning detection and quantification based on guided wave mode conversion features	Younho Cho	2006
5	Application of Hilbert-Huang signal processing to ultrasonic non-destructive testing of oil pipelines	Yi-Mei Mao	2006
6	Ultrasonic phased array inspection of electrofusion joints in polyethylene pipes	D. S. Caravaca	2007
7	Ultrasonic pig for submarine oil pipeline corrosion inspection	H. Lei	2009
8	A new type of wheeled intelligent ultrasonic thickness measurement system	Jin-Sheng Yang	2013
9	Automatic ultrasonic inspection of large-diameter pipes	N. P. Aleshin	2014
10	A survey on benchmark defects encountered in the oil pipe industries	Wissam Alobaidi	2015

6. Discussion

To summarize the objectives of our research reported in this paper are?
- Extending the service life of pipes through early quality control testing by finding more effective and appropriate ways to evaluate them.
- Demonstrating the limitations of two Ultrasonic testing techniques as covered in sections above.
- Employing the technique to extend the effectiveness of this technology in determining the location and depth of defects.

- Coupling ultrasonic technology with another NDT technique in an automated system to quickly determine the exact location and degree of the discontinuity.
- After designing a new system merging the Ultrasonic technique with another NDT technique, to build the actual system and employ it in real-world conditions to remotely detect and evaluate discontinuities.

The focus here will be on approaching straight beam and angle beam testing, with each considered as a different technique. The first is a longitudinal wave mode where the wave direction is parallel to the SP, and the second one is a shear wave mode where the wave direction is perpendicular to the SP. These techniques have different uses, as the straight beam is best for the pipe body and the angle beam for the pipe weld. As the review shows, the limitations for these techniques have been examined before. It should be mentioned that, while preparing a newly designed system, the limitations and advantages of the techniques must be considered. The resulting new system must be fit and appropriately in line with these technical conditions. As shown in **Figure 5**, the straight beam probe transmits an ultrasonic wave into the test object. In Case 1 there is no defect, thus the beam travels the full thickness of the material. The peaks for Case 1 represent the origination of the sonic signal, and the reflection of the signal from the back wall. In Case 2 probe S2 sends an unimpeded signal, which behaves like that in Case 1. But S3 sends a signal which is reflected by a defect, so the signal reflection arrives in less time, represented by the middle peak in the graph for Case 2. This yields the distance to defect (*DD*).

A consideration for angle beam testing is that the attenuation of the sonic waves varies depending on the material from which the object is made. As shown in **Figure 9** and **Figure 10**, β_R is adjusted according to the degree of penetration possible with a given material. For a more penetrable substance, β_R is greater, allowing a greater SKD, while a less penetrable substance requires a smaller β_R, with a respective reduction in SKD.

7. Conclusions

It is usual practice to base an NDT research on the various materials used in order to estimate and evaluate the suitability of ultrasonic NDE for testing the materials involved. The next step is to develop the needed devices, then to investigate the problem of using the ultrasonic NDE technique to assess the structural objects.

Through previous research, and the techniques that have been focused on, it can be seen that the straight probe works up to a limited thickness (such as the thickness of a pipe wall), and is employed to reveal subsurface defects, especially cracks that are parallel to the front and back surfaces of the test object. The example given in this article demonstrates a pipe wall with $D = 50.8$ mm, with eight cases of defect where *DD* ranges from 6.35 mm to 44.45 mm and presents the graphed results based on six different materials commonly used in the pipe construction industry (shown in **Figure 7**).

The angle beam is used to reveal discontinuities in the weld regions, with the limitation that the SKD cannot be increased unless the SP angle is increased. And that increased distance attenuates the sonic energy. The examples included in this article calculated the limitations of effective SD, SKD and SP using six reflection angles. Focusing on the 45° and 70° angles at the extremes for our example of $D = 25.4$ mm, we show that the 45° angle yields a ½ SKD (25.4 mm), full SKD (50.8 mm) and 1 ½ SKD (76.2 mm) with higher energy if we compare to the same material but a different angle, such as 70° angle which a greater test distance, but with more signal attenuation. Using the 70° angle we get ½ SKD (69.786 mm), full SKD (139.572 mm), and 1 ½ SKD (209.358 mm). The greater reach leads to reduced sensitivity because of the signal attenuation; operators must balance these factors when ultrasonic NDE is used. The general understanding is that 1 ½ SKD is the accepted limit of successful range for angle beam testing, all that according to the previous studies and technical websites. This requires continual attention by customer requirement and if there is no requirement, by the operator when assessing the integrity of pipes during manufacture.

It is recommended that a second, powerful NDE technique capable of remote detection would be useful alongside ultrasonic testing in order to direct the placement of the probe.

References

[1]   NDT Resource Center, The Collaboration for NDT Education, Iowa State University (2001-2014) Introduction to Ultrasonic Testing (Basic Principles of Ultrasonic Testing).
      https://www.nde-ed.org/EducationResources/CommunityCollege/Ultrasonics/Introduction/description.htm

[2]   Beard, M.D. and Lowe, M.J.S. (2003) Non-Destructive Testing of Rock Bolts Using Guided Ultrasonic Waves. *International Journal of Rock Mechanics & Mining Sciences*, **40**, 527-536.

http://dx.doi.org/10.1016/S1365-1609(03)00027-3

[3]   Chaki, S. and Bourse, G. (2009) Guided Ultrasonic Waves for Non-Destructive Monitoring of the Stress Levels in Pre-stressed Steel Strands. *Ultrasonics*, **49**, 162-171. http://dx.doi.org/10.1016/j.ultras.2008.07.009

[4]   Drinkwater, B.W. and Wilcox, P.D. (2006) Ultrasonic Arrays for Non-Destructive Evaluation: A Review. *NDT & E International*, **39**, 525-541. http://dx.doi.org/10.1016/j.ndteint.2006.03.006

[5]   Dutton, B., Clough, A.R., Rosli, M.H. and Edwards, R.S. (2011) Non-Contact Ultrasonic Detection of Angled Surface Defects. *NDT & E International*, **44**, 353-360. http://dx.doi.org/10.1016/j.ndteint.2011.02.001

[6]   Caravaca, D.S., Bird, C. and Kleiner, D. (2007) Ultrasonic Phased Array Inspection of Electrofusion Joints in Polyethylene Pipes. *Insight: Non-Destructive Testing and Condition Monitoring*, **49**, 83-86.
http://dx.doi.org/10.1784/insi.2007.49.2.83

[7]   Davis, J.R. (1992) Nondestructive Evaluation and Quality Control. *ASME Handbook*, **17**, 486-494.

[8]   Matz, V., Kreidl, M. and Šmíd, R. (2005) Classification of Ultrasonic Signals. *The 8th International Conference of the Slovenian Society for Non-Destructive Testing. Application of Contemporary Non-Destructive Testing in Engineering*, Portoroz, 27-33.

[9]   Mazeika, L., Kazys, R., Raisutis, R. and Sliteris, R. (2011) Ultrasonic Guided Wave Tomography for the Inspection of the Fuel Tanks Floor. *International Journal of Materials and Product Technology*, **41**, 128-139.
http://dx.doi.org/10.1504/IJMPT.2011.040291

[10] McNab, A. and Young, H.S. (1989) Knowledge-Based Approach to the Formulation of Ultrasonic Nondestructive Testing Procedures. *IEE Proceedings, Science*, **136**, 134-140.

[11] Nandi, A.K., Mampel, D. and Roscher, B. (1997) Blind Deconvolution of Ultrasonic Signals in Nondestructive Testing Applications. *IEEE Transactions on Signal Processing*, **45**, 1382-1390. http://dx.doi.org/10.1109/78.575716

[12] Mao, Y. and Que, P. (2006) Application of Hilbert-Huang Signal Processing to Ultrasonic Non-Destructive Testing of Oil Pipelines. *Journal of Zhejiang University SCIENCE A*, **7**, 130-134. http://dx.doi.org/10.1631/jzus.2006.A0130

[13] Ricci, M., Senni, L. and Burrascano, P. (2012) Exploiting Pseudorandom Sequences to Enhance Noise Immunity for Air-Coupled Ultrasonic Nondestructive Testing. *IEEE Transactions on Instrumentation and Measurement*, **61**, 2905-2915. http://dx.doi.org/10.1109/TIM.2012.2200409

[14] Salazar, A.J. and Rodríguez, E.C. (2011) Studies of the Effect of Surface Roughness in the Behaviour of Ultrasonic Signals in AISI-SAE-4340 Steel: Spectral and Wavelets Analysis. *International Journal of Microstructure and Materials Properties*, **6**, 224-235. http://dx.doi.org/10.1504/IJMMP.2011.043218

[15] Silva, C.E.R., Alvarenga, A.V. and Costa-Felix, R.P.B. (2012) Nondestructive Testing Ultrasonic Immersion Probe Assessment and Uncertainty Evaluation According to EN 12668-2:2010. *IEEE Transactions on Ultrasonics, Ferroelectrics, and Frequency Control*, **59**, 2338-2346. http://dx.doi.org/10.1109/TUFFC.2012.2459

[16] Zahran, O., Shihab, S. and Al-Nuaimy, W. (2002) Recent Developments in Ultrasonic Techniques for Rail-Track Inspection. Northampton British Institute of Non-Destructive Testing, Brownlow Hill, 55-60.

[17] Alobaidi, W., Sandgren, E. and Al-Rizzo, H. (2015) A Survey on Benchmark Defects Encountered in the Oil Pipe Industries. *International Journal of Scientific & Engineering Research*, **6**, 844-853.

[18] OLYMPUS Your Vision, Our Future (2015) Ultrasonic Flaw Detection Tutorial (What Is Ultrasound?).
http://www.olympus-ims.com/en/ndt-tutorials/flaw-detection/ultrasound/

[19] Lei, H., Huang, Z., Liang, W., Mao, Y. and Que, P.W. (2009) Ultrasonic Pig for Submarine Oil Pipeline Corrosion Inspection. *Russian Journal of Nondestructive Testing*, **45**, 285-291. http://dx.doi.org/10.1134/S106183090904010X

[20] OLYMPUS Your Vision, Our Future (2015) Ultrasonic Flaw Detection Tutorial (Wave Propagation).
http://www.olympus-ims.com/en/ndt-tutorials/flaw-detection/wave-propagation/

[21] Lucassen, J. and Tempel, M. (1972) Longitudinal Waves on Visco-Elastic Surfaces. *Journal of Colloid and Interface Science*, **41**, 491-498. http://dx.doi.org/10.1016/0021-9797(72)90373-6

[22] Gabriels, P., Snieder, R. and Nolet, G. (1987) *In Situ* Measurements of Shear-Wave Velocity in Sediments with Higher-Mode Rayleigh Waves. *Geophysical Prospecting*, **35**, 187-196.
http://dx.doi.org/10.1111/j.1365-2478.1987.tb00812.x

[23] Lysmer, J. and Waas, G. (1972) Shear Waves in Plane Infinite Structures. *Journal of the Engineering Mechanics Division*, **98**, 85-105.

[24] Aleshin, N.P., Gobov, Yu.L., Mikhailov, A.V., Smorodinskii, Ya.G. and Syrkin, M.M. (2014) Automatic Ultrasonic Inspection of Large Diameter Pipes. *Russian Journal of Nondestructive Testing*, **50**, 133-140.
http://dx.doi.org/10.1134/S1061830914030024

[25] Alleyne, D.N. and Cawley, P. (1992) The Interaction of Lamb Waves with Defects. *IEEE Transactions on Ultrasonics*,

Ferroelectrics, and Frequency Control, **39**, 381-397. http://dx.doi.org/10.1109/58.143172

[26] Leonard, K.R. and Hinders, M.K. (2005) Lamb Wave Tomography of Pipe-Like Structures. *Ultrasonics*, **43**, 574-583. http://dx.doi.org/10.1016/j.ultras.2004.12.006

[27] OLYMPUS Your Vision, Our Future (2015) Ultrasonic Flaw Detection Tutorial (General Concepts of Transducer Selection). http://www.olympus-ims.com/en/ndt-tutorials/flaw-detection/general-concepts/

[28] NDT Resource Center, The Collaboration for NDT Education, Iowa State University (2001-2014) Introduction to Ultrasonic Testing (Transducer Types). https://www.nde-ed.org/EducationResources/CommunityCollege/Ultrasonics/EquipmentTrans/transducertypes.htm

[29] Lach, M., Platte, M. and Ries, A. (1996) Piezoelectric Materials for Ultrasonic Probes. *NDT Net*, **1**, No. 9. http://www.ndt.net/article/platte2/platte2.htm

[30] Lowe, M.J.S., Alleyne, D.N. and Cawley, P. (1998) Defect Detection in Pipes Using Guided Waves. *Ultrasonics*, **36**, 147-154. http://dx.doi.org/10.1016/S0041-624X(97)00038-3

[31] Siqueira, M.H.S., Gatts, C.E.N., Da Silva, R.R. and Rebello, J.M.A. (2004) The Use of Ultrasonic Guided Waves and Wavelets Analysis in Pipe Inspection. *Ultrasonics*, **41**, 785-797. http://dx.doi.org/10.1016/j.ultras.2004.02.013

[32] Cho, Y., Oh, W.D. and Lee, J.H. (2006) A Wall Thinning Detection and Quantification Based on Guided Wave Mode Conversion Features. *Key Engineering Materials*, **321-323**, 795-798. http://dx.doi.org/10.4028/www.scientific.net/KEM.321-323.795

[33] NDT Resource Center, The Collaboration for NDT Education, Iowa State University (2001-2014) Introduction to Ultrasonic Testing (Normal Beam Inspection). https://www.nde-ed.org/EducationResources/CommunityCollege/Ultrasonics/MeasurementTech/beaminspection.htm

[34] Cleveland, D., Barron, A.R. and Mucciardi, A.N. (1980) Methods for Determining the Depth of Near-Surface Defects. *Journal of Nondestructive Evaluation*, **1**, 21-36. http://dx.doi.org/10.1007/BF00566229

[35] OLYMPUS Your Vision, Our Future (2015) Ultrasonic Flaw Detection Tutorial, Straight Beam Tests (Plates, Bars, Forgings, Castings, etc.). http://www.olympus-ims.com/en/ndt-tutorials/flaw-detection/straight-beam-tests/

[36] Yang, J.-S., Zhao, X.-G., Zhang, Y.-J. and Cao, Z. (2013) A New Type of Wheeled Intelligent Ultrasonic Thickness Measurement System. *Symposium on Piezoelectricity, Acoustic Waves and Device Applications* (*SPAWDA*), Changsha, 25-27 October 2013, 1-4. http://dx.doi.org/10.1109/spawda.2013.6841115

[37] NDT Resource Center, The Collaboration for NDT Education, Iowa State University (2001-2014) Introduction to Ultrasonic Testing (Angle Beams I). https://www.nde-ed.org/EducationResources/CommunityCollege/Ultrasonics/MeasurementTech/anglebeam1.htm

[38] NDT Resource Center, The Collaboration for NDT Education, Iowa State University (2001-2014) Introduction to Ultrasonic Testing (Angle Beams II). https://www.nde-ed.org/EducationResources/CommunityCollege/Ultrasonics/MeasurementTech/anglebeam2.htm

[39] NDT Resource Center, The Collaboration for NDT Education, Iowa State University (2001-2014) Introduction to Ultrasonic Testing (Angle Beam Inspection Calculations). https://www.nde-ed.org/GeneralResources/Formula/AngleBeamFormula/AngleBeamTrig.htm

18

Factor and Cluster Analysis as a Tool for Patient Segmentation Applied to Hospital Marketing in Jordan

Lamees M. Al-Durgham, Mahmoud A. Barghash

Industrial Engineering Department, Faculty of Engineering and Technology, The University of Jordan, Amman, Jordan
Email: l.aldurgham@ju.edu.jo, mabargha@ju.edu.jo

Abstract

Hospital marketing is becoming important for the survival and the prosperity of the health service. In addition, it indirectly acts as a formal feedback channel for the customer requirements, preferences, suggestions and complaints. In this work we have undertaken a survey based marketing study for two main objectives: The first being to better understand the patient clusters through k-means clustering and the second to understand customer perception of the different known quality perspectives through factor rotated and unrotated analysis. All of the questionnaires were designed according to international studies. Based on general descriptive statistics, items classified with higher variance but important, are: clean environment, doctors and nurses capabilities, and specialized doctors. Items that are less important with low variance are: food type, lighting and insurance. Also, items classified as more important with low variance are: recommended, no mistakes, and the cost. Using factor analysis rotated and unrotated reduced the variables into five main variables described as: medical aspects, psychological aspects, cost aspects, hospital image and ease of access and procedures. Using k-means clustering, the customers can be clustered into four main clusters with two of them described as general patient with wide variety of interest, serious cases interested in specialized doctors and food, and very serious case with high stress on equipment, no mistakes.

Keywords

Hospital Marketing, Factor Analysis, Clustering, Patient Segmentation, Customer Satisfaction

1. Introduction

It is argued that in the health care industry, increase in competitive pressures results in hospitals competing on the quality dimension [1]. In that respect, hospital competition might lead to better service and better efficiency. The improvement of efficiency over time is an indicator that hospitals have succeeded in their strategies to decrease the level of operational wastage in health care operations [2]. There are two consequences of competition among hospitals. One, in the absence of price competition, hospitals would perhaps compete on the quality dimension. This may lead to more resource consumption and hence lower efficiency. Two, increased competition should result in changes in demand, which would result in lower efficiency for those who are not able to attract patients, and higher efficiency for those who do [3]. Customer satisfaction is extremely important in highly competitive marketplace. Satisfaction is a person's feeling of pleasure or disappointment resulting from campaigning a product's perceived performance (or outcome) in relation to expectation. If the performance or expectations fall short, the customer is dissatisfied; if the performance matches the expectations, the customer is satisfied; if the performance exceeds expectation, the customer is highly satisfied or delighted [4]. Patients' satisfaction may drive for both changes in hospitals and for a better understanding of customer perception of quality through marketing studies.

It is becoming difficult for hospitals these days to depend on mere word of mouth promotion to attract patients, so hospital managements are putting extra effort in carving a brand image of the hospital and improving hospitals' visibility. In other words, many would agree that hospital marketing has evolved from being subtle to aggressive. Marketing is the management process that seeks to maximize returns to shareholders through developing relationships with valued customers and creating a competitive advantage [5]. The marketing concept came into use at the end of the 1940s (e.g., [6]-[13]). Marketing duality with respect to customers and businesses is also referred to in various definitions of marketing [14]. Marketing can be looked as the art of attracting and keeping profitable customers [15]. A company should not try to pursue and satisfy every customer [16].

Discussions of the role of marketing within organizations generally revolve around two perspectives (e.g., [17]-[20]). The first is a functional group perspective, which views marketing as an individual and distinct organizational entity (e.g., the marketing department); the second is an activity-based perspective, where marketing is treated as a set of activities undertaken by different people throughout the whole organization (e.g., market orientation as the responsibility of everyone). Research has typically separated the two approaches; for example, Hunt (1976) [21] took an activity based perspective, while Walker *et al.* (1987) [22] provided an example of the application of a functional group approach. More recently, the two perspectives have also been integrated [23]-[27].

A main component of marketing research is market segmentation and customer clustering. Clustering is an important mathematical tool for customer segmentation. Clustering is the problem of grouping objects on the basis of a similarity measure among them [28]. Relational clustering methods can be employed when a feature-based representation of the objects is not available, and their description is given in terms of pairwise (dis)similarities [28]. Clustering focuses on grouping objects on the basis of a similarity measure among them. It occurs very often in different disciplines and research areas. In some clustering applications, it is not possible to have a feature-based representation of the objects, and the description is given in terms of pairwise (dis)similarities. Some approaches have been proposed to cluster objects represented in this way, and are referred to as relational clustering methods [28].

Factor analysis is a method of exploring relationships among observed parameters. Several research works utilized principle component factor analysis and varimax rotating methods to extract factors related to hospital service quality [29]. The recommended first step in developing the measurement model is to examine several potentially meaningful structures in the data collected from former patients. Exploratory factor analysis (EFA) is commonly viewed as the best analytical tool for this purpose [30] [31]. International market segmentation has become an important issue in developing, positioning, and selling products. It helps companies to target potential customers at the international-segment level and to obtain an appropriate positioning [32]. Segmentation is therefore particularly important in enterprises that wish to develop and implement successful global marketing strategies [32]. Despite the obvious importance of international market segmentation for marketing as a discipline in general and international marketing in particular, it has received relatively little attention. In the literature a small percentage of papers dealt directly with international market segmentation [32] [33]. Segmentation is important to choose the most appropriate marketing strategies that better fit the interests of each segment [34] especially if segments can be characterized in terms of demographic characteristics [35] or even quality label and perceptions [36]. The importance of the segmentation of consumers has been proved in several studies [34] [36]-[42].

In this work we have applied customer segmentation to patients in Jordan through questionnaire result analysis for the purpose of identifying separate customer clusters and their requirements in hospitals. We also applied factor analysis to identify the important factors in customer requirements.

2. Mathematical Model

2.1. Factor Analysis

Factor analysis is related to the analysis of observable. If X_i is an observable trait. These $X = \begin{bmatrix} X_1 & X_2 & \cdots & X_p \end{bmatrix}^T$ with mean vector $\mu = \begin{bmatrix} \mu_1 & \mu_2 & \cdots & \mu_p \end{bmatrix}^T$. Let $f_1, f_2, \cdots, f_m$ be the unobservable common factors.

$$X = \mu + Lf + \varepsilon \quad \text{or} \quad X_i = \mu_i + l_{i1}f_1 + l_{i2}f_2 + \cdots + l_{im}f_m + \varepsilon_i \tag{1}$$

or

$$X_{ij} = \varepsilon_{ij} + \sum_{r=1}^{k} F_{ir} L_{rj}$$

and

$$X_i = \varepsilon_i + F_i L$$

or

$$X = \varepsilon + FL$$

where l_{ij}: factor loadings;

ε_i: independently distributed error terms with zero mean and finite variance;

F: the matrix $\begin{bmatrix} f_1, f_2, \cdots, f_m \end{bmatrix}$.

The communalities for the ith variable are computed by taking the sum of the squared loadings for that variable. This is expressed as

$$\hat{h}_i = \sum_{j=1}^{m} \hat{l}_{ij}^2 \tag{2}$$

The sample variance-covariance matrix and is expressed as

$$nV = X^T X \tag{3}$$

applying from (1) into (3)

$$nV = (\varepsilon + FL)^T (\varepsilon + FL)$$

$$nV = (\varepsilon^T + L^T F^T)(\varepsilon + FL)$$

$$nV = \varepsilon^T \varepsilon + \varepsilon^T FL + L^T F^T \varepsilon + L^T F^T FL$$

$$nV = n\psi + 0 + 0 + L^T IL$$

$$nV = n\psi + nL^T L$$

Then

$$V = \psi + L^T L \tag{4}$$

We can't actually calculate U until we know, or have to guess as to ψ. A reasonable and common starting-point is to do a linear regression of each feature j on all the other features, and then set ψ_j to the mean squared error for that regression.

$$U = V - \psi = L^T L \tag{5}$$

Through eigenvalue representation

$$U = e\lambda e^{\mathrm{T}} \tag{6}$$

We can represent the matrices as:

$$\lambda_q = \lambda_q^{1/2} \lambda_q^{1/2}$$

And thus

$$e_q \lambda_q e_q^{\mathrm{T}} = e_q \lambda_q^{1/2} \lambda_q^{1/2} e_q^{\mathrm{T}} = \left(e_q \lambda_q^{1/2}\right)\left(e_q \lambda_q^{1/2}\right)^{\mathrm{T}}$$

Applying into (5) gives

$$\hat{L} = \left(e_q \lambda_q^{1/2}\right)^{\mathrm{T}} \tag{7}$$

In summary, we have collected our eigenvectors into a matrix, but for each column of the matrix we will multiply it by the square root of the corresponding eigenvalue. This will now form our matrix L of factor loading in the factor analysis.

Factor rotation is motivated by the fact that these factor models are not unique. The rotated factor model is expressed as

$$X = \mu + L^* f^* + \varepsilon \tag{8}$$

where $L^* = LT$ and $f^* = fT$ where $T * T' = T'T = I$.

Varimax rotation is the most common of the rotations that are available which maximizes this quantity:

$$V = \frac{1}{p} \sum_{j=1}^{p} \left\{ \sum_{j=1}^{p} \left(\tilde{l}_{ij}^*\right)^4 - \frac{1}{p} \left(\sum_{j=1}^{p} \left(\tilde{l}_{ij}^*\right)^2\right)^2 \right\}. \tag{9}$$

We can then re-estimate the rotated loadings and matrices. Following to that we can find the vector of common factors for subject i, or $\hat{f}_i$ by minimizing the sum of the squared residuals:

$$\sum_{j=1}^{p} \varepsilon_{ij}^2 = \sum_{j=1}^{p} \left(y_{ij} - \mu_i - l_{j1} f_1 - l_{j2} f_2 - \cdots - l_{jm} f_m\right)^2 = \left(Y_i - \mu - L f_i\right)' \left(Y_i - \mu - L f_i\right) \tag{10}$$

$$\hat{f}_i = \left(\hat{L}'\hat{L}\right)^{-1} \hat{L}' \left(Y_i - \bar{y}\right) \tag{11}$$

2.2. *k*-Means Clustering Analysis

Clustering uses a set of input variables $\left(x_1, x_2, \cdots, x_n\right)$ for example columns in a survey to classify them into clusters. k-means clustering aims to partition this data and following to that the population filling the questionnaire into k ($\leq n$) sets $S = \{S_1, S_2, \cdots, S_k\}$ so as to minimize the within-cluster sum of squares namely:

$$\arg\min_{S} \sum_{i=1}^{k} \sum_{X \in S_i} \|X - \mu_i\|^2 \tag{12}$$

where μ_i is the mean of points in S_i.

We assume first a random k means $m_1^{(i)}, \cdots, m_k^{(i)}$ as centers of the clusters at the *i*th iteration.

Then we assign each point (customer) to one of the means Eucliden distance.

$$S_i^{(t)} = \left\{x_p : \left\|x_p - m_i^{(t)}\right\|^2 \leq \left\|x_p - m_j^{(t)}\right\|^2 \ \forall j, 1 \leq j \leq k\right\} \tag{13}$$

We then re-estimate the mean of the clusters

$$m_i^{(t+1)} = \frac{1}{\left|S_i^{(t)}\right|} \sum_{x_j \in S_i^{(t)}} x_j \tag{14}$$

The algorithm has converged when the assignments no longer change.

3. Results and Discussion

This section includes description of the questionnaire used (Section 3.1), general quantitative analysis is included in Section 3.2, factor analysis is described in Section 3.3, finally the clustering is in Section 3.4.

3.1. Questionnaire

The design of the questionnaire used in this study is as in **Table 1**, the first column in the questionnaire contains question number, the next column contains the questions, the questionnaire contains 19 questions, the questions cover different variables related to the location, the calmness and cleanness of the surrounding environment, the availability of parking, the cost and quality of the service, the accuracy of the medical procedures, the availability of specialized doctors and nurses, the popularity of the hospital among people, etc. the last column in the questionnaire is divided into ten columns with scales from 1 to 10, and it is required to fill only one from the 10 choices for each question.

We have distributed questionnaires to different people in different geographical location, and we got the results presented in this work.

3.2. General Quantitative Analysis

Table 2 contains the descriptive statistics for all the variables under investigation. It seems that all average values are around 5 which shows conflicting responses for all of the patients. High variance shows more disputed

Table 1. Marketing questionnaire used in this work.

Qu. No	Question	1 worst, 10 best									
		1	2	3	4	5	6	7	8	9	10
1	The location of the hospital (easy to reach).	1	2	3	4	5	6	7	8	9	10
2	The cleanness of the environment surrounding the hospital.	1	2	3	4	5	6	7	8	9	10
3	The calmness for the environment surrounding the hospital.	1	2	3	4	5	6	7	8	9	10
4	Availability of parking for visitors.	1	2	3	4	5	6	7	8	9	10
5	The cost compared to the health care.	1	2	3	4	5	6	7	8	9	10
6	The type of food (appearance, color, cleanness).	1	2	3	4	5	6	7	8	9	10
7	The quality of food service (on time, proper for the patient).	1	2	3	4	5	6	7	8	9	10
8	The quality of the overall hospital service.	1	2	3	4	5	6	7	8	9	10
9	How this hospital is known among people (good, bad).	1	2	3	4	5	6	7	8	9	10
10	The ability to avoid medical errors.	1	2	3	4	5	6	7	8	9	10
11	Qualified and well trained doctors.	1	2	3	4	5	6	7	8	9	10
12	Qualified and well trained nurses.	1	2	3	4	5	6	7	8	9	10
13	The quality of lightness (sun, suitable lamps, etc.).	1	2	3	4	5	6	7	8	9	10
14	Interior design (furniture, curtains, etc.).	1	2	3	4	5	6	7	8	9	10
15	Entertainment (TV, magazines, medical brochure, …).	1	2	3	4	5	6	7	8	9	10
16	Availability of specialist doctor.	1	2	3	4	5	6	7	8	9	10
17	The relationship between your doctor and the hospital.	1	2	3	4	5	6	7	8	9	10
18	You have health insurance for this hospital.	1	2	3	4	5	6	7	8	9	10
19	The availability of laboratory, certain medical tests, etc.	1	2	3	4	5	6	7	8	9	10

Table 2. Variance explained for each attribute/object in 3-dimensional map.

Attribute/object	Mean	Variance
Location	4.838	1.187
Clean environment	4.850	1.502
Calmness	4.512	1.390
Parking	4.475	1.372
Cost	5.200	0.903
Food type	4.875	0.774
Food service	5.050	1.016
Quality	5.000	1.160
Reputation	4.950	1.196
No mistakes	5.037	0.989
Doctors capability	5.225	1.379
Nurses capability	5.287	1.342
Lighting	4.912	0.782
Interior design	5.287	1.342
Entertainment	5.062	1.062
Specialized doctor	5.300	1.258
Recommended	5.600	0.894
Insurance	4.738	0.834
Special X-rays and test	5.375	1.201

categories, clean environment, doctors capability, and nurses capability, and specialized doctors. The explanation for the higher variance of these variables might be explained as a difference in responses between patient with critical cases for example the patient who wants to do an operation, and the patients doing routine hospital visits *i.e.* patients with simple procedure.

Low variances reflect low dispute, there is less dispute that the food type, the lighting, and the insurance are relatively less important to patients, on the other hand there is less dispute that the recommended, no mistakes, and the cost are relatively more important.

3.3. Factor Analysis

SPSS software was used to conduct factor analysis. Communality shows how much each attribute is explained by the factors. **Table 3** is the table of communalities which shows how much of the variance in the variables has been accounted for by the extracted factors. **Table 3** illustrates that all communalities are considered to be of high values for the 19 attributes, so the 19 attributes will be taken and considered for the next explanations.

According to Equation (1) communalities are calculated as the sum of the loadings of the variables which are calculated after removing the static mean (μ) and the noise factor (ε). Thus it reflects better the importance of the different variables to the patients. Variables with higher communalities are considered more important to the patients, namely no mistakes, interior design, the environment of the hospital, and the cost of service. Insurance, parking, and recommended are considered less important by the patients.

Table 4 shows the rotated matrix factor analysis. Column 1 is the factor where the initial number of factors is the same as the number of variables used in the factor analysis. Column 2 is the initial eigenvalues, eigenvalues are the variances of the factors. Column 2 contains three columns the first one is the total, the second one is the % of variance, and the last one is the cumulative %. The first one which is the total contains the eigenvalues. The

Table 3. Communalities generated by SPSS factor analysis.

	Communalities	
	Initial	Extraction
Location	1.000	0.812
Environment	1.000	0.817
Calmness	1.000	0.721
Parking	1.000	0.633
Cost	1.000	0.816
Food type	1.000	0.686
Food service	1.000	0.800
Quality	1.000	0.779
Reputation	1.000	0.708
Mistakes	1.000	0.838
Doctors	1.000	0.743
Nurses	1.000	0.700
Lighting	1.000	0.780
Interior	1.000	0.819
Entertainment	1.000	0.640
Specialized	1.000	0.760
Insurance	1.000	0.604
Recommended	1.000	0.667
Special test	1.000	0.764

Extraction method: Principal component analysis.

first factor will always account for the most variance (and hence have the highest eigenvalue), and the next factor will account for as much of the left over variance as it can, and so on. The second one which is the % of variance this column contains the percent of total variance accounted for by each factor. The last column is % of variance this column contains the cumulative percentage of variance accounted for by the current and all preceding factors. Next the table contains the rotation sums of squared loadings column the values in this panel of the table represent the distribution of the variance after the varimax rotation.

Table 4 shows that 74.138% of variance is explained by the 5 factors from the whole data, which is a representative result considering the sample size of 1000 respondents. The first factor explains about 20.274%, the second explains 15.308%. The third factor has a variance of 14.16%, the forth and the fifth factors explain the rest of the data.

The Scree plot graphs the eigenvalue against the factor number. **Figure 1** shows the Scree plot generated by the SPSS. From the Scree plot, it is known that SPSS software as a default takes the components which have an Eigen values above 1.

Component 5 has Eigen value of 0.992, so it can be included also. Then that bring us with 5 components/ factors which is reliable with 74.138%.

Table 4 shows the rotated component matrix generated by SPSS software. The 19 attributes are reduced into 5 major factors as discussed before.

Using the marketing engineering software we obtained the results in **Table 5**. Based on the largest absolute values in each column we defined the factors in **Table 6**. For example the largest values in the first column in

Table 4. Variance of five factors.

Component	Total variance explained					
	Initial eigenvalues			Rotation sums of squared loadings		
	Total	% variance	Cumulative %	Total	% variance	Cumulative %
1	8.839	46.519	46.519	3.852	20.274	20.274
2	1.869	9.839	56.359	2.908	15.308	35.582
3	1.343	7.068	63.427	2.690	14.160	49.742
4	1.043	5.489	68.916	2.461	12.954	62.696
5	0.992	5.222	74.138	2.174	11.441	74.138
6	0.734	3.866	78.004			
7	0.642	3.379	81.383			
8	0.605	3.186	84.569			
9	0.533	2.804	87.373			
10	0.486	2.560	89.932			
11	0.389	2.046	91.978			
12	0.315	1.660	93.638			
13	0.275	1.445	95.083			
14	0.234	1.234	96.317			
15	0.200	1.055	97.371			
16	0.162	0.855	98.227			
17	0.119	0.626	98.853			
18	0.115	0.606	99.460			
19	0.103	0.540	100.000			

Extraction method: Principal component analysis.

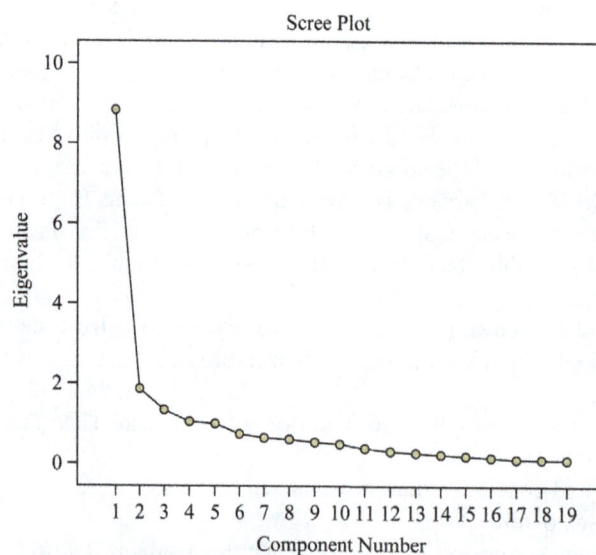

Figure 1. Scree plot for factor analysis.

Table 5. Unrotated factor analysis results.

Calmness	−0.7785	−0.03165	0.05774	−0.2707	−0.03227	−0.09609
Parking	−0.6825	−0.1989	−0.04391	−0.07001	0.3144	−0.2852
Cost of treatment	−0.6329	0.06477	0.4393	0.365	−0.1926	0.253
Type of food	−0.6726	−0.2475	0.2828	0.3363	0.04383	0.04354
Food service	−0.6553	−0.2232	0.2935	−0.468	0.1261	0.1479
Quality of service	−0.7833	0.1113	−0.07888	0.2283	0.3112	−0.0948
Reputation	−0.6823	0.1217	−0.1811	0.4052	0.2216	−0.2786
No medical mistakes	−0.8113	0.3225	0.1165	−0.2577	0.01397	−0.1828
Doctors capability	−0.8085	0.2051	0.1513	0.1068	−0.04818	−0.1571
Nurses capability	−0.7892	0.1284	0.1713	−0.1181	−0.1231	−0.2595
Lighting	−0.5567	−0.2816	−0.5236	−0.2737	−0.03765	0.1603
Interior design	−0.3982	−0.5432	−0.4766	0.1103	−0.3136	−0.2428
Entertainment	−0.4899	−0.6564	0.04552	0.17	−0.1453	0.07799
Specialized doctors	−0.7443	−0.364	0.2355	−0.01953	0.1111	0.231
Recommended by	−0.7733	−0.00921	−0.0456	0.03258	0.09037	0.1639
Health insurance	−0.4968	0.299	−0.4618	0.2368	0.2675	0.3898
Special X-ray	−0.7705	0.2394	−0.2141	−0.2444	0.1013	0.1589

Table 6. Factors and their major components.

Proposed factor name	Major components defining the factor
Factor 1: Medical aspects	• No medical mistakes • Doctors' capabilities • Nurses' capabilities • Calmness • Clean environment • Special tests and X-rays
Factor 2: Psychological aspect	• Entertainment • Interior design
Factor 3: Cost aspects	• Health insurance • Cost of treatment
Factor 4: Hospital image	• Food service • Reputation
Factor 5: Ease of access and easy and quality procedures	• Location • Clean environment • Parking • Quality of service

Table 5 are related to no medical mistakes, doctors' capabilities, nurses' capabilities, calmness, clean environment, and special tests and X-rays. These components are all related to medical aspects. The largest values in the next column are related to entertainment and interior design. We proposed that these two components can be defined as the psychological aspects. The largest absolute values in the third column are from health insurance, and cost of treatment, both are components of cost aspects. From the forth column the values are related to the food service and the reputation of the hospital, we proposed that these two components are defining the image of the hospital, from the last column in **Table 5**, location, clean environment, parking, and quality of service are the

components that have the highest absolute values, and all these components are related to ease of access and easy and quality procedures.

Table 7 shows the rotated components matrix, the idea of rotation is to reduce the number of factors on which the variables under investigation. Rotation does not actually change anything but makes the interpretation of the analysis easier.

From Table 7 we gathered the components with the maximum absolute values in each column, and defined the factor for these components in Table 8, from column 1 in Table 7. It is clear that food service, calmness, and no medical mistakes are the components with the highest values, all of these components are related to medical mistakes, quality of service, and psychological aspects.

The second column shows that reputation, and quality have the maximum values, these two components reflect the image of the hospital, from the third column it is clear that the cost of the service and the food type are the components with the highest values, they are related to the quality and the cost. The forth column in Table 7 reveals that the location of the hospital and the environment are the components with the highest absolute values, these components are related to ease of access and the psychological aspects. From the last column in Table 7 it is clear that the lighting and the interior design are the components with the highest values, these two components are related to the psychological aspects.

3.4. Clustering

The clustering analysis study are shown in Table 9 resulted in a best number of clusters to be four, taking into account the five factors resulting from the factor analysis. The confidence level was set to be 95%.

Table 7. Rotated component matrix.

| | Rotated component matrix[a] | | | | |
| | Component | | | | |
	1	2	3	4	5
Location	0.040	0.221	0.181	0.840	0.153
Environment	0.498	0.205	0.088	0.702	0.163
Calmness	0.712	0.216	0.172	0.320	0.188
Parking	0.569	0.423	0.227	−0.053	0.277
Cost	0.115	0.151	0.786	0.401	−0.044
Food type	0.270	0.257	0.700	0.091	0.222
Food service	0.849	−0.010	0.269	0.037	0.070
Quality	0.357	0.706	0.341	0.152	0.121
Reputation	0.153	0.728	0.319	0.184	0.137
Mistakes	0.657	0.351	0.192	0.478	−0.131
Doctors	0.390	0.397	0.471	0.459	−0.033
Nurses	0.562	0.230	0.342	0.463	0.024
Lighting	0.062	0.157	0.094	0.057	0.860
Interior	0.085	0.096	0.063	0.160	0.879
Entertainment	0.259	−0.026	0.539	−0.043	0.529
Specialized	0.562	0.171	0.586	0.016	0.265
Insurance	0.408	0.440	0.388	0.255	0.168
Recommended	0.054	0.789	−0.020	0.191	0.069
Special test	0.593	0.511	−0.014	0.380	0.086

Extraction method: Principal component analysis; rotation method: Varimax with Kaiser Normalization; [a]Rotation converged in 7 iterations.

Table 8. Proposed factors and its major components.

Proposed factor name	Major components defining the factor
Medical aspects, quality, and psychological aspects	• Food service • Calmness • No medical mistakes
The image of the hospital	• The hospital is recommended • The reputation of the hospital • The quality
Quality and cost	• The cost of the service • The food type
Ease of access	• The location of the hospital • The environment
Psychological aspects	• Lighting • Interior design

Table 9. Clustering results.

	Overall	CL1	CL2	CL3	CL4
Location	8.21	9.00	4.00	7.09	7.72
Environment	8.41	9.16	4.00	7.45	8.06
Calmness	8.02	8.91	3.57	7.09	7.17
Parking	7.27	8.19	3.43	6.18	6.17
Cost	8.06	8.81	5.14	4.55	8.67
Types of food	8.32	9.05	6.00	5.91	8.11
Food service	8.17	8.77	5.00	6.82	8.11
Quality	8.26	9.28	4.14	6.91	7.06
Reputation	8.02	8.80	4.00	7.18	7.33
Mistakes	8.90	9.70	3.57	6.91	9.33
Doctors capability	8.90	9.64	5.29	6.64	9.06
Nurses capability	8.56	9.44	4.29	6.18	8.56
Lighting	7.68	8.28	5.71	7.82	6.22
Interior design	6.75	7.34	5.29	7.45	4.78
Entertainment	7.49	8.22	5.86	6.73	6.00
Specialized doctor	8.58	9.41	6.43	6.55	7.72
Recommended	7.95	9.12	4.00	6.09	6.44
Insurance	7.87	8.61	4.00	7.82	6.78
X-ray	8.67	9.47	2.86	8.73	8.06

Cluster 1 this cluster represents 64% of the sample, this cluster represents general patient while concerned about all features equally. They will be selecting a hospital with no medical mistakes is very important to them, the presence of specialized doctors, capable doctors and capable nurses in the hospital is also very important. This group seeks hospital with certain tests or X-rays. Hospital locations are of great importance to them, they also seek a calm and clean surrounding area around the hospital. It is important for them to find a parking area for their cars as well as their visiting relatives'. From their point of view, it is important that hospital offers good quality food to its in-patients, the overall quality of services is of much importance to them. They prefer hospitals with considerably good reputation. It is important to have entertainment features in the hospital room (T.V., magazines, medical brochures, etc.). It is also preferred for them if the treating physician of a patient recommends the hospital. They prefer a hospital where they are health insured. Cost of treatment is a major factor for selecting hospitals for these people. Good Food service in the cafeteria of the hospital is also important to them. The lighting of the room is very important to them, *i.e.* proper sunshine and good electrical lighting is important for these people.

Cluster 2, this cluster represents (7%) of the sample, it is clear that this cluster represent the patients with critical cases, since the availability of Specialized doctor in the hospital is of the highest importance for them. They also require the hospital to offer a good type food to its patients.

Cluster 3, this cluster represents (11%) of the sample, specialized equipment are of major importance to them, this cluster shares a high interest with cluster 1 in some of the attribute. The lighting of the room is very important to them, *i.e.* proper sunshine and good electrical lighting is important for these people. They prefer a hospital where they are health insured. Their major interest is in the interior design of the hospital room. Comfortable interior design is of high importance to them. They prefer hospitals with considerably good reputation.

The last cluster is Cluster 4 with 18% from the sample. It is clear that this cluster is for the patients with the very serious cases, a hospital known for its accurate and reliable results and no medical mistakes committed during patients' treatment is very important to them. The presence of capable doctors and nurses in the hospital is very important to them. This cluster also share common interest with cluster 1 in some of the attributes. The cost of treatment and the presence of specialized equipment are major factors for selecting hospitals for the people in this cluster. Good Food service in the cafeteria of the hospital is also important to them.

The results acquired from the Clustering analysis model shows that the majority of the sample (64%) are interested in almost all of the attributes. That means that the chosen attributes are of great importance to many people and that they highly affect their hospitals' choice. Clusters 2, 3 and 4 are also interested in some of the attributes which confirms the importance of these attributes to people.

4. Conclusions

From the analysis of the results we concluded the following:

1) Based on general descriptive statistics, items classified with higher variance but important, are: clean environment, doctors and nurses capabilities, and specialized doctors. Items that are less important with low variance are: food type, lighting and insurance. Also, items classified as more important with low variance are: recommended, no mistakes, and the cost.

2) Factor analysis rotated and unrotated shows that we can sum up the variable into five main variables described as: medical aspects, psychological aspects, cost aspects, hospital image and procedures.

3) Using *k*-means clustering: the customers can be clustered into four main clusters:

Cluster 1 (64% of the sample): this cluster represents general patients who usually select a hospital with high emphasis on no medical mistakes, specialized doctors, capable doctors and nurses, availability of certain tests, hospital locations, calm and clean surrounding area around the hospital, acceptable cost of treatment.

Cluster 2 (7% of the sample): critical cases consider availability of specialized doctor in the hospital and a good type food.

Cluster 3 (11% of the sample) considers specialized equipment, interior design, health insured and good reputation. This cluster shares a high interest with cluster 1 in some of the attribute.

Cluster 4 (18% from the sample) may represent very serious cases and consider no medical mistakes, capable doctors and nurses, cost of treatment, and the presence of specialized doctors.

References

[1]   Nyman, J. and Bricker, D. (1989) Profit Incentives and Technical Efficiency in the Production of Nursing Home Care. *Review of Economics and Statistics*, **71**, 586-594.

[2]   Assaf, A.G. and Josiassen, A. (2012) Time-Varying Production Efficiency in the Health Care Foodservice Industry: A Bayesian Method. *Journal of Business Research*, **65**, 617-625. http://dx.doi.org/10.1016/j.jbusres.2011.02.049

[3]   Chang, H., Chang, W.-J., Das, S. and Li, S.-H. (2004) Health Care Regulation and the Operating Efficiency of Hospitals: Evidence from Taiwan. *Journal of Accounting and Public Policy*, **23**, 483-510. http://dx.doi.org/10.1016/j.jaccpubpol.2004.10.004

[4]   Kotler, P. and Armstrong, G. (2004) Principles of Marketing. 10th Edition, Pearson-Prentice Hall, New Jersey.

[5]   Kotler, P. and Lane, K.K. (2008) Marketing Management. 12th Edition, Pearson-Prentice Hall, New Jersey.

[6] Ames, B.C. (1970) Trappings vs. Substance in Industrial Marketing. *Harvard Business Review*, **48**, 93-102.

[7] Anderson, E.W. and Mittal, V. (2000) Strengthening the Satisfaction-Profit Chain. *Journal of Service Research*, **3**, 107-120. http://dx.doi.org/10.1177/109467050032001

[8] Bagozzi, R.P. (1975) Marketing as Exchange. *Journal of Marketing*, **39**, 32-39. http://dx.doi.org/10.2307/1250593

[9] Felton, A.P. (1959) Making the Marketing Concept Work. *Harvard Business Review*, **37**, 55-65.

[10] Grønroos, C. (1990) Marketing Redefined. *Management Decision*, **28**, 5-9. http://dx.doi.org/10.1108/00251749010139116

[11] Kohli, A.K. and Jaworski, B.J. (1990) Market Orientation: The Construct, Research Propositions, and Managerial Implications. *Journal of Marketing*, **54**, 1-18. http://dx.doi.org/10.2307/1251866

[12] Kohli, A.K. and Jaworski, B.J. (1993) Market Orientation: Antecedents and Consequence. *Journal of Marketing*, **57**, 50-70.

[13] Narver, J.H. and Slater, S.F. (1990) The Effect of Market Orientation on Business Profitability. *Journal of Marketing*, **54**, 20-35. http://dx.doi.org/10.2307/1251757

[14] Helgesen, Ø. (2007) Customer Accounting and Customer Profitability Analysis for the Order Handling Industry—A Managerial Accounting Approach. *Industrial Marketing Management*, **36**, 757-769. http://dx.doi.org/10.1016/j.indmarman.2006.06.002

[15] Kotler, P. and Armstrong, G. (1996) Principles of Marketing. 7th Edition, Prentice-Hall, Englewood Cliffs.

[16] Berger, P.D. and Nasr, N.I. (1998) Customer Lifetime Value: Marketing Models and Applications. *Journal of Interactive Marketing*, **12**.

[17] Piercy, N.F. (1986) The Role and Function of the Chief Marketing Executive and the Marketing Department: A Study of Medium-Sized Companies in the UK. *Journal of Marketing Management*, **1**, 265-289. http://dx.doi.org/10.1080/0267257X.1986.9963990

[18] Varadarajan, R.P. (1992) Marketing's Contribution to the Strategy Dialogue: The View from a Different Looking Glass. *Journal of the Academy of Marketing Science*, **20**, 335-344. http://dx.doi.org/10.1007/BF02725210

[19] Webster Jr., F.E. (1992) The Changing Role of Marketing in the Corporation. *Journal of Marketing*, **56**, 1-17.

[20] Workman Jr., J.P. (1993) Marketing's Limited Role in New Product Development in One Computer Systems Firm. *Journal of Marketing Research*, **30**, 405-421.

[21] Hunt, S.D. (1976) The Nature and Scope of Marketing. *Journal of Marketing*, **40**, 17-28. http://dx.doi.org/10.2307/1249990

[22] Walker Jr., O.C. and Ruekert, R. (1987) Marketing's Role in the Implementation of Business Strategies: A Critical Review and Conceptual Framework. *Journal of Marketing*, **51**, 15-34.

[23] Workman Jr., J.P., Homburg, C. and Gruner, K. (1998) Marketing Organization: An Integrative Framework of Dimensions and Determinants. *Journal of Marketing*, **62**, 21-41.

[24] Moorman, C. and Rust, R.T. (1999) The Role of Marketing. *Journal of Marketing*, **63**, 180-197. http://dx.doi.org/10.2307/1252111

[25] Merlo, O. and Auh, S. (2009) The Effects of Entrepreneurial Orientation, Market Orientation, and Marketing Subunit Influence on Firm Performance. *Marketing Letters*, **20**, 295-311. http://dx.doi.org/10.1007/s11002-009-9072-7

[26] Verhoef, P.C. and Leeflang, P.S.H. (2009) Understanding Marketing Department's Influence within the Firm. *Journal of Marketing*, **73**, 14-37. http://dx.doi.org/10.1509/jmkg.73.2.14

[27] Merlo, O. and Auh, S. (2010) Marketing's Strategic Influence in Australian Firms: A Review and Survey. *Australasian Marketing Journal*, **18**, 49-56. http://dx.doi.org/10.1016/j.ausmj.2010.01.002

[28] Filippone, M. (2009) Dealing with Non-Metric Dissimilarities in Fuzzy Central Clustering Algorithms. *International Journal of Approximate Reasoning*, **50**, 363-384. http://dx.doi.org/10.1016/j.ijar.2008.08.006

[29] Lee, W.-I., Shih, B.-Y. and Chung, Y.-S. (2008) The Exploration of Consumers' Behavior in Choosing Hospital by the Application of Neural Network. *Expert Systems with Applications*, **34**, 806-816. http://dx.doi.org/10.1016/j.eswa.2006.10.020

[30] Briggs, S.R. and Cheek, J.M. (1986) The Role of Factor Analysis in the Development and Evaluation of Personality Scales. *Journal of Personality*, **54**, 106-148. http://dx.doi.org/10.1111/j.1467-6494.1986.tb00391.x

[31] Koerner, M.M. (2000) The Conceptual Domain of Service Quality for Inpatient Nursing Services. *Journal of Business Research*, **48**, 267-283. http://dx.doi.org/10.1016/S0148-2963(98)00092-7

[32] Steenkamp, J.B.E.M. and Ter Hofstede, F. (2002) International Market Segmentation: Issues and Perspectives. *International Journal of Research in Marketing*, **19**, 185-213. http://dx.doi.org/10.1016/S0167-8116(02)00076-9

[33] Aulakh, P.S. and Kotabe, M. (1993) An Assessment of Theoretical and Methodological Development in International Marketing: 1980-1990. *Journal of International Marketing*, **1**, 5-28.

[34] Naes, T., KubberØd, E. and Sivertsen, H. (2001) Identifying and Interpreting Market Segments Using Conjoint Analysis. *Food Quality and Preference*, **12**, 133-143. http://dx.doi.org/10.1016/S0950-3293(00)00039-2

[35] Andrews, R.L. and Currin, I.S. (2003) Recovering and Profiling the True Segmentation Structure in Markets: An Empirical Investigation. *International Journal of Research in Marketing*, **20**, 177-192. http://dx.doi.org/10.1016/S0167-8116(03)00017-X

[36] Sepúlveda, W.S., Maza, M.T. and Mantecón, A.R. (2010) Factors Associated with the Purchase of Designation of Origin Lamb Meat. *Meat Science*, **85**, 167-173. http://dx.doi.org/10.1016/j.meatsci.2009.12.021

[37] Carbonell, L., Izquierdo, L., Carbonell, I. and Costell, E. (2008) Segmentation of Food Consumers According to Their Correlations with Sensory Attributes Projected on Preference Spaces. *Food Quality and Preference*, **19**, 71-78. http://dx.doi.org/10.1016/j.foodqual.2007.06.006

[38] Fonti Furnols, M., San Julián, R., Guerrero, L., Sañudo, C., Campo, M.M., Olleta, J.L., *et al.* (2006) Acceptability of Lamb Meat from Different Producing Systems and Ageing Time to German, Spanish and British Consumers. *Meat Science*, **72**, 545-554. http://dx.doi.org/10.1016/j.meatsci.2005.09.002

[39] Oliver, M.A., Nute, G.R., Fonti Furnols, M., San Julián, R., Campo, M.M., Sañudo, C., *et al.* (2006) Eating Quality of Beef from Different Production Systems, Assessed by German, Spanish and British Consumers. *Meat Science*, **74**, 422-435. http://dx.doi.org/10.1016/j.meatsci.2006.03.010

[40] Realini, C.E., Fonti Furnols, M., Guerrero, L., Campo, M.M., Sañudo, C., Nute, G.R., *et al.* (2009) Effect of Finishing Diet on Consumer Acceptability of Uruguayan Beef in the European Market. *Meat Science*, **81**, 499-506. http://dx.doi.org/10.1016/j.meatsci.2008.10.005

[41] Verbeke, W., Pérez-Cueto, F.J.A., de Barcellos, M.D., Krystallis, A. and Grunert, K.G. (2009) European Citizen and Consumer Attitudes and Preferences regarding Beef and Pork. *Meat Science*, **84**, 284-292. http://dx.doi.org/10.1016/j.meatsci.2009.05.001

[42] Fonti Furnols, M., Realini, C., Montossi, F., Sañudo, C., Campo, M.M., Oliver, M.A., Nute, G.R. and Guerrero, L. (2011) Consumer's Purchasing Intention for Lamb Meat Affected by Country of Origin, Feeding System and Meat Price: A Conjoint Study in Spain, France and United Kingdom. *Food Quality and Preference*, **22**, 443-451. http://dx.doi.org/10.1016/j.foodqual.2011.02.007

19

A Comprehensive Evaluation of Eye Surgery Performance by Sigma Quality Level for Eye Care Hospitals in Turkey

İbrahim Şahbaz[1], Mehmet Tolga Taner[2], Gamze Kağan[3], Engin Erbaş[4]

[1]Department of Opticianry, Üsküdar University, Istanbul, Turkey
[2]Department of Business Administration, Doğuş University, Istanbul, Turkey
[3]Department of Occupational Health and Safety, Üsküdar University, Istanbul, Turkey
[4]Institute of Health Sciences, Üsküdar University, Istanbul, Turkey
Email: ibrahim.sahbaz@uskudar.edu.tr, mtaner@dogus.edu.tr, gamze.kagan@uskudar.edu.tr, enginerbas78@hotmail.com

Abstract

Sigma Quality Levels are statistical indices that can play an important role in establishment of more reliable and robust surgical processes. This article aims to highlight the value of interpreting sigma levels as a modern means of both the efficiency and success rate of nine different type of eye surgery processes involved. Data were compiled from a comprehensive study of multiple sources—all based on medical practices—carried out in a range of eye care hospitals throughout Turkey. It is found that the eye surgery processes in Turkey operates at an average of 3.4816 sigma level. This corresponds to a DPMO of 23,725 and a yield of approximately 97.63% in 7292 surgeries. Thus, Turkish eye care hospitals have to make significant changes to drive to a 6σ level performance.

Keywords

Six Sigma, Ophthalmology, Surgeries, Complications

1. Introduction

During the past three decades, profound tecnological advances such as lasers, molecular genetics and immunology, have occurred in the diverse fields of medicine [1]. This progress has encouraged similar advances in almost every aspect of ophthalmic practice. The implementation and adaptation of so much information lead to more robust practices resulting in far fewer types of complication [2].

Pertaining to both ocular and visual health, maintenance and restoration are complimentary parts of the overall management of patients with eye diseases [3]. However, eye surgeries are often further complicated by results which may lead to a major cost burden on the healthcare systems and even endanger the patient's life resulting in the loss of well-being and the quality of life by undergoing the risk. This burden may include direct costs of treating these complications, opportunity costs and lost productivity costs due to disability as well as the indirect costs related to extended periof of stay in hospital. In some cases, due to ophthalmic drugs being of high costs, financial burden can also play a key role in treatment.

Each surgical procedure carries certain risks with it. Therefore, it is impossible to guarantee that it will produce perfect results without complications [2]. While most eye surgeries have become reliably safe procedures with the vast amount of surgical experience accumulated and the presence of high technology equipment, complications are an expected part of surgical practice [3]. Since these acute, sub-acute and/or chronic conditions take serious toll on human health, economic and emotional pressures at times leading to acute life-threatening complications, their elimination by continuously improved ophthalmic surgical processes is self evident.

Most surgeries are complicated processes where at least a standard outcome is desired by the independent skills but collective work of ophthalmic surgeon, nurse, assistant surgeon and technician who are also reliant on equipment and materials. Some examples for desired outcomes of ophthalmic surgeries may include ocular alignment, high visual acuity, high vision quality, corrected refractive error, emmetropia, orthophoria, fusion, clear cornea and excellent prognosis. Certain complications do not usually occur if the surgery was not performed in a manner below this standard of care. On the other hand, certain complications may be difficult for even the most experienced surgeon to protect against while others might be complications which only seldom occur due to surgeon negligence.

Eye research has grown exponentially in recent years requiring the personnel involved to maintain a consistent and updated level of knowledge and technique so as to remain abreast of all modern clinical procedures and the scientists to revise their individual fields of expertise [1]. Coupled with this dichotomy is the wide range of complications to establish their root causes and their prevention depends on many variables. These variables necessitate a reference work that provides concise, easily understandable information on a wide range of surgeries.

This study presents and compares the complication rates of nine different types of eye surgeries, namely trabulectomy, IntraLase surgery, intravitreal injections, cataract surgery in patients with pseudoexfoliation syndrome, pars plana vitrectomy, phacoemulsification cataract surgery, penetrating keratoplasty and LASIK surgery and strabismus surgery [4]-[12]. To achieve this, a Six Sigma metric named Sigma Quality Level is employed.

2. Methodology

In recent years, Six Sigma has become the most popular quality and process improvement methodology which strives for elimination of defects in the processes whose origin is traced back to the pioneering and innovation made at Motorola Company and its adoption by many manufacturing and service companies worldwide [13]. It is improving the outcomes of modern healthcare processes today [14].

Six Sigma Methodology is a statistical approach to measure variance incorporating normal distribution. The more standard deviations (σ), i.e. an indicator of the variation of the process, the more capable is the process. A Six Sigma process means that 6σ equates in percentage terms to 99.99966% accuracy or to 3.4 defects per million opportunities (DPMO) to make a defect (**Figure 1**).

Sigma Quality Level is a measure used to indicate how often the defects are likely to occur. It is a simple statistic that puts a given defect rate on a "six-sigma" scale and measures the quality maturity of the process. Here, sigma is a mathematical term and the key measure of variability. It emphasizes need to control both the average and variability of a process. According to [15], most companies produce a defect rate of between 35,000 and 50,000 of DPMO which equates to a sigma quality level of 3σ to 3.5σ.

Table 1 shows different Sigma levels and associated defects per million opportunities. The statistical representation of Six Sigma describes quantitatively how a process is performing. For example, 1σ level indicates that it tolerates 690,000 defects per million opportunities with 30.9% yield. 6σ level allows only 3.4 defects per million opportunities with 99.99966 yield. This means that 6σ is not the same as zero defects, however, it is a significant threshold of performance (**Table 1**).

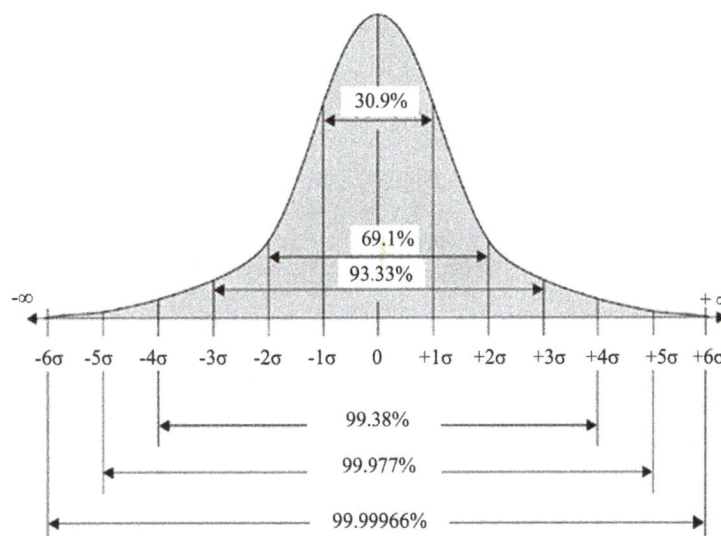

Figure 1. Six sigma (6σ) and normal distribution.

Table 1. Sigma performance levels.

Sigma Level (σ)	DPMO	Defects (%)	Yield (%)
6	3.4	0.00034	99.99966
5	233	0.023	99.977
4	6.210	0.62	99.38
3	66.807	6.68	93.33
2	308.537	30.8	69.2
1	691.462	69.1	30.9

3. Six Sigma's DMAIC

3.1. Define Phase

The primary objective of Six Sigma is to reduce variations, in products and processes, to achieve high quality levels. To identify weaknesses and eliminate defects in the process, Six Sigma makes use of a structured sequence of steps called DMAIC (Define, Measure, Analyze, Improve and Control) and finds the root causes behind problems to reach near perfect processes.

In each hospital under study, a Six Sigma team was gathered from a surgeon, nurse and technician. As the desired outcomes, the teams first *defined* the objectives of their surgical processes (**Table 2**).

The team defined a "complication" as a defect, *i.e.* any unwanted outcome inhibiting the patient to be cured and stable which compounds the illness and decreases the patient's quality of life or prolongs the planned hospital stay [13]. They noted the complications as they occurred. They stated that the surgical process with less amount of complications has higher success [16]. Thus, they used DMAIC to analyse, improve and modify the complicated ophthalmic processes by identifying and eliminating root causes of the complications.

Table 2 records the results of a range of eye surgeries, their length of duration and their success rates measured on the sigma level. While the duration of surgeries varies between surgery types, it may also be different within the same surgery type depending on the amount of time it takes to achieve the objective.

3.2. Measure Phase

The team calculated the sigma level of each complication and type of surgery from DPMO (**Table 2**). DPMO is calculated from Equation (1) as follows [13]:

Table 2. Number, duration, yield, sigma level and objective per surgery type.

Surgery	Surgery Name	Objective	Number of Surgeries	Duration (mins)	Yield (%)	Sigma Level
I	LASIK Surgery	Emmetropia; Corrected Refractive Error	2357	5 - 10	98.7	3.7135
II	Strabismus Surgery	Orthophoria; Binocular Function; Ocular Alignment; High Visual Acuity; Fusion	473	30 - 60	95.6	3.2025
III	Intravitreal Injection	Improvement of Vision Acuity; Prevention of Worsening of Visual Acuity	229	2 - 3	96.1	3.2657
IV	Phacoemulsification Cataract Surgery	High Visual Acuity	1050	10 - 30	99.4	3.9580
V	Cataract Surgery in Patients with Pseudoexfoliation Syndrome	High Visual Acuity	151	20 - 30	98.6	3.7030
VI	Pars plana vitrectomy	High Visual Acuity; High Vision Quality	2272	60 - 120	99.1	3.8559
VII	IntraLase Surgery	Emmetropia; Corrected Refractive Error	448	5 - 10	96.9	3.3547
VIII	Penetrating keratoplasty	Improved Visual Acuity; Excellent Prognosis; Clear Cornea	59	60 - 90	95.0	3.1418
IX	Trabeculectomy	IOP under control; Long-term Survival of Optic Nerve; Preservation of Visual Acuity; Preservation of Visual Field	253	25 - 35	94.9	3.1391

$$DPMO = 1,000,000 \times (B/A) \tag{1}$$

where A is the total number of surgeries performed and B is the total number of complications occurred. Then, DPMO is translated to a Sigma level of Z-statistic. For a given sigma level, the DPMO is provided, assuming normality and a 1.5 sigma shift [16]. The higher level of sigma indicates a lower rate of complications and a more efficient process [16].

A total of 7292 procedures have already been carried out. Among the nine types of surgeries, phacoemulsification cataract surgeryyielded the highest sigma level (3.9580), whereas trabulectomy did the lowest (3.1391).

Alternate methods (e.g. Lasik surgery vs Intralase surgery; phacoemulsification cataract surgery vs cataract surgery in patients with pseudoexfoliation syndrome) demonstrate that lasik surgery (3.7135) and phacoemulsification cataract surgery (3.9580) result in having more successful surgerical processes than IntraLase surgery (3.3547) and cataract surgery in patients with pseudoexfoliation syndrome (3.7030), respectively.

Sigma levels for all surgery types were found to be under 4.00 sigma level indicating that all processes need improvement and preventative/corrective actions. From the sigma levels per surgery type in **Table 2**, the average is measured and found that the eye surgery processes in Turkey operates at a 3.4816 sigma level. If it were assumed that each surgery yielded at most one complication, this level would correspond to a DPMO of 23,725 and a yield of approximately 97.63% in 7292 surgeries. This means that the Turkish eye care hospitals have to make significant changes to drive to a 6σ level performance.

Then, the team prioritized the complications according to how serious their consequences were (*i.e.* severity score) and how frequently they occurred (*i.e.* occurrence rate). They determined the severity of each complication and assigned scores for them. The severity of each complication was scored from 1 to 4 (*i.e.* a complication with no harm = 1; temporary harm = 2; permanent harm = 3; death = 4).

Severity score for all the observed complications per surgery type are listed in **Table 3** along with their occurence rates (%) and sigma levels.

3.3. Analyze Phase

The teams evaluated each process while attempting to assess the multipleCritical-to-Quality (CTQ) factors that cause each acute, sub-acute and chronicpre-, intra- and post-surgical complication (**Table 4**). By brainstorming, the team identified the quality factors affecting each process. They determined that the complications were caused by the surgeon, assistant surgeon, nurse, technician, anaesthesiologist, patient, donor cornea, equipment, materials and hygiene. Then, they *analyzed* the processes depending on the effect frequency of each quality factor. Then, they separated these factors according to how frequently they affect the process and cause a

Table 3. Severity score of complications (Occurrence Rate %; Sigma Level) per Surgery Type.

	Severity Score			
	1	**2**	**3**	**4**
LASIK	Subconjunctival haemorrhage (55.49%; 1.36), Interface debris (0.17%; 4.43).	Dry eye syndrome (100%; −6.26), Limbal Haemorrhage (0.81%; 3.91), Epithelial erosion (0.25%; 4.30), Decentered flap (0.25%; 4.30), Free flap (0.04%; 4.84).	Epilethial ingrowth (0.47%; 4.10), Thin flap (0.47%; 4.10), Undercorrection (0.47%; 4.10), Overcorrection (0.30%; 4.25), Shifting sands of the Sahara (0.08%; 4.64), Wrinkles (0.08%; 4.64), Small flap (0.04%; 4.84).	Flap edge melt (2.84%; 3.40), Incomplete flapv (0.21%; 4.36), Buttonhole (0.08%; 4.64), Wrong insertion of patient's biometric data (0.08%; 4.64), Inadequate suction (0.04%; 4.84), Sliding flap (0.04%; 4.84).
Strabismus	Haemorrhage (100%; −6.35), Neusea and vomiting (28.96%; 2.05), Corneal abrasions (0.63%; 3.99).	Bleeding (1.48%; 3.68), Prolapsed Tenon's capsule (0.21%; 4.36).	Partial scleral damage (0.85%; 3.89), Surgery to the wrong muscle (0.63%; 3.99).	Undercorrection (4.86%; 3.16), No change in vertical strabismus degree (1.48%; 3.68), Slipped muscle (1.27%; 3.74), Orbital fat penetration (0.63%; 3.99), Wrong site surgery (0.63%; 3.99), Overcorrection (0.63%; 3.99), Lost muscle (0.21%; 4.36), Wrong type of intervention tothe muscle (0.21%; 4.36), Apnea (0.21%; 4.36).
Intravitreal Injection	Subconjunctival Haemorrhage (45.85%; 1.60).	Increase in IOP (18.34%, 2.40), Anterior uveitis (6.99%; 2.98), Anterior/Posterior Inflammation (2.18; 3.52).	Cataract (0.44%; 4.12)	Endophthalmitis (0.44%; 4.12), Retinal detachment (0.44%; 4.12).
Phacoemulsification	Capsular tension ring Implantation (0.57%; 4.03), Retained cortex material (0.76%; 3.93), Iridodonesis (1.44%; 3.78).	Damage to the IOL (1.33%; 3.72), Iris prolapse (0.19%; 4.39), Pupillary irregularity, Iris atrophy (2.86%; 3.40), Fibrin reaction (0.76%; 3.93).	Radial tears in the anterior capsule (0.57%; 4.03), Iridodialysis (0.095%; 4.60), Zonular disinsertion (0.095%; 4.60), Intraocular pressureelevation and Glaucoma (1.91%; 3.20), Irvine-Gass syndrome (1.52%; 3.66).	Posterior capsular tear (2.09%; 3.53), Endophthalmitis (0.38%; 4.17), Corneal edema and bullous keratopathy (0.095%; 4.60).
Pseudoexfoliation	Iris retraction hooks (2.64%; 3.44), Retained cortex material (0.66%; 3.98), Pupillar membrane (0.66%; 3.98).	Glaucoma (1.98%; 3.56), Pupillary irregularity (0.66%; 3.98), Iris sphincter tears (0.66%; 3.98).	Zonular dialysis (2.64%; 3.44), Iridodialysis (0.66%; 3.98).	Posterior capsule rupture (11.25%; 2.71), IOL dislocation (0.66%; 3.98).
Pars plana vitrectomy	Posterior synechiae (5.24%; 3.12), Conjunctival haemorrhage (30.24%; 2.02).	Cystoid macular edema (2.82%; 3.41), Glaucoma (1.94%; 3.57), Hypertony (1.67%; 3.63), Iatrogenic retinal hemorrhage (0.75%; 3.93), Hypotony (0.66%; 3.98), Silicone oil into anterior Chamber (0.53%; 4.06), Migration of gas into anterior chamber (0.48%; 4.09), IOL dislocation (0.31%; 4.24), Complete dislocation of the IOLinto the vitreous cavity (0.18%; 4.42), Posterior capsular tear (0.18%; 4.42), Endophthalmitis (0.18%; 4.42), Expulsive haemorrhage (0.08%; 4.63).	Intraocular haemorrhage (4.97%; 3.15), Iatrogenic retinal breaks (4.62%; 3.18), Epiretinal and subretinal membrane (1.98%; 3.56), Cataract (0.92%; 3.86), Iatrogenic retinal detachment (0.18%; 4.42).	Retinal redetachment (3.04%; 3.38), Choroidal detachment (0.04%; 4.83), Escape of gas (0.13%; 4.51).

Continued

IntraLase	Subconjunctival haemorrhage (79.91%; 0.66), Chemosis (2.67; 3.43).	Decenteralised flap (1.33%; 3.71), Suction loss (0.89%; 3.87), Interface debris (0.66%; 3.97), Punctate Epitheliopathy (9.82%; 2.79).	Photophobia (13.83%; 2.59), Sidecut inefficiency (use of scissors) (2.23%; 3.51), Sidecut inefficiency (use of syringe) (2%, 3.55), Lamellar Keratitis at peripheral flap borders (1.78%; 3.60), Dry eye (1.78%; 3.60).	Overcorrection (3.57%; 3.30), Incomplete flap (2.67%; 3.43), Undercorrection (2.67; 3.43), Flap wrinkling (1.78%; 3.60), Diffuse lamellar keratitis (1.56%; 3.65), Tear in flap (0.22%; 4.34).
Penetrating keratoplasty	Posterior capsule opafication (5.08%; 3.14), Wound leaks and iris tissue prolapse (1.69%; 3.62).	Glaucoma (27.12%; 2.11), Anterior Synechiae (3.39; 3.33), Cystoid macular edema (3.39; 3.33).	Infectious keratitis and suture abscesses (6.78%; 2.99), Persistent epithelial defects (3.39; 3.33), Corneal Abscess (3.39; 3.33).	Primary Graft Failure (10.17%; 2.77), Graft Rejection (6.78%; 2.99), Acute choroidal detachment (1.69%; 3.62).
Trabeculectomy	None	Conjunctival Incontinence and Perforation (1.79%; 3.56), Hyphema (19.36%; 2.36), Fibrin reaction (12.64%; 2.64).	Hypotony (44.66%; 1.63%), Shallow anterior chamber (29.64%; 2.03), Choroidal detachment (10.67%; 2.74), Positive Seidel Test (3.16%; 3.36), Early cataract formation (1.18%; 3.76), Malignant Glaucoma (0.79%; 3.91).	Expulsive haemorrhage (0.39%; 4.16), Chronic hypotony and maculopathy (3.16%; 3.36), Retinal detachment (0.39%; 4.16).

complication: "Vital few" CTQs are those factors that affect the process the most and are often the main sources of the complications (**Table 4**).

3.4. Improve Phase

The Six Sigma team recommended the following *improvements* in the eye units:

The demand of everyday life places an ever increasing burden not only on the active surgeon but also on all members of his inherent team. The team must be trained on Six Sigma; understand Six Sigma philosophies and principles, including the supporting systems and tools and understands all aspects of the DMAIC model in accordance with Six Sigma principles.

The burden and pace of ophthalmic techniques places an obligation on both the surgeon and his team to make every effort to adapt, whenever possible, to ensure that all modern techniques are well known, practised and when required revised and improved by his team through constant consultation and feedback.

Being the black belt who leads problem-solving projects, the surgeon has to demonstrate team leadership. The focus of all success in this ophthalmic endeavour falls on the shoulders of well-qualified, updated, experienced and efficient surgeon. A surgeon needs to be knowledgeable, experienced, procedurally skilled and have good judgment. These include the knowledge of the tools, instruments and materials used, anaesthesia, the knowledge to treat injury and understanding of the techniques and indications of the types of surgical procedures. The absence of any of the above qualities will impair his surgical procedure.

Any ophthalmic surgery demands qualifications that are obtained by special training and skills. However, no matter how carefully planned and supervised a training program, some complications cannot be prevented. Thus, a required part of surgical training must be experience with complications and unexpected events. These criteria are self evident to a proficient surgeon. Without required knowledge and experience, they will be unable to cope with such situations as will arise. Thus, an essential component of a fully developed surgeon comes from diligent experience in situations where complications and unexpected events arise.

The surgeon must maintain a highly flexible approach to learning as many of the details even though some of the principles may change with time. Refinements and improvements develop at such a remarkable rate that every surgeon must realize that techniques widely used today may be outdated in the very near future. In fact, this flexible approach to learning is a requirement for reducing the number and type of complications and thus,

Table 4. CTQs per surgery type.

Surgery	Vital Few CTQ	Trivial Many CTQ
I	Experience of refractive surgeon; Type of microkeratome; Hygiene of microkeratome	Cleaning of Patient's Cornea by Refractive Surgeon; Patient's Eye Anatomy
II	Patient's eye anatomy; experience of anaesthesiologist; experience and attention of strabismus surgeon	Type and Quality of Suture; Attention of assistant surgeon; Attention of nurse
III	Experience of retina specialist; Attention of retina specialist; Patient's ocular pathology	Sterilization and hygiene; Dosage of drug/agent; Chemical properties of drug/agent
IV	Experience of ophthalmic surgeon; Patient's anatomy; Cooperation of patient during surgery	Sterilization and hygiene; Attention of assistant surgeon; Calibration of equipment; Quality/chemical composition of intraocular material
V	Experience of ophthalmic surgeon; Patient's anatomy; Cooperation of patient during surgery	Sterilization and hygiene; Attention of assistant surgeon; Calibration of equipment; Quality/chemical composition of intraocular material
VI	Experience of vitreoretinal surgeon; Attention of vitreoretinal surgeon; Patient's anatomy	Sterilization and hygiene; Amount of silicone oil; Amount of gas
VII	Experience of refractive surgeon; Patient's anatomy; Calibration of laser power	Patient's psychology; Sterilization and hygiene; Suction-ring's pressure
VIII	Patient's eye anatomy; Suitability of donor cornea	Experience of ophthalmic surgeon; Sterilization and hygiene; Performance of equipment
IX	Patient's eye anatomy; Experience of ophthalmic surgeon	Quality of surgical equipment; Quality and type of suture; Experience of staff

improving the overall surgical process.

The surgeon should be well-prepared for the surgery. This requires knowledge of the desired goals, the patient, the materials used, the techniques employed and oneself. In addition, the surgeon must be in control of the whole procedure, including what happens prior to the actual surgical procedure, what happens in the operating room, including the surgical team and the appropriate surgical technique.

Even before the surgery starts, the qualifications of separate individuals and how well they function together as a team is paramount to the success of the procedure about to be undertaken. No individual is too important or unimportant simply because the procedure does not flow freely. However, each surgical procedure, while independent, still depends on the separate efficiencies of the individuals involved.

Because a successful team functions throughout surgery as a medical entity, each member is in effect a subset of a larger unit. As each team is comprised of the sum of its parts, the burden of success is shared by each team member to the degree that the other members of the team rely on them. No individual can be considered as too important or unimportant, high and low of the attempt of the impending surgery is to be a success.

Recent technological developments has dramatically changed surgeon's roles by providing the means for surgeons to be more effective [17]. What is more, very convincing improvements in the surgical instruments and materials have resulted due to these open-ended advances. However, the techniques used are being adopted so rapidly and extensively that a single surgeon cannot possibly be fully competent in all aspects.

The surgeon's success will depend in no small part on the quality and type of the range of materials (e.g. suture, intraocular material) he has at hand. He is required to be fully satisfied and perform regular checks on the materials he uses as they will no longer be visible on completion of surgery. Examples of these are a comprehensive knowledge of the range of intraocular material, their restricted usage and their mean life time. A surgeon's updated knowledge and comprehension of the best dosage and frequency of e.g. drugs, agents, silicone oil, gas, is always required dealing with the age, condition and seriousness of the condition being treated.

In the midst of the surgical procedure, a surgeon must not divert his attention. The sole purpose of his work is efficiency in achieving as best he possibly can a successful outcome to the patient he is presently attending. To achieve this success, a proficient and professional atmosphere must be maintained (e.g. no external distraction, no annoyances within the workplace, no potential disturbances from outside during functional hours) where everybody knows the sequence, the dangers and the precautions which are required on completion of the surgery.

A clear indicator of a surgeon's efficiency and care for his patient will be decided by the quality of the instrumentation he uses everyday in his surgery. While the best equipment may not always guaranteee the perfect result, it is a statement by the surgeon that he approaches all surgery in the most professional manner.

Ensuring good visualization of the surgical field is an essential principle. Seeing clearly requires proper direction and intensity of lighting, proper positioning of the patient, skillful assistants, proper positioning of the surgeon's hands and competence in the use of appropriate optical aids.

Every professional category learns best from within its own circle. For this reason, assistant surgeons require exposure frequently on the present and new techniques being practised within the operating room. It can also be of great value to have readily written reference available in the operating room so that junior team members have access to them at all functional times.

For people who encounter ocular health problems resulting in e.g. poorer vision; the treatment and quality of eye surgery available is of special importance. Ophthalmic patients often require excellent medical, surgical and emotional care. However, complete success is not an invariable outcome of surgical treatment. Thus, patients must be prepared to accept results desirable than hoped for. Each patient is unique with his or her nature, needs and wants. Thus, surgeons must understand the unique qualities of each patient at each interaction and patient's response to disease. It is also vital that the patient must have confidence in his or her surgeon, and that the confidence is deserved.

Nurses are trained to become the green belts of Six Sigma projects. They must assist the black belt surgeon with data collection; analyzing and solving quality problems. Assistant surgeons are the yellow belts as theyhave a small role, interest, or need to develop foundational knowledge of Six Sigma, whether as an entry level employee..

Nurses must make sure that noone in the staff breaks the hygiene rules since the success of any medical procedure must begin with strict adherence to hygenic standards. In addition, those higher up in contact with a patient throughout surgery bear a heavy burden to ensure success as maintaining hygenic standards and practices. Infections are a major cause of preventable ocular morbidity. The surgeon and his team should be alert to the possibility that ophthalmic instruments are improperly sterilized and contaminated ophthalmic solutions are infected. Hands also play a major role in the transmission of infection. They should be washed or disinfected before and after the examination of every patient.

The advent of improved products such as lenses requires the surgeon to pay close and consistent reference to the new products available so as to ensure the ocular health of his patients. Constant reference to catalogue results will help him to choose the proper item or range of items.

Prior to undergoing any surgical procedure, a detailed examination of the overall health and specific ailment of the patient must be prepared by the most proficient professional. Any additional conditions (e.g. diabetes) must be clearly brought to the teams' attentions and in the weeks before surgery occurs. The same condition should be checked by both the local doctor and the patient.

Throughout surgery, the patient will respond to its alien surrounding if he has clearly been informed of the stages required in the simplest possible language and if required, this knowledge can be repeated in a friendly manner. A clear contrast exists here between a team who are functioning together daily and a patient who is experiencing this procedure for the first time in foreign surroundings. The success of the patient is more important than the success of the procedure although the patient's health has without doubt dependent on the procedure just carried.

As soon as a patient leaves surgery, he is at his vulnerable requiring regular and friendly checkups to ensure the patient is getting better. Contact with members of family can help to ensure the patient is relaxed after surgery and also gives the surgeon an occasion to mention any potential complications which may result from the procedure just finished.

Subsequent examination of the patient's recovery visits must be carried out by the best qualified individual appointed by the surgeon with steady contact being available to surgeon to report and record any potential complications.

Follow-up examinations by the surgeon or his team also require a realistic assessment of the patients' post-operational recouperation with any complications being dealt with as quickly and efficiently as possible. A few words of encouragement by the medical team to describe the procedure carried out on the patient as having been successful will go a long way to remove the natural worries a patient will always have. The patient needs to know how best he can part take in his own recovery. Such aids as eye drops and regular eye washing with a reputed eye lotion and informing the surgeon of any potential infection will be necessary.

3.5. Control Phase

To keep the surgical processes under *control*, the team initiated and implemented the improvement plan. Although it was subject to revision with a view of standardising the tested improved surgical processes, knowledge gained from corrective/preventative recommendations was shared and institutionalised.

4. Conclusions

This study has shown potential use of Sigma Quality Level as a statistical index that provides a quantitative measure of the capability of specialised eye care processes. Sigma level can both be used as an organizational or industrial metric in healthcare.

A sigma level is a measure of the error rate of a process, based on the DPMO estimate. It offers an indicator of how often complications are likely to occur, where a higher sigma level indicates a process that is less likely to create complications. Consequently, as sigma level of quality increases, reliability of the surgery improves, the need for reoperation diminishes, follow-up time declines, surgery duration length goes down, costs go down, and patient satisfaction goes up. In addition, sigma level can be used to measure the occurence of a complication or the success rate of a surgery. A surgery that yields a low number of complications will have a high sigma level.

By determining the sigma level, practitioners can make use of Six Sigma's DMAIC to prevent the occurrence of complications. First, CTQs should be determined. Then, preventative measures should be undertaken for each complication. This will remove variation from ophthalmic processes and fewer complications inherently result. A reduction in complications can, in turn, help eliminate waste from processes in the following ways: fewer complications decrease the number of materials that must be scrapped; and fewer complications also mean that energy, costs (e.g. insurance cost, re-operation cost, psychological burden, cost of labour, materials cost, training cost) and time to follow-up and re-work the patient to fix the complications are eliminated.

To achieve such continuous improvement by Six Sigma, leadership commitment is necessary. Eye care center's top management (and surgeon as a black belt in Six Sigma projects) is in a position to initiate Six Sigma deployment and play an active role in the whole deployment cycle. Six Sigma starts by providing senior leadership with training in the Six Sigma principles and tools, and it needs to direct the development of a management infrastructure to support it. This involves reducing the organizational hierarchy levels and removing procedural barriers to change.

The return on investment gained from the information and knowledge the DMAIC tool creates can be substantial. This tool requires a great deal of coordination within the team. If achieved, it can greatly improve a eye care processes' ability to be controlled and analyzed during process improvement projects.

Throughout this study, attention has been given to the best available information at hand and is presented in a factual manner, to be of use to the greatest possible range of medical personnel.

The proposed method is versatile enough to incorporate any further clinical developments which take place into the future. The medical world plays no small part in the updating of their operational results so that all involved can have a realistic scale of medical procedures. A databank, reviewing and updating sigma levels, needs to be established and directed through a range of medical practitioners with thte most updated qualifications in their varied range of eye surgeries.

A variety of different headings would be required to ensure that this method will have the success rate in highlighting and correcting any complications which result from the surgery technique used.

The Six Sigma methodology has proved both informative and versatile in addressing the core facts and it is our belief that it can easily been adapted to suit a range of studies in other related or distinct fields of medicine— dentistry, anaesthesia and orthopedics to name a few.

References

[1]   Probst, L.E., Tsai, J.H. and Goodman, G. (2012) Ophthalmology: Clinical and Surgical Principles. Slack Incorporated, Thorofare, NJ.

[2]   Boyd, S. and Wu, L. (2009) Management of Complications in Ophthalmic Surgery. Jaypee Highlights Medical Publishers Inc., Panama.

[3]   Agarwal, A. and Jacob, S. (2012) Complications in Ocular Surgery: A Guide to Managing the Most Common Challenges. Slack Incorporated, Thorofare, NJ.

[4]   Oztürker, C., Sahbaz, I., Oztürker, Z.K., Taner, M.T., Bayraktar, S. and Kagan, G. (2014) Development of a Six Sigma Infastructure for Trabulectomy Process. *American Journal of Operations Research*, **4**, 246-254. http://dx.doi.org/10.4236/ajor.2014.44024

[5]   Sahbaz, I., Taner, M.T., Sahandar, U.T., Kagan, G. and Erbas, E. (2014) Elimination of Post-Operative Complications in Penetrating Keratoplasty by Deploying Six Sigma. *American Journal of Operations Research*, **4**, 189-196. http://dx.doi.org/10.4236/ajor.2014.44018

[6]   Sahbaz, I., Taner, M.T., Eliacik, M., Kagan, G. and Erbas, E. (2014) Adoption of Six Sigma's DMAIC to Reduce Complications in IntraLase Surgeries. *International Journal of Statistics in Medical Research*, **3**, 126-133. http://dx.doi.org/10.6000/1929-6029.2014.03.02.6

[7]   Sahbaz, I., Taner, M.T., Eliacik, M., Kagan, G., Erbas, E. and Enginyurt, H. (2014) Deployment of Six Sigma Methodology to Reduce Complications in Intravitreal Injections. *International Review of Management and Marketing*, **4**, 160-166.

[8]   Sahbaz, I., Taner, M.T., Kagan, G., Sanisoglu, H., Durmus, E., Tunca, M., Erbas, E., Kagan, S.B., Kagan, M.K. and Enginyurt, H. (2014) Development of a Six Sigma Infrastructure for Cataract Surgery in Patients with Pseudoexfoliation Syndrome. *Archives of Business Research*, **2**, 15-23. http://dx.doi.org/10.14738/abr.22.173

[9]   Sahbaz, I., Taner, M.T., Kagan, G., Sanisoglu, H., Erbas, E., Durmus, E., Tunca, M. and Enginyurt, H. (2014) Deployment of Six Sigma Methodology in Phacoemulsification Cataract Surgeries. *International Review of Management and Marketing*, **4**, 123-131.

[10]   Sahbaz, I., Taner, M.T., Sanisoglu, H., Kar, T., Kagan, G., Durmus, E., Tunca, M., Erbas, E., Armagan, I. and Kagan, M.K. (2014) Deployment of Six Sigma Methodology to Pars Plana Vitrectomy. *International Journal of Statistics in Medical Research*, **3**, 94-102. http://dx.doi.org/10.6000/1929-6029.2014.03.02.3

[11]   Taner, M.T., Kagan, G., Sahbaz, I., Erbas, E. and Kagan, S.B. (2014) A Preliminary Study for Six Sigma Implementation in Laser *in Situ* Keratomileusis (LASIK) Surgeries. *International Review of Management and Marketing*, **4**, 24-33.

[12]   Taner, M.T., Sahbaz, I., Kagan, G., Atwat, K. and Erbas, E. (2014) Development of Six Sigma Infrastructure for Strabismus Surgeries. *International Review of Management and Marketing*, **4**, 49-58.

[13]   Taner, M.T., Kagan, G., Celik, S., Erbas, E. and Kagan, M.K. (2013) Formation of Six Sigma Infrastructure for the Coronary Stenting Process. *International Review of Management and Marketing*, **3**, 232-242.

[14]   Taner, M.T., Sezen, B. and Antony, J. (2007) An Overview of Six Sigma Applications in Healthcare Industry. *International Journal of Health Care Quality Assurance*, **20**, 329-340. http://dx.doi.org/10.1108/09526860710754398

[15]   Conlin, M. (1998) Revealed at Last: The Secret of Jack Welch's Success. *Forbes*, **16**, 44.

[16]   Taner, M.T. (2013) Application of Six Sigma Methodology to a Cataract Surgery Unit. *International Journal of Health Care Quality Assurance*, **26**, 768-785. http://dx.doi.org/10.1108/IJHCQA-02-2012-0022

[17]   Spath, G.L., Danesh-Meyer, H., Goldberg, I. and Kampik, A. (2012) Ophthalmic Surgery: Principles and Practice. 4th Edition, Elsevier Health Sciences, Edinburgh.

20

On the Gas Routing via Game Theory

Irinel Dragan

University of Texas, Mathematics, Arlington, TX, USA
Email: dragan@uta.edu

Abstract

The delivery of the natural gas obtained by drilling, fracking and sending the product to consumers is done usually in two phases: in the first phase, the gas is collected from all wells spread on a large area, and belonging to several companies, and is sent to a depot owned by the city; then, in the second phase, another company is taking the gas on a network of ducts belonging to the city, along the streets to the neighborhoods and the individual consumers. The first phase is managed by the gas producing companies on the ducts owned by each company, possibly also on some public ducts. In this paper, we discuss only this first phase, to show why the benefits of these companies depend on the cooperation of the producers, and further, how a fair allocation of the total gas obtained, to the drilling companies, is computed. Following the model of flow games, we generate a cooperative transferable utilities game, as shown in the first section, and in this game any efficient value gives an allocation of benefits to the owners of ducts in the total network. However, it may well happen that the chosen value is not coalitional rational, in the game, that is, it does not belong to the Core of the game. By using the results obtained in an earlier work of the author, sketched in the second section, we show in the last section how the same allocation may be associated to a new game, which has the corresponding value a coalitional rational value. An example of a three person flow game shows the game generation, as well as the procedure to be used for obtaining the new game in which the same value, a Shapley Value, will give a coalitional rational allocation.

Keywords

Cooperative TU Game, Core, Shapley Value, The Inverse Problem, Coalitional Rationality

1. The Flow Game, a Model for Gas Routing

The flow games have been introduced by E. Kalai and E. Zemel (1982), in [1], as a class of cooperative transferable utilities games generated on a graph by means of a max-flow-min-cut algorithm. We use this model to show how this may be used for gas routing, in case that the gas extracted by several companies is sent to a common depot, while the distribution is done by a public company. An example with three companies and a graph of individual ducts belonging to each company would help in the understanding of any similar situation.

Example 1. Consider the case of a city which has wells belonging to three companies, K_1, K_2, K_3, at the nodes A, B, C, of a graph, respectively. The companies are able to extract natural gas as follows: company K_1, 11 tones at A; company K_2, 13 tones at B; and company K_3, 9 tones at C. The set of nodes in the graph is $N = \{S, A, B, C, D, E, F, G, T\}$; where S is an auxiliary common source, T is an auxiliary common sink, and the arcs are belonging to the companies as follows: $A_1 = \{[B, F], [C, E], [E, G]\}$, $A_2 = \{[A, E], [C, F], [F, G]\}$, $A_3 = \{[A, D], [B, D], [D, G]\}$, with capacities shown in the matrix below. There is also a public duct, belonging to the city, connecting the node G to the depot node T, from which a distribution company is allocating the deliveries to customers throughout the city, on a public network. Let this arc be $[G, T]$, with the capacity 25. We added an auxiliary node S and the auxiliary arcs $[S, A]$, $[S, B]$, $[S, C]$, with capacities equal to the production power of each company, that is 11, 13, 9, respectively, for computational purposes. By inspection we easily determine that no gas producing company can send to the depot alone any amount of gas, so that if we denote the amounts provided by the singletons by $P(\{i\})$, $\forall i \in N$, we have $P(\{i\}) = 0$, $\forall i \in N$. Instead, for each pair of companies, $\{i, j\}$, by using all the arcs belonging to the pair (K_i, K_j), we obtain the maximal amounts provided by the corresponding coalitions, listed as

$$P(K_1, K_2) = 22, P(K_1, K_3) = 18, P(K_2, K_3) = 18,$$

while the cooperation of all companies will give

$$P(K_1, K_2, K_3) = 25.$$

The maximal amounts for coalitions of size two are obvious from the figure; the maximal amount provided by the grand coalition should be computed in the network by a max-flow-min-cut algorithm, for example the Ford-Fulkerson algorithm (see [2]). If there are more companies producing natural gas, the game is larger and the computation is more difficult, but the approach may be the same.

The capacities of the ducts represented by the arcs of the graph are shown in the following figure:

	A	B	C	D	E	F	G
S	11	13	9	-	-	-	-
A	-	-	-	9	13	-	-
B	-	-	-	9	-	11	-
C	-	-	-	-	11	13	-
D	-	-	-	-	-	-	9
E	-	-	-	-	-	-	11
F	-	-	-	-	-	-	13

A graph with capacities associated to the above model.

The graph has the arcs for which there are positive capacities in the table, plus the entering and exiting auxiliary arcs, and one public arc discussed below. Now, note that the columns for S and T and the rows for G and T have been omitted, because they are empty, except a public duct, belonging to the city, $[G, T]$, with a capacity of 25. The graph and the capacities are taken arbitrarily by the author.

As shown by the numbers that give the characteristic function of the game, the three companies should cooperate in order to get a maximum benefit. Now, the three-person cooperative TU game and any fair schedule of allocations, are obtained by using an efficient value of the game:

$$N = \{1, 2, 3\}$$

$$P(1) = P(2) = P(3) = 0, P(1, 2) = 22, P(1, 3) = P(2, 3) = 18, P(1, 2, 3) = 25,$$

where we replaced the names of the companies by the indices and the brackets by parantheses, to keep the notation simple.

As the worth of singletons is zero, a simpler value like the Center of the Imputation Set, expressed by the formula:

$$CIS_i(N,P) = v(i) + \frac{1}{n}\left[v(N) - \sum_{j \in N} v(j)\right], \forall i \in N,$$

gives

$$CIS(N,P) = \left(\frac{25}{3}, \frac{25}{3}, \frac{25}{3}\right),$$

which is not fair. Therefore, we shall use the Shapley Value, given by the formula:

$$SH_i(N,P) = \sum_{S:i \in S} \frac{(s-1)!(n-s)!}{n!}\left[P(S) - P(S-\{i\})\right], \forall i \in N,$$

which has more properties due to its axiomatic definition (see [2]). By using the formula, we obtain $SH(N,P) = (9,9,7)$. The Core of the game is given by the conditions

$$x_i \geq 0, \forall i \in N, x_1 + x_2 \geq 22, x_1 + x_3 \geq 18, x_2 + x_3 \geq 18, x_1 + x_2 + x_3 = 25.$$

Obviously, the Shapley Value does not belong to the Core, as

$$SH_1(N,P) + SH_2(N,P) = 18 \geq 22 = P(1,2),$$

and all the other inequalities for coalitions of size two do not hold, that is the Shapley Value is not coalitional rational, even though it is efficient, as the sum of components makes 25. Of course, any pair of companies may break the grand coalition and make a two person coalition which may give more gain to each player, while the third player is left out of the deal, an unfair solution. This instability is due exactly to the fact that the solution is not coalitional rational, which could be a motivation for the present work.

To be able to give a fair allocation, we shall solve a connected problem that has been discussed in an earlier work of the author (see [3]). To get the paper self contained we shall remind some concepts and results from the mentioned work. The new problem introduced there is: find out a new game, with the same Shapley Value, which is coalitional rational in the new game. This problem is connected to the Inverse Problem for the Shapley Value, introduced also in some work of the author (see [4]).

2. The Inverse Problem and the Coalitional Rationality

In [4], the Inverse Problem for the Shapley Value has been introduced and solved. For the Shapley Value the problem may be stated: let $L \in R^n$ be the Shapley Value of a cooperative TU game with n players; find out the set of all TU games with n players, for which the Shapley Value equals L. In [3], a connected problem has been introduced and solved: let $L \in R^n$ be the Shapley Value of a cooperative TU game; find out a new game, with the same Shapley Value, in which the Shapley Value is coalitional rational. In other words, in the Inverse Set relative to the Shapley Value, find out a game in which the Shapley Value is in the Core of the game. To sketch the use of the results from [3] and [4] in the gas routing problem, we give here some earlier results.

It is well known that the set of TU games with the set of players N, forms a vector space of dimension $2^n - 1$, (see [2]). Hence, any game may be written as a vector in that space, for example in R^7 for our game of Example 1. The results of Linear Algebra will be useful in this respect. A basis for the vector space was introduced in [4], by the formulas

$$W = \left\{w_S \in R^{2^n - 1} : S \subseteq N, S \neq \emptyset\right\},$$

where the basic vectors are defined for $S \subset N$ by $w_S(S) = |S|$, $w_S(T) = -1$, $\forall T = S \cup \{i\}$, $i \in N - S$, $w_S(T) = 0$, otherwise, while $w_N(N) = n$, $w_N(T) = 0$, otherwise.

As mentioned above all these vectors, invented in [4], are linearly independent, hence they form a basis; an easy exercise is to write the seven basic vectors for our example 1. Now, any game (N,w) in the vector space may be written as a linear combination of the basic vectors as

$$w = \sum_{S:|S| \le n-2} c_S w_S + \sum_{i \in N} c_{N-\{i\}} w_{N-\{i\}} + c_N w_N,$$

where the coefficients are constants which may be determined from this vector equation written in scalar form. To solve the Inverse Problem we used the results:

$$SH(N, w_S) = 0, \forall S, |S| \le n-2, \; SH_i\left(N, w_{N-\{j\}}\right) = -\delta_i^j, \forall i, j \in N, \; SH_i\left(N, w_N\right) = 1, \forall i \in N,$$

and the linearity of the Shapley Value. By applying these results in the above expansion of (N, w), we get

$$SH_i(N, w) = c_N - c_{N-\{i\}}, \forall i \in N.$$

After eliminating the constants $c_{N-\{i\}}$, we obtain an explicit formula for the games in the Inverse Set relative to the Shapley Value L:

$$w = \sum_{S:|S| \le n-2} c_N w_N + c_N \left(w_N + \sum_{i \in N} w_{N-\{i\}} \right) - \sum_{i \in N} L_i w_{N-\{i\}},$$

(see [4], Theorem 3.5).

As mentioned above, we shall be looking for a new game in the Inverse Set relative to the Shapley Value L, which is coalitional rational. We shall confine ourselves to find such a game in the subfamily of the Inverse Set defined by all $c_S = 0, \forall S, |S| \le n-2,$ that is in the set of games

$$w = c_N \left(w_N + \sum_{i \in N} w_{N-\{i\}} \right) - \sum_{i \in N} L_i w_{N-\{i\}}.$$

This subset of the Inverse Set will be called the almost null subfamily. Taking into account the expressions of the basic games shown above, the games in the family given by this vector formula, may be rewritten in scalar form as

$$w(N-\{i\}) = (n-1)(c_N - L_i), \forall i \in N, \quad w(N) = \sum_{i \in N} L_i, \quad w(S) = 0, \text{ otherwise (*)}$$

Now, the Core conditions written together give an inequality determining the values of parameter c_N for those games in which the Shapley Value is coalitional rational:

$$c_N \le \frac{1}{n-1} MIN \left\{ \sum_{j \in N-\{i\}} L_j + (n-1)L_i \right\}, \quad (**)$$

where the minimum is taken over the index i.

The results already shown provide a procedure for solving our above stated problem:

- Compute the Shapley Value of the given game. Check whether, or not, the Shapley Value is in the Core. If yes, stop, the problem is solved; otherwise.
- Compute the game in the Inverse Set given by (*), for the Shapley Value L.
- Take a value for c_N in formula (*). to satisfy the inequality (**), and find out the desired game in the almost null subfamily.

Now, we intend to apply the results to the gas routing problem.

3. An Application to the Gas Routing Problem

Return to the Example 1 given in the first section, in which we associated to the network the cooperative TU game obtained by finding for each coalition the maximum flow that could be delivered by using the ducts owned by each company in the coalition and the final public arc. The TU game is

$$v(1) = v(2) = v(3) = 0, v(1,2) = 22, v(1,3) = v(2,3) = 18, v(1,2,3) = 25,$$

and we can compute the Shapley Value: $SH(N, v) = (9, 9, 7)$.

Obviously, this is an efficient vector, but it is not belonging to the Core, because

$$SH_1(N, v) + SH_2(N, v) \ge v(1, 2), \quad SH_1(N, v) + SH_3(N, v) \ge v(1, 3), \quad SH_2(N, v) + SH_3(N, v) \ge v(2, 3),$$

are not satisfied. Then, start the procedure described above, by deriving the inequality (**):

$$c_N \leq \frac{1}{2} MIN(34, 34, 32) = 16.$$

We choose the largest value of the parameter that satisfies the inequality, $c_N = 16$, and plug into the formulas (*):

$$w(1,2) = 2(c_N - 7), w(1,3) = 2(c_N - 9), w(2,3) = 2(c_N - 9), w(1,2,3) = 25,$$

where the null worth of the singletons is omitted. We obtain

$$w(1) = w(2) = w(3) = 0, w(1,2) = 18, w(1,3) = w(2,3) = 14, w(1,2,3) = 25.$$

Further, we can compute again the Shapley Value and find out the same vector as before. This is not only efficient, but also coalitional rational, because the Core conditions hold. A similar situation and a similar strategy may be used for the Shapley Value in the general case of games with any number of players. Moreover, the procedure will work also for any other value which is efficient. A value which is not efficient is requiring a new definition of the coalitional rational value.

The procedure may be extended to this case by using the results following from the ideas developed in the joint paper [5], which were used in the paper [6].

Note that the present paper is an application of the results obtained in the previous paper [3], while for more general cases the Inverse set defined in [6] is needed.

References

[1]   Kalai, E. and Zemel, E. (1982) On Totally Balanced Games and Games of Flow. *Mathematics of Operations Research*, **7**, 476-478. http://dx.doi.org/10.1287/moor.7.3.476

[2]   Owen, G. (1995) Game Theory. 3rd Edition, Academic Press, San Diego.

[3]   Dragan, I. (2014) On the Coalitional Rationality of the Shapley Value and Other Efficient Values. *American Journal of Operations Research*, **4**, 328-334. http://dx.doi.org/10.4236/ajor.2014.44022

[4]   Dragan, I. (1991) The Potential Basis and the Weighted Shapley Value. *Libertas Mathematica*, **11**, 139-146.

[5]   Dragan, I. and Martinez-Legaz, J.E. (2001) On the Semivalues and the Power Core of Cooperative TU Games. *International Game Theory Review*, **3**, 127-139. http://dx.doi.org/10.1142/s0219198901000324

[6]   Dragan, I. (2005) On the Inverse Problem for Semivalues of Cooperative TU Games. *International Journal of Pure and Applied Mathematics*, **22**, 545-561.

21

A Production Inventory Model of Constant Production Rate and Demand of Level Dependent Linear Trend

Shirajul Islam Ukil, Md. Sharif Uddin

Jahangirnagar University, Savar, Bangladesh
Email: shirajukil@yahoo.com

Abstract

The proposed model considers the products with finite shelf-life which causes a small amount of decay. The market demand is assumed to be level dependent and in a linear form. The model has also considered the constant production rate which stops attaining a desired level of inventories and that is the highest level of inventories. Production starts with a buffer stock and without any sort of backlogs. Due to the market demand and product's decay, the inventory reduces to the level of buffer stock where again the production cycle starts. With a numerical search procedure the proof of the proposed model has been shown. The objective of the model is to obtain the total average optimum inventory cost and optimum ordering cycle.

Keywords

Production Inventory, Level Dependent Linear Trend, Constant Production Rate

1. Introduction

With a view to solving the inventory problems, it is highly essential for the business institutions to obtain the economic order quantity (EOQ) and obtaining this quantity leads to reduce the total average inventory cost. This is why the business institutions emphasize on inventory management and solving inventory problems. The problem can only be solved if a suitable inventory model could be established which is fit for all the parameters concerned like, market demand, production rate, product's life, etc. The innovative EOQ model is, therefore, a highly demand on regular basis and when required in spite of having existence of huge number of inventory models. Inventory, indeed, is a stock of materials. Inventory problems are mainly related to the proper management of this inventory which can lead to minimize the inventory cost. Generally, we have two kinds of materials in our daily needs as far as damage, wastage, deterioration or decay is concerned. Items like radioactive substances, food grains, fashionable items, pharmaceuticals, etc. are the items of finite life and the items like electronic goods, steels, woods, etc. are the items of ling life. Due to the limited shelf-life and market demand, the stock level or inventory continuously decreases and the items in the inventory deplete or deteriorate. This deterioration affects the inventory and inventory cost increases. To make the inventory cost at optimum level *i.e.* to get the minimum inventory cost, a suitable inventory model is required to be formulated. An inventory model with linear demand, small amount of decay and constant production rate has been proposed in this paper to minimize inventory cost. Keeping the buffer stock as a reserve, the production is assumed to start and after certain periods at the highest level of inventory, it stops. In this model, we have considered constant production rate along with the deterioration, whereas the classical inventory models and many researchers use the instantaneous replenishment. Finally, by proving the convex property and using a numerical searcher procedure, the paper justified the correctness of the model.

2. Literature Review

Since long the researchers had been focusing on obtaining inventory models suitable to the needs in real life with a view to solving inventory problems. The problems are related with what will be the pattern of demand in the market, what may be the production rate of the business institutions, whether there will be finite life of the products, whether backlogs or shortages and delay in payments are allowed etc. Many researchers have structured various types of inventory model basing on the situation or the market demand. It may arise different types of demands in the market. Demand may be linear, quadratic, exponential, time dependent, level or stock dependent, price dependent etc. Considering all the parameters the inventory model is designed. There are two types of models in this field which covers all the parameters mentioned above. One is deterministic model which deals with the constant demand and lead time; the other one is stochastic or probabilistic model which deals randomly with the variable demand and lead time. In this review of the literature, mostly the inventory models with deterministic demand have been discussed. Determining EOQ is one of the most important factors to formulate the inventory model. The ultimate aim of formulating the model is to minimize the inventory cost by finding the EOQ. Harris [1] was the first researcher to study the inventory model. The new horizon is opened in the field of the inventory control management since he presented the famous EOQ formula. Whitin [2] developed the inventory model for the first time which was suitable for fashionable goods considering its little decay in the inventory. Ghare and Schrader [3] first pointed out the effect of decay in inventory analysis and discovered the economic order quantity (EOQ) model. They showed the nature of the consumption of the deteriorating items. Skouri and Papachristos [4] discussed a continuous review inventory model considering the five costs as deterioration, holding, shortage, opportunity cost due to the lost sales and the replenishment cost due to the linearly dependency on the lot size. Chund and Wee [5] developed an integrated two stages production inventory deteriorating model for the buyer and the supplier on the basis of stock dependent selling rate considering imperfect items and just in time (JIT) multiple deliveries. Applying inventory replenishment policy Cheng and Wang [6] expressed the inventory model for deteriorating items with trapezoidal type demand rate, where the demand rate is a piecewise linearly functions. Hassan and Bozorgi [7] developed the location of distribution centers with inventory. Sarker *et al.* [8] explained an inventory model where demand was a composite function consisting of a constant component and a variable component proportional to inventory level in a period in which decay was exponential and inventory was positive. Tripathy and Mishra [9] discuss the inventory model with ordering policy for weibull deteriorating items, quadratic demand, and permissible delay in payments. Khieng *et al.* [10] presented a production model for the lot-size, order level inventory system with finite production rate and the effect of decay. Ekramol [11] [12] considered various production rates assuming the demand is constant. Mishra *et al.* [13] explained an inventory model for deteriorating items with time dependent demand and time varying holding cost under partial backlogging. Ukil [14] discussed the effect of just in time manufacturing system on EOQ model. Sivazlian and Stenfel [15] determined the optimum value of time cycle by using the graphical solution of the equation to obtain the economic order quantity model. Shah and Jaiswal [16] and Dye [17] established an inventory model by considering demand as a function of selling price and three parameters of Weibull rate of deterioration. Billington [18] discussed classic economic production quantity (EPQ) model without backorders or backlogs. Pakkala and Achary [19] established a deterministic inventory model for deteriorating items with two warehouses, while the replenishment rate was finite, demand was uniform and shortage was allowed. Abad [20] discussed regarding optimal pricing and lot sizing under conditions of perish ability and partial backordering. Sing and Pattanak [21]-[23] developed the model for deterioration and time dependent quadratic demand under permissible delay in payment, whereas we have used the demand of linear trend but ignoring the payment aspect. Amutha and Chandrasekaran [24] formulated the inventory model with deterioration items, quadratic demand and time dependent holding cost, but in our proposed model, we have emphasized on the production rate, linear type of demand and constant holding cost. Ouyang and Cheng [25] explained the inventory model for deteriorating items with exponential declining demand and partial backlogging. Dave and Patel [26] introduced an inventory model for deteriorating Items with time proportional demand, but we have considered the demand which is level dependent and a type of linear trend. Teng *et al.* [27] developed the model with deteriorating items and shortages assuming that the demand function was positive and fluctuating with respect to time, but in the proposed model, the demand was considered as a linear function and production starts with a buffer stock as a reserve. The previous model established various types of inventory models considering several parameters which did not consider the production rate, linear demand along with the buffer stock. Here comes the necessity to build the proposed model.

3. Assumptions and Notations

- Production rate λ is constant and greater than demand rate at any time.
- $a + bI(\theta)$ = Demand rate at any instant θ, where "$a = 0, 1, 2, \cdots$" and $0 < b < 1$ satisfying the condition $\lambda > a + bI(\theta)$.
- For unit inventory, amount of decay rate μ is very small and constant.
- Production starts with a few amounts of items in the inventory as a buffer stock.
- Inventory level is highest at a specific level and after this point, the inventory depletes quickly due to demand and deterioration.
- Shortages are not allowed.
- $I(\theta)$ = Inventory level at instant θ.
- I_1 = Un-decayed inventory at $T = 0$ to t_1.
- I_2 = Un-decayed inventory at $T = t_1$ to T_1.
- D_1 = Deteriorating inventory at $T = 0$ to t_1.
- D_2 = Deteriorating inventory at $T = t_1$ to T_1.
- Q, Q_1 and Q = Inventory level at time $T = 0$, t_1 and T_1 respectively. Here, Q is the buffer stock.
- $d\theta$ = Vary small portion of instant θ.
- K_0 = Set up cost.
- h = Average holding cost.
- $TC = TC(T_1)$ = Total average inventory cost in terms of T_1.
- t_1 = Time when inventory gets maximum level.
- T_1 = Total time cycle.
- Q_1^* = Optimum order quantity.
- t_1^* = Optimum time at maximum inventory.
- T_1^* = Optimum order interval.
- TC^* = Total average optimum inventory cost.

4. Formulation of the Model

Basing on the demand pattern, the business institution decides the structure of the model (**Figure 1**). This demand very often changes because of various reasons. In reality, the demand may at times the demand be dependent on the level or the stock on hand in the inventory. To meet this type of situation, this model is developed. The model is suitable for those kinds of products which have finite shelf-life and ultimately causes the products decay due to its limited life. At the beginning, while time $T = 0$, the production λ starts with Q inventories and the production remains constant for entire production cycle.

The inventory increases at the rate of $\lambda - a - bI(\theta) - \mu I(\theta)$ during $T = 0$ to t_1. The market demand is $a + bI(\theta)$ and $\mu I(\theta)$ is the decay of $I(\theta)$ inventories at instant θ where, μ is the decay of unit inventory in the mentioned period. With the help of the above criteria we can formulate the differential equations as

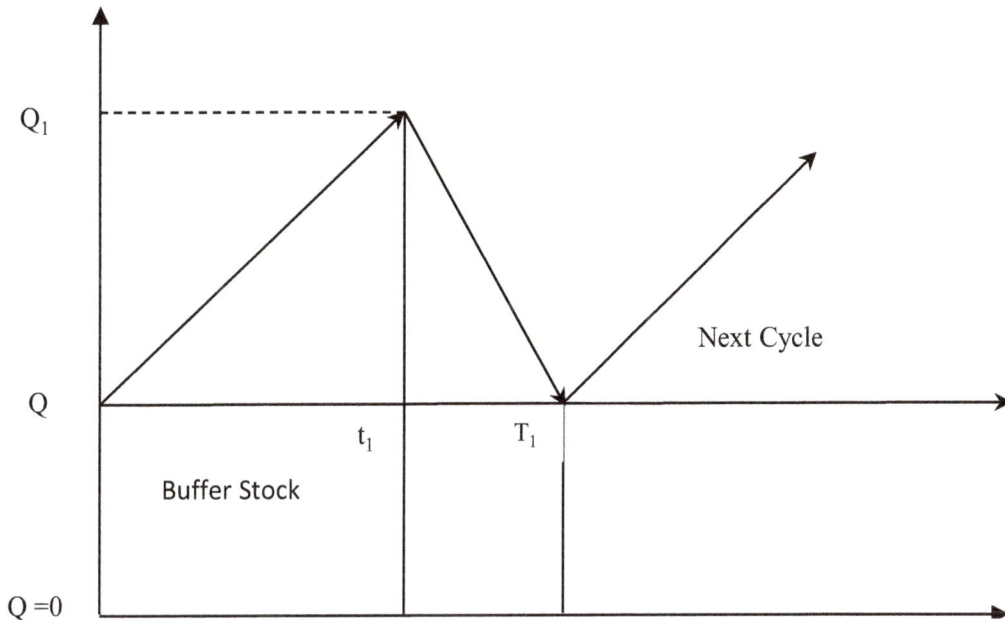

Figure 1. Inventory model with linear demand.

below:

$$I(\theta + d\theta) = I(\theta) + \{\lambda - a - bI(\theta)\}d\theta - \mu I(\theta)d\theta$$

$$\text{or, } I(\theta + d\theta) - I(\theta) = \{\lambda - a - bI(\theta) - \mu I(\theta)\}d\theta$$

$$\text{or, } \lim_{d\theta \to 0} \frac{I(\theta + d\theta) - I(\theta)}{d\theta} = \{\lambda - a - bI(\theta)\} - \mu I(\theta)$$

$$\text{or, } \frac{d}{d\theta}I(\theta) + \mu I(\theta) = \lambda - a - bI(\theta)$$

The general solution of the differential equation is $I(\theta) = \dfrac{\lambda - a}{\mu + b} + Ae^{-(\mu+b)\theta}$.

Applying the following boundary condition, we get $I(\theta) = Q$ at $\theta = 0$, Solving these equations, we get,

$$A = Q - \frac{\lambda - a}{\mu + b},$$

$$\text{Therefore, } I(\theta) = \frac{\lambda - a}{\mu + b} + \left(Q - \frac{\lambda - a}{\mu + b}\right)e^{-(\mu+b)\theta} \tag{1}$$

From the other boundary condition, *i.e.* at $\theta = t_1$, $I(\theta) = Q_1$, taking up to first degree of μ, we get the following equation:

$$\begin{aligned}
Q_1 &= \frac{\lambda - a}{\mu + b} + \left(Q - \frac{\lambda - a}{\mu + b}\right)e^{-(\mu+b)t_1} \\
&= \frac{\lambda - a}{\mu + b} + \left(Q - \frac{\lambda - a}{\mu + b}\right)\{1 - (\mu + b)t_1\} \\
&= Q + Q(\mu + b)t_1 - (\lambda - a)t_1 \\
&= Q + (\lambda - a - Q\mu - Qb)t_1
\end{aligned} \tag{2}$$

Using the Equation (1) and considering up to second degree of μ for our convenience, the total undecayed inventory during $\theta = 0$ to t_1 we get,

$$
\begin{aligned}
I_1 &= \int_0^{t_1} I(\theta)\,\mathrm{d}\theta = \int_0^{t_1}\left[\frac{\lambda-a}{\mu+b}+\left(Q-\frac{\lambda-a}{\mu+b}\right)e^{-(\mu+b)\theta}\right]\mathrm{d}\theta \\
&= \left[\frac{\lambda-a}{\mu+b}+\left(Q-\frac{\lambda-a}{\mu+b}\right)\left\{\frac{e^{-(\mu+b)\theta}}{-(\mu+b)}\right\}\right]_0^{t_1} \\
&= \frac{(\lambda-a)t_1}{\mu+b}-\left(Q-\frac{\lambda-a}{\mu+b}\right)\left\{\frac{e^{-(\mu+b)t_1}-1}{\mu+b}\right\} \\
&= \frac{(\lambda-a)t_1}{\mu+b}-\left(Q-\frac{\lambda-a}{\mu+b}\right)\left(\frac{1}{\mu+b}\right)\left\{-(\mu+b)t_1+\frac{1}{2}(\mu+b)^2 t_1^2\right\} \\
&= Qt_1-\frac{1}{2}Q(\mu+b)t_1^2+\frac{1}{2}(\lambda-a)t_1^2
\end{aligned} \tag{3}
$$

We calculate the deteriorating items during the period considering the decay of the items as below:

$$
\begin{aligned}
D_1 &= \int_0^{t_1}\mu I(\theta)\,\mathrm{d}\theta = \mu\int_0^{t_1}\left[\frac{\lambda-a}{\mu+b}+\left(Q-\frac{\lambda-a}{\mu+b}\right)e^{-(\mu+b)\theta}\right]\mathrm{d}\theta \\
&= Q\mu t_1-\frac{1}{2}Q(\mu+b)\mu t_1^2+\frac{1}{2}(\lambda-a)\mu t_1^2
\end{aligned} \tag{4}
$$

On the other hand, the inventory decreases at the rate of $a+bI(\theta)+\mu I(\theta)$ during $T=t_1$ to T_1 as there is no production after time t_1. The inventory depletes due to market demand and the deterioration of the items. Similar approach as used before can be applied to get another differential equation which is as follows:

$$
\frac{\mathrm{d}}{\mathrm{d}\theta}I(\theta)+\mu I(\theta)=-a-bI(\theta)
$$

The general solution of the differential equation is defined below:

$$
I(\theta)=\frac{-ab^n}{\mu}+Be^{-\mu\theta}
$$

Applying the boundary condition at $\theta = T_1$, we get, $I(\theta)=Q$.

By solving we get, $B=\left(Q+\dfrac{a}{\mu+b}\right)e^{(\mu+b)T_1}$.

$$
\text{Therefore, } I(\theta)=-\frac{a}{\mu+b}+\left(Q+\frac{a}{\mu+b}\right)e^{(\mu+b)(T_1-\theta)} \tag{5}
$$

Substituting another boundary condition, *i.e.* at $\theta = t_1$, $I(\theta)=Q_1$, taking up to first order of μ, we get the following equation:

$$
\begin{aligned}
Q_1 &= -\frac{a}{\mu+b}+\left(Q+\frac{a}{\mu+b}\right)e^{(\mu+b)(T_1-t_1)} \\
&= -\frac{a}{\mu+b}+\left(Q+\frac{a}{\mu+b}\right)\left\{1+(\mu+b)(T_1-t_1)\right\} \\
&= Q+\left\{a+Q(\mu+b)\right\}(T_1-t_1)
\end{aligned} \tag{6}
$$

Now, using Equation (5) and considering up to the first degree of μ we get the un-decayed inventory during $T=t_1$ to T_1 as:

$$I_2 = \int_{t_1}^{T_1} I(\theta) \, d\theta = \int_{t_1}^{T_1} \left\{ -\frac{a}{\mu+b} + \left(Q + \frac{a}{\mu+b} \right) e^{(\mu+b)(T_1-\theta)} \right\} d\theta$$

$$= \left[-\frac{a}{\mu+b} + \left(Q + \frac{a}{\mu+b} \right) \left\{ \frac{e^{(\mu+b)(T_1-\theta)}}{-(\mu+b)} \right\} \right]_{t_1}^{T_1}$$

$$= -\frac{a}{\mu+b}(T_1-t_1) - \left(Q + \frac{a}{\mu+b} \right) \left(\frac{1}{\mu+b} \right) \left\{ 1 - e^{(\mu+b)(T_1-t_1)} \right\} \qquad (7)$$

$$= -\frac{a}{\mu+b}(T_1-t_1) + \left(Q + \frac{a}{\mu+b} \right)(T_1-t_1)$$

$$= Q(T_1-t_1)$$

Considering the decay of the items, we calculate the deteriorating items during the period as below:

$$D_2 = \int_{t_1}^{T_1} \mu I(\theta) \, d\theta = \int_{t_1}^{T_1} \mu \left\{ -\frac{a}{\mu+b} + \left(Q + \frac{a}{\mu+b} \right) e^{(\mu+b)(T_1-\theta)} \right\} d\theta = Q\mu(T_1-t_1) \qquad (8)$$

From Equations (2) and (6), we get,

$$Q + (\lambda - a - Q\mu - Qb)t_1 = Q + \{a + Q(\mu+b)\}(T_1-t_1)$$

Or, $\quad t_1 = \dfrac{a + Q(\mu+b)}{\lambda} T_1 \qquad (9)$

Considering the value as $\quad \dfrac{a + Q(\mu+b)}{\lambda} = v \qquad (10)$

We construct the following equation with the help of Equation (9),

$$t_1 = vT \qquad (11)$$

Total Cost Function: The cost function can be described in the following form,

$$TC = \frac{K_0 + h(I_1 + D_1 + I_2 + D_2)}{T_1} \qquad (12)$$

By substituting the Equations (3), (4), (7), (8) and (11) in (12), we get the value of total average inventory cost as below,

$$TC = \frac{1}{T_1} \left[K_0 + hQt_1 - \frac{h}{2} Q(\mu+b)t_1^2 + \frac{h}{2}(\lambda-a)t_1^2 + hQ\mu t_1 - \frac{h}{2}Q(\mu+b)\mu t_1^2 + \frac{h}{2}(\lambda-a)\mu t_1^2 \right]$$

$$= \frac{1}{T_1} \left[K_0 + hQ(1+\mu)t_1 + \frac{h}{2}(\lambda-a-Q\mu-Qb)(1+\mu)t_1^2 + hQ(1+\mu)(T_1-t_1) \right] \qquad (13)$$

$$= \frac{K_0}{T_1} + hQ(1+\mu)v + \frac{h}{2}(\lambda-a-Q\mu-Qb)(1+\mu)v^2 T_1 + hQ(1+\mu)(1-v)$$

$$= \frac{K_0}{T_1} + hQ(1+\mu) + \frac{h}{2}(\lambda-a-Q\mu-Qb)(1+\mu)v^2 T_1$$

Now with a view to obtaining the total time cycle T_1 that minimizes the total average inventory cost for the inventory system we shall adopt the convex property. The total average inventory cost is depicted by the equation no (13). To obtain the optimum time cycle and verify Equation (13) as convex in T_1, we must satisfy the following well established convex property,

(i) $\dfrac{d}{dQ_1} TC = 0 \quad$ and

(ii) $\dfrac{d^2}{dQ_1}TC > 0$

Now differentiating Equation (13) with respect to T_1 we get the following equation,

$$\frac{dTC}{dT} = -\frac{K_0}{T_1^2} + \frac{h}{2}(\lambda - a - Q\mu - Qb)(1+\mu)v^2 \tag{14}$$

Putting the value of Equation (14) in the convex property (i) and then using (10), we get

$$\frac{K_0}{T_1^2} = \frac{h}{2}(\lambda - a - Q\mu - Qb)(1+\mu)v^2$$

Or, $\quad T_1^2 = \dfrac{2K_0\lambda^2}{h(\lambda - a - Q\mu - Qb)(1+\mu)(a + Q\mu + Qb)^2}$

Or, $\quad T_1 = \sqrt{\dfrac{2K_0\lambda^2}{h(\lambda - a - Q\mu - Qb)(1+\mu)(a + Q\mu + Qb)^2}} \tag{15}$

Now with the help of Equations (11) and (15), we get the value of t_1 as below,

$$t_1 = \sqrt{\frac{2K_0}{h(\lambda - a - Q\mu - Qb)(1+\mu)}} \tag{16}$$

Again differentiating Equation (14) with respect to T_1, we get,

$$\frac{d^2}{dT_1^2}TC = \frac{2K_0}{T_1^3} \tag{17}$$

From Equation (17) we come to an end that the convex property (ii) is satisfied, *i.e.* $\dfrac{d^2}{dT_1^2}TC > 0$, as K_0 and

T_1 both is positive. Finally, we conclude that total cost function (13) is convex in T_1. Hence, there is an optimal solution in T_1 for which the total average inventory cost must be minimal.

5. Numerical Search Procedure

According to the result in section 5, we give an example that may illustrate how the numerical search procedure works. Suppose that there is a product which is a linear function in the inventory system and adopts the following parameters:

$$K_0 = 100, Q = 10, \lambda = 50, h = 2, a = 5, b = 0.8, \text{ and } \mu = 0.01.$$

We now put all the values in Equations (15), (16), (2) and (13) and then we get the results as optimum order interval $T_1^* = 6.252$ units, optimum time $t_1 = 1.638$ units at maximum inventory level, optimum order quantity $Q_1^* = 70.442$ units and total average optimum inventory cost $TC^* = 52.19$ units respectively. Substituting the values of T_1 arbitrarily either bigger or lesser than T_1^*, we get the inventory cost gradually increased from the minimum inventory cost at optimum level, which is show in **Table 1** and **Figure 2**. The table and figure justify the total average optimum inventory cost. If we analyze the table and figures we can observe that in a particular point total inventory cost is minimal and the order interval is optimum.

6. Sensitivity Test

Now, how the inventory system or the solution is affected by even a little changes of parameters Q, λ, a, b, h and

Table 1. Order interval (T_1) verses total cost (TC).

Order Interval (T_1)	5.250	5.500	5.750	6.000	**6.252**	6.500	6.750	7.000	7.250
Total Cost (TC)	52.67	52.45	52.30	52.21	**52.19**	52.21	52.28	52.39	52.54

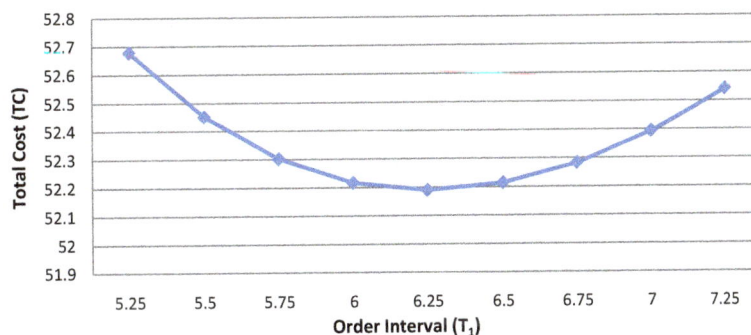

Figure 2. Time verses total cost.

Table 2. Effects of the changes of parameters.

Parameters	Change in %	Value of			
		t_1^*	T_1^*	Q_1^*	TC^*
Q	+50	1.736	5.061	68.808	70.696
	+25	1.685	5.570	69.625	61.395
	+10	1.656	5.954	70.115	55.852
	−10	1.620	6.592	70,769	48.562
	−25	1.595	7.200	71.259	43.205
	−50	1.555	8.591	72.076	34.569
λ	+50	1.265	7.241	111.392	48.121
	+25	1.416	6.754	90.918	49.900
	+10	1.537	6.454	78.632	51.205
	−10	1.762	6.052	62.252	53.265
	−25	2.014	5.765	49.967	54.996
	−50	2.884	5.505	29.492	56.826
a	+50	1.697	5.438	66.347	57.338
	+25	1.667	5.807	68.395	54.747
	+10	1.649	6.063	69.623	53.200
	−10	1.627	6.457	71.261	51.192
	−25	1.611	6.797	72.490	49.727
	−50	1.582	7.477	74.537	47.378
b	+50	1.735	5.072	63.890	60.491
	+25	1.684	5.577	67.166	56.293
	+10	1.656	5.957	69.132	53.812
	−10	1.621	6.588	71.753	50.602
	−25	1.595	7.186	73.7187	48.302
	−50	1.556	8.549	76.994	44.752
h	+50	1.337	5.105	70.442	70.284
	+25	1.465	5.592	70.442	61.237
	+10	1.562	5.961	70.442	55.808
	−10	1.727	6.590	70.442	48.570
	−25	1.891	7.219	70.442	43.142
	−50	2.317	8.842	70.442	34.095
μ	+50	1.635	6.217	70.360	52.469
	+25	1.636	6.238	70.409	52.301
	+10	1.637	6.245	70.426	52.245
	−10	1.638	6.259	70.459	52.133
	−25	1.639	6.266	70.475	52.078
	−50	1.640	6.287	70.524	51.910

μ on the optimal time t_1^* at maximum inventory level, optimum length of ordering cycle T_1^*, optimal order quantity Q_1^* and the total average optimum inventory cost TC^* per unit time in the model, will be shown in the following table. If the parameters change the values mentioned in **Table 2** by adding and subtracting respectively, we see the effect on the on the optimal time t_1^* at maximum inventory level, optimum length of ordering cycle T_1^*, optimal order quantity Q_1^* and the total average optimum inventory cost TC^* per unit time. While the change of one parameter take place, the other parameter must remain unchanged. **Table 2** shows the effect or the sensitivity.

Table 2 shows that small amount of a particular parameter may affect on the values of t_1^*, T_1^*, Q_1^* and TC^* even on great extent. On the basis of the results obtained in **Table 2**, the following observations can be highlighted:

- t_1^* and TC^* decrease while T_1^* and Q_1^* increase with increase in the value of the parameter λ. Here λ is highly sensitive to Q_1^* and moderately sensitive to other values.
- t_1^* and TC^* increase while T_1^* and Q_1^* decrease with increase in the value of the parameter Q, a and b. Here, Q, a and b all are moderately sensitive to the values of t_1^*, T_1^*, Q_1^* and TC^*.
- t_1^*, T_1^* and Q_1^* decrease, while TC^* increases with increase in the value of the parameter μ. Here, μ is moderately sensitive to all the values of t_1^*, T_1^*, Q_1^* and TC^*.
- t_1^* and T_1^* decrease and TC^* increases, while Q_1^* remain unchanged with increase in the value of the parameter h. Here, h is highly sensitive to the value of TC^* and moderately sensitive to all other values.

7. Conclusion

Because of the development of inventory management in the present age, the business institution cannot think its cost minimization without the proper use of it. By the proper use, management and thereby developing the suitable inventory models, the business enterprise can save its huge inventory cost. Before using model the enterprise needs to know the actual pattern of demand in the market. This demand always fluctuates. The suitable model is developed by considering the actual demand. The inventory model we have proposed in this paper is dependent on the stock, even we have considered buffer stock. Hence, the stock goes out due to any unavoidable circumstances, demand could still be met. The model also considers the deterioration, so due to the finite shelf-life of the items this model gives the correct. In the proposed model, the production rate and the decay have been considered constant through. The model develops an algorithm to determine the optimum ordering cost, total average optimum inventory cost, optimum time at maximum inventory level and optimum time cycle. The model could establish that with a particular order level $Q_1^* = 70.442$, the total average optimum inventory cost $TC^* = 52.189$ units.

Acknowledgements

The authors thank the editor and the reviewers for their valuable comments which could play a significant role to improve the standard of the manuscript.

References

[1] Harris, F.W. (1957) Operations and Costs. A. W. Shaw Company, Chicago, 48-54.

[2] Whitin, T.M. (1957) Theory of Inventory Management. Princeton University Press, Princeton, NJ, 62-72.

[3] Ghare, P.M. and Schrader, G.F. (1963) A Model for an Exponential Decaying Inventory. *Journal of Industrial Engineering*, **14**, 238-243.

[4] Skouri, K. and Papachristos, S. (2002) A Continuous Review Inventory Model, with Deteriorating Items, Time Varying Demand, Linear Replenishment Cost, Partially Time Varying Backlogging. *Applied Mathematical Modeling*, **26**, 603-617. http://dx.doi.org/10.1016/S0307-904X(01)00071-3

[5] Chund, C.J. and Wee, H.M. (2008) Scheduling and Replenishment Plan for an Integrated Deteriorating Inventory Model with Stock-Dependent Selling Rate. *International Journal of Advanced Manufacturing Technology*, **35**, 665-679. http://dx.doi.org/10.1007/s00170-006-0744-7

[6] Cheng, M.B. and Wang, G.Q. (2009) A Note on the Inventory Model for Deteriorating Items with Trapezoidal Type Demand Rate. *Computers and Industrial Engineering*, **56**, 1296-1300. http://dx.doi.org/10.1016/j.cie.2008.07.020

[7] Shavandi, H. and Sozorgi, B. (2012) Developing a Location Inventory Model under Fuzzy Environment. *International Journal of Advanced Manufacturing Technology*, **63**, 191-200. http://dx.doi.org/10.1007/s00170-012-3897-6

[8]   Sarker, B.R. Mukhaerjee, S. and Balam, C.V. (1997) An Order-Level Lot Size Inventory Model with Inventory-Level Dependent Demand and Deterioration. *International Journal of Production Economics*, **48**, 227-236. http://dx.doi.org/10.1016/S0925-5273(96)00107-7

[9]   Tripathy, C.K. and Mishra, U. (2010) Ordering Policy for Weibull Deteriorating Items for Quadratic Demand with Permissible Delay in Payments. *Applied Mathematical Science*, **4**, 2181-2191.

[10] Khieng, J.H., Labban. J. and Richard, J.L. (1991) An Order Level Lot Size Inventory Model for Deteriorating Items with Finite Replenishment Rate. *Computers Industrial Engineering*, **20**, 187-197. http://dx.doi.org/10.1016/0360-8352(91)90024-Z

[11] Ekramol, I.M. (2004) A Production Inventory Model for Deteriorating Items with Various Production Rates and Constant Demand. *Proceedings of the Annual Conference of KMA and National Seminar on Fuzzy Mathematics and Applications*, Payyanur, 8-10 January 2004, 14-23.

[12] Ekramol, I.M. (2007) A Production Inventory with Three Production Rates and Constant Demands. *Bangladesh Islamic University Journal*, **1**, 14-20.

[13] Mishra, V.K., Singh, L.S. and Kumar, R. (2013) An Inventory Model for Deteriorating Items with Time Dependent Demand and Time Varying Holding Cost under Partial Backlogging. *Journal of Industrial Engineering International*, **9**, 1-4. http://dx.doi.org/10.1186/2251-712x-9-4

[14] Ukil, S.I., Ahmed, M.M., Sultana, S. and Sharif, U.M. (2015) Effect on Probabilistic Continuous EOQ Review Model after Applying Third Party Logistics. *Journal of Mechanics of Continua and Mathematical Science*, **9**, 1385-1396.

[15] Sivazlin, B.D. and Stenfel, L.E. (1975) Analysis of System in Operations Research. 203-230.

[16] Shah, Y.K. and Jaiswal, M.C. (1977) Order Level Inventory Model for a System of Constant Rate of Deterioration. *Opsearch*, **14**, 174-184.

[17] Dye, C.Y. (1915) Joint Pricing and Ordering Policy for a Deteriorating Inventory with Partial Backlogging. *Omega*, **35**, 184-189. http://dx.doi.org/10.1016/j.omega.2005.05.002

[18] Billington, P.L. (1987) The Classic Economic Production Quantity Model with Set up Cost as a Function of Capital Expenditure. *Decision Series*, **18**, 25-42. http://dx.doi.org/10.1111/j.1540-5915.1987.tb01501.x

[19] Pakkala, T.P.M. and Achary, K.K. (1992) A Deterministic Inventory Model for Deteriorating Items with Two Warehouses and Finite Replenishment Rate. *European Journal of Operational Research*, **57**, 71-76. http://dx.doi.org/10.1016/0377-2217(92)90306-T

[20] Abad, P.L. (1996) Optimal Pricing and Lot Sizing under Conditions of Perish Ability and Partial Backordering. *Management Science*, **42**, 1093-1104. http://dx.doi.org/10.1287/mnsc.42.8.1093

[21] Singh, T. and Pattnayak, H. (2013) An EOQ Model for Deteriorating Items with Linear Demand, Variable Deterioration and Partial Backlogging. *Journal of Service Science and Management*, **6**, 186-190. http://dx.doi.org/10.4236/jssm.2013.62019

[22] Singh, T. and Pattnayak, H. (2012) An EOQ Model for a Deteriorating Item with Time Dependent Exponentially Declining Demand under Permissible Delay in Payment. *IOSR Journal of Mathematics*, **2**, 30-37. http://dx.doi.org/10.9790/5728-0223037

[23] Singh, T. and Pattnayak, H. (2013) An EOQ Model for a Deteriorating Item with Time Dependent Quadratic Demand and Variable Deterioration under Permissible Delay in Payment. *Applied Mathematical Science*, **7**, 2939-2951.

[24] Amutha, R. and Chandrasekaran, E. (2013) An EOQ Model for Deteriorating Items with Quadratic Demand and Tie Dependent Holding Cost. *International Journal of Emerging Science and Engineering*, **1**, 5-6.

[25] Ouyang, W. and Cheng, X. (2005) An Inventory Model for Deteriorating Items with Exponential Declining Demand and Partial Backlogging. *Yugoslav Journal of Operation Research*, **15**, 277-288. http://dx.doi.org/10.2298/YJOR0502277O

[26] Dave, U. and Patel, L.K. (1981) (T, Si) Policy Inventory Model for Deteriorating Items with Time Proportional Demand. *Journal of the Operational Research Society*, **32**, 137-142.

[27] Teng, J.T., Chern, M.S. and Yang, H.L. (1999) Deterministic Lot Size Inventory Models with Shortages and Deteriorating for Fluctuating Demand. *Operation Research Letters*, **24**, 65-72. http://dx.doi.org/10.1016/S0167-6377(98)00042-X

22

A Quantitative Factorial Component Analysis to Investigate the Recent Changes of Japan's Weight-Based Food Self-Sufficiency Ratio

Kunihisa Yoshii[1], Tatsuo Oyama[2]

[1]Ministry of Agriculture, Forestry and Fisheries (MAFF), Tokyo, Japan
[2]National Graduate Institute for Policy Studies (GRIPS), Tokyo, Japan
Email: oyamat@grips.ac.jp

Abstract

We investigate the weight-based food self-sufficiency ratio (*WSSR*) for Japan over a 50-year period (1961-2011) by applying factorial component analysis technique in order to measure the changes of the *WSSR* quantitatively. Quantitative data analysis is employed to determine the drivers of those changes. Numerical results show that Japan experienced a drastic decline in its food self-sufficiency ratio (*FSSR*) during the above period. The factorial component analysis shows that such a decline was caused by the changes in the *FSSR* of the food groups/items, not in the quantity of the food supply. A number of characteristics of those changes are presented and a list of major food groups that have major impacts on the changes is constructed. The findings in this paper reiterate the alarming food security problem in Japan and provide clear insight into the causes of this problem. The findings in this study pick up where previous studies have left off, aid the food-related policy-making process and identify new ideas for future food research.

Keywords

Food Self-Sufficiency Ratio, Food Security, Factorial Component Analysis

1. Introduction

"Food security" is an important issue in Japan. This is partly due to the surge in world food and agricultural commodity prices in 2007, but it also reflects the food supply insufficiency the country has experienced over the past few decades. "Food security" is a multi-faceted and complex term, with varied applications. Indeed, among scholars and more broadly in the government, "food security" is used to refer to different concepts and contexts. The most widely used definition that was agreed upon by the Food and Agriculture Organization (FAO) of the United Nations was adopted in 1996 by representatives from over 180 countries, including Japan, with the Rome Declaration on World Food Security: *"Food security exists when all people, at all times, have access to sufficient, safe and nutritious food to meet their dietary needs and food preferences for an active and healthy life"*[1]. However, this definition allows for a range of interpretations and applications, depending on the purpose of the user. In developing countries, food insecurity may be a matter of life or death. In developed countries, food insecurity may create political riots and pressure national security and sovereignty. The rice replacement program in Japan in the after-war period is but one example.

Living in the world's third largest economy, Japanese enjoy a standard of living that most people in the world can only dream of. However, even though Japan is a prosperous nation, and Japanese have enough food to lead "an active and healthy life", it is widely believed in Japan that the country has a problem that is theoretically faced only by the world's poorer countries: food insecurity. Although this may sound ironic, even contradictory, some Japanese worry about an excessive nutritional intake of animal fats and unbalanced diet, many worry about Japan's future food supply as its food self-sufficiency ratio (*FSSR*) has been declining over the past half century.

The *FSSR* is used to represent the magnitude of domestic production as a proportion of domestic utilization (including consumption). It is defined as the percentage of domestic production against domestic utilization. On a calorie basis, Japan's *FSSR* has drastically dropped from 79% in 1960 to 39% in 2005, a drop of 40 percentage points in 45 years. This trend was publicly noted in 1973 and has been documented by Ogura [1], Higuchi [2] and Saeki [3], and more recently by, among others, Kako [4], Tanaka and Hosoe [5], Mashimo [6], Yoshii and Oyama [7], Trung, *et al.* [8], Hayami and Godo [9], Hayami [10], and in various reports issued by Japan's Ministry of Agriculture, Forestry and Fisheries(MAFF) (2000-2008) [11].

These studies focus on the exogenous elements that are factors of the decline of Japan's *WSSR* and economic implications both for supply and demand of food. Many have given attention to domestic and international economic policies. However, few have examined the endogenous factors that account for the downward trend in the *FSSR*. It is important, though, to investigate the drastic change in Japan's *FSSR* in more detail, from the inside out in order to gain an overall understanding of the situation. Such research would help policy makers in planning new directions for the country's future food supply and food policy.

We seek to enrich the literature by performing quantitative factorial component analysis to investigate, in detail, the endogenous elements that have affected Japan's *WSSR* over the 50 years from 1961 to 2011. In the next section the paper starts by constructing and presenting a weight-based *WSSR* for Japan. Then in Section 3 the ratio will then be decomposed into factorial components for quantitative analysis. The characteristics of the changes will be analyzed to provide insights into the most important factors that have driven down Japan's *WSSR*. The findings, which are presented in Section 4, will have implications for policy directions. Summary and conclusions are given in the final Section 5.

2. Japan's Food Self-Sufficiency Ratio and Its Past Trend

Calorie intake is often used in Japan as the basis for calculating the nation's *FSSR*. The *FSSR* on a calorie basis is defined as the percentage of the net calorie intake per capita per day supplied by domestic production over the total net calorie intake per capita per day. Japan's *FSSR* on a calorie basis drastically fell from 79% in 1960 to 39% in 2005 (Kako [4]), and has stayed around 39% thereafter until now. These figures, with minor adjustments, were widely used in academic papers and policy discussions in MAFF (various reports/papers 2007-2009), Kako [4], Tanaka and Hosoe [5], Mashimo [6], Yoshii and Oyama [7], Trung, *et al.* [8].

There are other methods to measure a country's *FSSR*. One measure using weight is called the "*WSSR* on a weight basis" (*WSSR*). Similarly, the *FSSR* measure based on money is called the "*FSSR* on a monetary basis" (*MSSR*). Japanese make little use of the *MSSR* while the *WSSR* is rarely given official attention in Japan. Japanese government, however, has announced the target for the *FSSR* on a calorie basis as 45% in 2025 even though it seems to be very difficult to be attained as the current value is still low as of 39% in 2014.

---

[1]FAO, "Rome Declaration on World Food Security and World Food Summit Plan of Action", World Food Summit, Rome, 13-17 November 1996.

This paper, however, opts to use a more internationally recognized approach to calculate Japan's *FSSR* so that a comparative perspective may be obtained. This paper employs FAO definitions and utilizes its Food Balance Sheets to reassess Japan's data from 1961 to 2011 so that a comprehensive picture of the country's food supply patterns during that period can be obtained. Data for those sheets were standardized and updated in December 2009. We have used the most recent data set FAO-STAT as the 2011 data is the latest one at this stage.

Using the FAO's internationally acknowledged methods of data classification for representing actual agricultural production, a measure of Japan's *FSSR* on a weighted basis (hereinafter, the "*WSSR*") is developed to estimate the magnitude of domestic production in relation to overall domestic utilization. Let N be the set of food groups/items concerned. Then Japan's *FSSR*, on a weight basis, denoted by *WSSR*, is formulated as follows:

$$WSSR = \frac{\sum_{i \in N} DP_i}{\sum_{i \in N} DP_i + \sum_{i \in N} IM_i + \sum_{i \in N} SV_i - \sum_{i \in N} EX_i} \tag{1}$$

where

DP_i	:	Domestic production of food group/item i (ton), $i \in N$;
IM_i	:	Imports of food group/item i (ton), $i \in N$;
SV_i	:	Stock variation (increase or decrease) of food group/item i (ton), $i \in N$;
EX_i	:	Exports of food group/item i (ton), $i \in N$,

While (1) is used to obtain a value for the *FSSR* of all food groups/items, the self-sufficiency ratio of each food group/item i on a weight basis (%), denoted by $WSSR_i$, can also be calculated in a similar manner.

$$WSSR_i = \frac{DP_i}{DP_i + IM_i + SV_i - EX_i}, \quad i \in N. \tag{2}$$

Likewise, a food import dependency ratio for Japan (the "*FIDR*") on a weight basis (%) is developed to assess the importance of imported food in the country. The *FIDR* expresses the magnitude of imports in relation to domestic utilization and is formulated in a similar manner to its *WSSR* counterpart except for the difference in the numerator, where domestic production is replaced by the value for import, as follows.

$$FIDR = \frac{\sum_{i \in N} IM_i}{\sum_{i \in N} DP_i + \sum_{i \in N} IM_i + \sum_{i \in N} SV_i - \sum_{i \in N} EX_i}. \tag{3}$$

The numerical results for *WSSR* and *FIDR* were derived by computing the values for (1), (2) and (3) using the FAO 2010 data for Japan's Food Balance Sheets from 1961 to 2011. The values are given in **Table 1** and graphed in **Figure 1**. Twenty major food groups (*MFG*) were used to compute the ratio. The group labeled "Miscellaneous" is omitted due to missing data throughout the reference period[2]. The use of all 20 *MFG* allows for a consistent measure of the total food supply and utilization in the country. These *MFG* and the food items included in each grouping are given in **Table 2**.

The results show that the values for both the *WSSR* and the *FIDR* changed drastically over time in ways consistent with the views of those concerned about Japan's future food supply and import dependency. It is obvious from the graph in **Figure 1** that the values for the *WSSR* fell sharply from 87% in 1961 (or an average of 80% in the 60s) to an average of 54% in the first five years of the new millennium. It is a drop of nearly 34 percentage points in 45 years, or a loss of 39% in value. The values for *FIDR* rose from just above 14% in 1961 to more than 49% in 2005. That is an increase of 35 percentage points over the same period, or a 2.5-fold gain in value. In the meantime, food supply in Calories (kcal) continued to rise as Japanese sought a higher living standard and the economy grew. The calorie intake rose from 2468 Calories per capita per day in 1961 to 2752 Calories in

---

[2]The major food groups were: (1) Cereals-Excluding Beer; (2) Starchy Roots; (3) Sugar Crops; (4) Sugar & Sweeteners; (5) Pulses; (6) Tree Nuts; (7) Oil Crops; (8) Vegetable Oils; (9) Vegetables; (10) Fruits-Excluding Wine; (11) Stimulants; (12) Spices; (13) Alcoholic Beverages (14) Meat; (15) Offals; (16) Animal Fats; (17) Eggs; (18) Milk-Excluding Butter; (19) Fish, Seafood; and (20) Aquatic Products, Other.

Table 1. *WSSR* and *FIDR* on a weight basis (%), and Food supply (kcal/capita/day) (kcal), Japan 1961-2011.

	1961	1962	1963	1964	1965	1966	1967	1968	1969	1970
WSSR (%)	86.4	84.8	82.2	79.9	80.0	77.5	77.7	77.2	74.7	72.3
FIDR (%)	15.0	15.9	19.2	21.6	23.2	24.6	25.5	26.1	27.4	30.4
Food supply (kcal)	2525	2572	2609	2632	2621	2643	2690	2700	2699	2738
	1971	**1972**	**1973**	**1974**	**1975**	**1976**	**1977**	**1978**	**1979**	**1980**
WSSR (%)	71.2	70.7	68.8	67.9	68.6	66.2	66.0	65.7	64.9	63.8
FIDR (%)	29.6	30.5	33.2	34.5	33.0	35.7	36.0	36.5	37.4	36.8
Food supply (kcal)	2729	2782	2773	2743	2717	2752	2774	2791	2808	2799
	1981	**1982**	**1983**	**1984**	**1985**	**1986**	**1987**	**1988**	**1989**	**1990**
WSSR (%)	64.9	65.6	64.8	65.2	64.0	63.7	62.5	61.4	61.4	60.8
FIDR (%)	36.3	35.9	36.7	37.7	38.4	38.1	38.9	40.8	40.0	40.3
Food supply (kcal)	2750	2814	2829	2827	2862	2874	2896	2942	2969	2949
	1991	**1992**	**1993**	**1994**	**1995**	**1996**	**1997**	**1998**	**1999**	**2000**
WSSR (%)	58.8	58.5	56.9	57.2	55.6	55.3	55.4	54.7	54.0	53.6
FIDR (%)	41.9	42.4	43.6	44.7	45.9	45.5	46.0	45.7	47.3	48.0
Food supply (kcal)	2934	2943	2926	2932	2921	2963	2939	2895	2898	2900
	2001	**2002**	**2003**	**2004**	**2005**	**2006**	**2007**	**2008**	**2009**	**2010**
WSSR (%)	53.1	52.5	52.2	52.1	52.2	51.7	52.8	53.5	52.7	51.4
FIDR (%)	47.7	48.4	48.3	49.1	49.5	49.7	48.4	48.4	47.4	49.8
Food supply (kcal)	2890	2853	2842	2843	2828	2777	2816	2732	2674	2692
	2011									
WSSR (%)	51.3									
FIDR (%)	49.9									
Food supply (kcal)	2719									

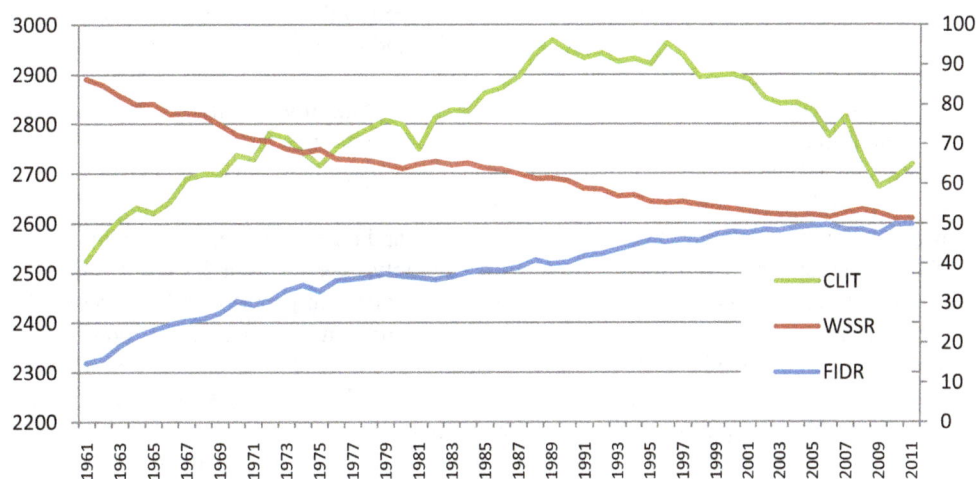

Figure 1. Japan's per capita per day calorie intake (left axes) and *WSSR*, *FIDR* (right axes).

1972, peaking at 2859 Calories in 1996 before slightly decreasing to 2743 Calories in 2005. That was a 11% increase in daily calorie intake over the last half century. Nevertheless, the 2005 level was just about that of the early 1970s, implying no net improvement in calorie intake over the 35-year period.

Taken altogether, the three trends indicate that while the quality of life is improving, Japan produces only half of its food needs and has come to depend on foreign imports for the other half. Setting aside differences in the

Table 2. 20 *MFG* and the items included.

MFG	Items included
Cereals-Excluding Beer	Wheat; Rice (Milled Equivalent); Barley; Maize; Rye; Oats; Millet; Sorghum; Cereals, Other.
Starchy Roots	Cassava; Potatoes; Sweet Potatoes; Yams; Roots, Other.
Sugar Crops	Sugar Cane; Sugar Beet.
Sugar & Sweeteners	Sugar, Non-Centrifugal; Sugar (Raw Equivalent); Sweeteners, Other; Honey.
Pulses	Beans; Peas; Pulses, Other.
Tree Nuts	
Oil Crops	Soyabeans; Groundnuts (Shelled Equation); Sunflowerseed; Rape and Mustardseed; Cottonseed; Coconuts-Incl Copra; Sesameseed; Palmkernels; Olives; Oilcrops, Other.
Vegetable Oils	Soyabean Oil; Groundnut Oil; Sunflowerseed Oil; Rape and Mustard Oil; Cottonseed Oil; Palmkernel Oil; Palm Oil; Coconut Oil; Sesameseed Oil; Olive Oil; Ricebran Oil; Maize Germ Oil; Oilcrops Oil, Other.
Vegetables	Tomatoes; Onions; Vegetables, Other.
Fruits-Excluding Wine	Oranges, Mandarines; Lemons, Limes; Grapefruit; Citrus, Other; Bananas; Plantains; Apples; Pineapples; Dates; Grapes; Fruits, Other
Stimulants	Coffee; Cocoa Beans; Tea.
Spices	Pepper; Pimento; Cloves; Spices, Other.
Alcoholic Beverages	Wine; Beer; Beverages, Fermented; Beverages, Alcoholic; Alcohol, Non-Food.
Meat	Bovine Meat; Mutton & Goat Meat; Pig Meat; Poultry Meat; Meat, Other.
Offals	
Animal Fats	Butter, Ghee; Cream; Fats, Animals, Raw; Fish, Body Oil; Fish, Liver Oil.
Eggs	
Milk-Excluding Butter	
Fish, Seafood	Freshwater Fish; Demersal Fish; Pelagic Fish; Marine Fish, Other; Crustaceans; Cephalopods; Molluscs, Other.
Aquatic Products, Other	Meat, Aquatic Mammals; Aquatic Animals, Others; Aquatic Plants.
Miscellaneous	

statistics used, the results correspond to the findings described in the MAFF papers and in the previous studies by Kako [4], Tanaka and Hosoe [5], Mashimo [6], Yoshii and Oyama [7], Trung, *et al.* [8], Hayami and Godo [9], and Hayami [10].

Whether the trends are cause for "alarm" depends on one's point of view. Many other variables also shape the political decision-making process in this globalized era. One could say that these trends might raise much concerns in both domestic and international arenas, given their impact both on Japan's food production, prices and trade, and the world's needs for overall global sufficiency.

In addition to realizing the changes in both *WSSR*, *FIDR* and calorie intake, one should also note that the pattern of food consumption in Japan has changed considerably over the past 50 years. Mashimo [6] has commented on the reasons for such changes[3], and studied three different patterns of Japanese food consumption in 2005. The first pattern was the *status quo* in 2005. The second pattern was a set of food intake and nutritional data recommended by the Ministry of Health, Welfare and Labor (MHWL). This pattern has been created as the ideal for Japanese to maintain their health, avoiding lifestyle-related sickness (MHWL pattern). And the third pattern was a set of food requirements based on the daily meal menus organized by Setsuko Shirone, an expert of sustainable food consumption and organic agriculture. Mashimo [6] called this the Chisan-chisho pattern (LP-LC pattern), after a popular movement that encouraged local production and local consumption in Japan. In 2005, according to the MHWL pattern, the ingestion of grains, potatoes and vegetables would increase, while the consumption of meat, milk products, sugar and fat would drastically decrease. This tendency was even more radical in the LP-LC pattern. The difference was the high amount of marine product intakes that was still considered to be possible. All of this, however, consisted of small fish and coastal fish, as well as the continued con-

---

[3]Mashimo argued that dietary changes in Japan dated back to the 1960s and 1970s with the rice diversion program and the "westernization" of the Japanese dietary habit. See Mashimo (2008) for details.

sumption of other domestically available marine species (Mashimo [6]).

3. Method and Numerical Results for Factorial Component Changes

3.1. Method for Measuring Changes

This part of the paper employs a factorial component analysis to assess the weight of endogenous factors that might account for the decline in the *WSSR*. The annual total change in the *WSSR* is broken into two major factorial components: (1) the change due to the *WSSR* change of each *MFG* component, and (2) the change due to *MFG*'s quantity supply change. These two factors are then compared to examine the irrelative impact on the *WSSR*.

Recalling the definition of food self-sufficiency ratio, *WSSR* is defined as the magnitude of domestic production in relation to overall domestic utilization. In other words, *WSSR* is the fraction of the total domestic utilization times its own self-sufficiency ratio (which equals domestic production) over the total domestic utilization. In mathematical terms, *WSSR*—hereinafter denoted by *R*—is defined as follows:

$$R = \frac{\sum_{i \in N} p_i w_i}{\sum_{i \in N} w_i} \tag{4}$$

where,

p_i	:	$MFGi$'s weight-based self-sufficiency ratio, $i \in N$;
w_i	:	$MFGi$'s quantity (weight) supply, $i \in N$,

The first derivative of (4) with respect to time shows that changes in *R* consists of both changes in p_i and w_i. In general terms, the annual change in Japan's food self-sufficiency ratio is the combination of a component change in the self-sufficiency ratios of the concerned *MFG* and a component change in the quantity of those *MFG*'s supply. Specifically, a small change in the value of *R*, denoted as ΔR, can be decomposed into two components corresponding to the SSR change and the quantity change respectively, as follows:

$$\frac{dR}{dt} = \sum_{i \in N} \frac{\partial R}{\partial p_i} \frac{dp_i}{dt} + \sum_{i \in N} \frac{\partial R}{\partial w_i} \frac{dw_i}{dt} \tag{5}$$

Denoting the above "annual" changes corresponding to the changes for *R*, p_i and w_i as ΔR, Δp_i and Δw_i, respectively, we can rewrite the expression in (5) as follows.

$$\Delta R = \sum_{i \in N} \frac{\partial R}{\partial p_i} \Delta p_i + \sum_{i \in N} \frac{\partial R}{\partial w_i} \Delta w_i \tag{6}$$

Let $W = \sum_{i \in N} w_i$ be the total supply for domestic utilization, then the right-hand-side (RHS) terms in (6) could be rewritten as:

$$\frac{\partial R}{\partial p_i} = \frac{w_i}{W}, \quad i \in N \tag{7}$$

$$\frac{\partial R}{\partial w_i} = \frac{1}{W}\left(p_i - \frac{\sum_{i \in N} p_i w_i}{W} \right) = \frac{\sum_{j \in N, j \neq i} (p_i - p_j) w_j}{W^2} = \frac{p_i - R}{W}, \quad i \in N \tag{8}$$

$$\Delta p_i = p_i^t - p_i^{t-1} \quad \text{and} \quad \Delta w_i = w_i^t - w_i^{t-1}, \quad i \in N \tag{9}$$

Replacing (7) and (8) in (6) yields

$$\Delta R = \sum_{i \in N} \left(\frac{w_i}{W} \right) \left(p_i^t - p_i^{t-1} \right) + \sum_{i \in N} \left(\frac{p_i - R}{W} \right) \left(w_i^t - w_i^{t-1} \right) \tag{10}$$

$$\Delta R = \Delta R_p + \Delta R_w \tag{11}$$

Note that the left-hand-side (LHS) in (9) is the total annual change in Japan's food self-sufficiency ratio. The first term of the RHS in (9) is the factorial change in the major food groups' self-sufficiency ratios, and the second term is the factorial change of their supply quantities.

3.2. Numerical Results for Factorial Component Changes

The numerical results for the *WSSR* for each of the 20 *MFG*, and the values for all the terms in (6) were computed through (9) using the 2010 FAO data for Japan's Food Balance Sheets from 1961 to 2011. The results are presented in the **Table A** in the Appendix and in **Table 3** below.

Table 3. Numerical results for W, R, ΔR, ΔR_p and ΔR_w.

	1961	1962	1963	1964	1965	1966	1967	1968	1969	1970
W	495.0	515.7	537.1	541.3	548.2	563.1	577.8	599.2	593.3	601.5
R	86.4	84.8	82.2	79.9	80.0	77.5	77.7	77.2	74.7	72.3
ΔR		−1.355	−2.217	−2.161	0.710	−1.502	−0.130	−0.154	−1.486	−1.642
ΔR_p		−1.606	−2.616	−2.001	0.333	−1.604	−0.011	−0.658	−1.151	−1.432
ΔR_w		0.251	0.399	−0.160	0.377	0.102	−0.119	0.503	−0.335	−0.210
	1971	1972	1973	1974	1975	1976	1977	1978	1979	1980
W	604	623.9	619.1	609.7	612.7	611.8	634	639.1	646.3	637.5
R	71.2	70.7	68.8	67.9	68.6	66.2	66.0	65.7	64.9	63.8
ΔR	−0.316	−0.083	−1.484	−0.848	0.550	−2.331	0.224	0.453	−0.237	−0.542
ΔRp	−0.864	−0.434	−1.412	−0.683	0.186	−1.976	−0.467	0.298	−0.256	−0.220
ΔR_w	0.548	0.351	−0.073	−0.165	0.363	−0.355	0.691	0.155	0.019	−0.322
	1981	1982	1983	1984	1985	1986	1987	1988	1989	1990
W	627.8	642.3	643.3	636	649.9	656.8	667.3	674.3	683.3	674.1
R	64.9	65.6	64.8	65.2	64.0	63.7	62.5	61.4	61.4	60.8
ΔR	0.753	0.686	−0.544	0.068	−1.089	−0.018	−0.665	−1.690	0.001	−0.431
ΔR_p	0.707	0.414	−0.501	0.145	−1.143	−0.410	−1.055	−1.930	0.132	−0.470
ΔR_w	0.046	0.272	−0.043	−0.077	0.054	0.391	0.390	0.240	−0.131	0.039
	1991	1992	1993	1994	1995	1996	1997	1998	1999	2000
W	670.4	678.9	669.5	677	679.6	677.5	670.6	652.1	658.9	655.2
R	58.8	58.5	56.9	57.2	55.6	55.3	55.4	54.7	54.0	53.6
ΔR	−1.955	−0.166	−1.230	−0.846	−1.834	−0.017	0.455	−1.200	−0.709	−0.974
ΔR_p	−1.921	−0.475	−1.076	−1.120	−1.847	0.133	0.566	−0.784	−0.640	−0.736
ΔR_w	−0.034	0.309	−0.155	0.274	0.013	−0.151	−0.110	−0.416	−0.069	−0.238
	2001	2002	2003	2004	2005	2006	2007	2008	2009	2010
W	649.9	640.6	631.8	627.7	628.4	609.8	616.5	598.8	585.8	581
R	53.1	52.5	52.2	52.1	52.2	51.7	52.8	53.5	52.7	51.4
ΔR	−0.657	−0.720	−0.235	−1.089	−0.128	−0.576	0.942	0.428	−0.265	−1.649
ΔR_p	−0.263	−0.155	−0.101	−0.924	0.066	−0.513	1.085	0.388	−0.082	−1.282
ΔR_w	−0.394	−0.565	−0.134	−0.165	−0.194	−0.063	−0.142	0.040	−0.183	−0.366
	2011									
W	591.7									
R	51.3									
ΔR	−0.727									
ΔR_p	−0.877									
ΔR_w	0.150									

Using the breakdown in formula (6), we can map the relations between *WSSR*'s annual factorial changes, ΔR_p and ΔR_w, and the total changes ΔR. Using the computed data, the graphs in **Figure 2** show those relations and the 3-period moving average trend for ΔR.

Figure 2 shows that a strong correlation between the factorial change in *MFG*'s *WSSR* (ΔR_p) and the total change in *WSSR* (ΔR) does exist, but we cannot find a relation between the factorial change in *MFG*'s quantity supply (ΔR_w) and ΔR. The graphs show that ΔR_w fluctuated in a narrow band close to the origin0.0 and registered mostly positive values except for the period after year 1995. On the other hand, ΔR_p and ΔR took on mostly negative values and moved in close tandem with each other in a much wider fluctuation, mainly below zero. Positive values and minor adjustments or changes in *MFG*'s supply tend to stabilize and mitigate the total changes in *WSSR*. Nonetheless, changes in *MFG*'s *WSSR* diminish that effect as sharp falls tend to cause a drastic fall in the total change in *WSSR*. The 3-year moving average trend line also shows that, on average, ΔR moved in a wide band far below zero in a way that reflected the broad negative fluctuation of ΔR_p, and cancelling out the positive effect brought about by ΔR_w. The moving average stays mostly below zero, implying and verifying the fact that *WSSR* declines most of the time.

This data analysis shows that the declining trend of *WSSR* is mainly due to the declining trend of the MFG's *WSSR* rather than due to the changes in *MFG*'s quantity supply.

3.3. Trend Analysis on the Factorial Change

We analyze the time series trend of factorial component changes in order to investigate the change in the *WSSR* in more detail. We divide the whole time span of 1961-2011 into four sub-periods, trying to find the specific characteristic of the three elements concerned (ΔR, ΔR_p and ΔR_w) in each of these four sub-periods, which we denote by I, II, III, and IV, respectively.

Sub-period I (1961-1976): This is the longest sub-period, characterized by large negative values for ΔR and ΔR_p. During the sub-period the rise in the values for ΔR_w kept the values for ΔR from falling more than they did. This period corresponds to the time when many food-related policies, such as the rice diversion program, which started in 1970 to reduce the domestic rice production by 30% - 40%, thus policy changes were adopted in Japan. The period also saw the "westernization" of Japanese diets. Although agricultural production in this period performed satisfactorily, new demands created huge shortages in food items which were not produced domestically. The outcome was a sharp fall in the values of ΔR_p for many food items (*i.e.*, *MFG*) which led to the sharp fall in the values of ΔR (*i.e.*, *WSSR*). Major falls occurred in rice, grain and wheat in the "Cereals" *MFG* and in potatoes and sweet potatoes in the "Starchy Roots" *MFG*. On the other hand, weights of vegetables, fish, seafood, milk and fruits increased in a large scale, which made the values of ΔR_w rather stable.

Sub-period II (1977-1984): This is the shortest sub-period, lasting only eight years, where the fluctuations in ΔR and ΔR_p were within a narrow band. Again in this sub-period, the values for ΔR_w remained mostly highly and strongly positive. ΔR remained only slightly negative through this period, and for some years even rose into positive territory. This period came after a long period of decline in the *WSSR* that had begun to alarm Japanese in the early 1970s, following the Oil Crisis that rocked Japan's economy. A mild "return to local foodstuff" saw the *WSSR* rise somewhat. However, Japan's relatively open agricultural trade policy tended to offset the return, and the *WSSR* fluctuated during this time. Major contributors to the fluctuation were grain and wheat in the "Cereals" *MFG*. New impacts came from wine and alcoholic beverages in the "Alcoholic Beverages" *MFG*, the "Vegetables" *MFG*, and a range of foreign fruits from the "Fruits-Excluding Wine" *MFG*.

Sub-period III (1985-1996): During this period, Japan returned to the pattern that had been found for the first sub-period. It was characterized by negative values for ΔR and ΔR_p (though with slightly smaller values). The values for ΔR_w were positive from 1985-1989. Then they turned negative, which, together with ΔR_p, amplified the negative values being registered for ΔR. This period marks the beginning of Japan being the world's largest net food importer. High volumes of imports in wheat, grain in the "Cereals" *MFG*, in potatoes in "Starchy Roots", in wine and alcoholic beverages in "Alcoholic Beverages", in tomatoes, onions and other vegetables in "Vegetables", in apples, bananas, oranges, pineapples, grapes in "Fruits-Excluding Wine" contributed mostly to low negative values in ΔR_p. The new commodities were marine fish and other seafood in the "Fish, Seafood" *MFG*, and bovine meat, mutton and goat meat in the "Meat" *MFG*.

Sub-period IV (1997-2005): This was a stable, but all negative period. The values for ΔR_p and ΔR_w were negative almost all time, thereby keeping those for ΔR in the negative zone. The moving average was quite

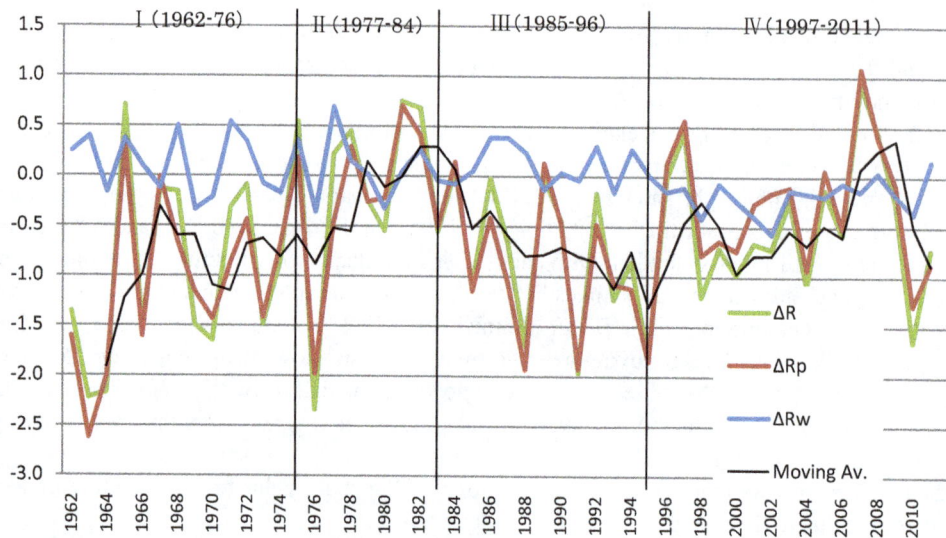

Figure 2. ΔR_p, ΔR_w, ΔR and its 3-year moving average of ΔR.

smooth and the negative scale was not that as much as that of the former three sub-groups. Thus the "stabilized period" is explainable. After three periods of sharp declines and fluctuation, the values for ΔR_p and ΔR_w seem to "adapt" themselves to the new domestic demands and dietary habits. This results in a more stable set of slightly negative values for ΔR. The commodities remain those noted above.

4. Investigation on the Relation among Factorial Component Changes

4.1. Impact of Factorial Component Changes on the Total Change

To analyze the impact of factorial component changes on the total change, the values for ΔR_p and ΔR_w are individually regressed against those for ΔR. **Figure 2** plots the two combinations (ΔR, ΔR_p) and (ΔR, ΔR_w) to represent the comparative relation between the two components and the total change. We note that, firstly, the values for the total change ΔR vary within an interval ranging from -3.0 to 1.0. Secondly, the values for the factorial component change ΔR_p range within the same interval $-3.0 < \Delta R_p < 1.0$. Thirdly, the values for the factorial component change ΔR_w range in a much narrower interval $-0.5 < \Delta R_w < 0.5$.

The graphs in **Figure 3** show that the data points for the (ΔR, ΔR_p) combinations, letting ΔR and ΔR_p expressed by x and y, respectively, are best explained by the single variable linear regression model $y = 0.874x - 0.124$ (with $R^2 = 0.897$). This means that the component change ΔR_p is generally explained by the total change ΔR with almost 90% goodness of fit. Moreover, only one-fifth (20%) of the (ΔR, ΔR_p) combinations, *i.e.*, 9 out of 50 combinations, are located in the first coordinate while almost 70%, *i.e.*, 37 out of 50 combinations, are in the third coordinate, meaning the negative change in the values for ΔR is mostly attributed to the negative change in the values for ΔR_p, not that of those for ΔR_w. In other words, the declining trend of *WSSR* is attributed to the decline in the self-sufficiency ratios of the foodstuff themselves, not that of the change in the food supply quantity.

On the other hand, the graph does not show a satisfactory correlation for the (ΔR, ΔR_w) combinations. These combinationsscatter along a narrow rectangle given by $-3.0 < \Delta R < 1.0$, and $-0.5 < \Delta R_w < 0.5$, meaning that the factorial component ΔR_w has not changed on a large scale in the past 45 years and has not contributed greatly to the total change ΔR. We also note that data points in the first coordinate (*i.e.*, positive values for both ΔR and ΔR_w) stay close to the origin, implying not much change and impact on ΔR. Positive changes in ΔR_w are mostly reflected in the increase in Calorie supply (kcal/capita/day) rather than in ΔR and *WSSR* as a whole.

4.2. Relation between Factorial Components

Table 4 presents the frequency of the values for ΔR_p and ΔR_w during the reference period. It shows how they are distributed, over 51 years as a whole (50 years of changes), and in each of the above-mentioned four sub-periods in particular. It can be seen from the table that the number of years when the values for ΔR_w are nonnegative and that when the values for ΔR_w are negative are equal as 23 and 27, respectively, thus they are not so different.

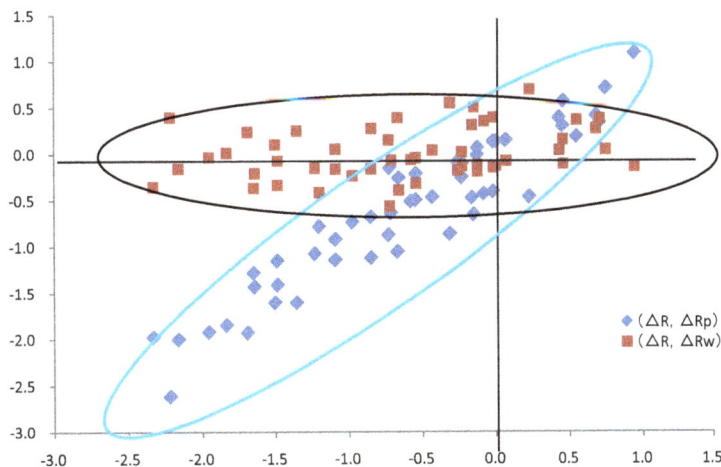

Figure 3. Plots of $(\Delta R, \Delta R_p)$ and $(\Delta R, \Delta R_w)$.

Table 4. ΔR_p and ΔR_w frequency.

Sub-period	ΔR_p				ΔR_w		
	$\Delta R_p < -2$	$-2 < \Delta R_p < -1$	$-1 < \Delta R_p < 0$	$0 < \Delta R_p$	$\Delta R_w < -1$	$-1 < \Delta R_w < 0$	$0 < \Delta R_w$
I (1962–1976)	2	6	5	2	0	7	8
II (1977–1984)	0	0	4	4	0	3	5
III (1985–1996)	0	7	3	2	0	4	8
IV (1997–2011)	0	1	10	4	0	13	2
Total	2	14	22	12	0	27	23

On the other hand, regarding the values of ΔR_p, they are positive only 12 years among 50 years while other 38 years show negative values.

In the sub-period I only 2 years out of 15 are in the 1st coordinate, *i.e.*, ΔR_p and ΔR_w are both positive while other 6 years are in the 2nd coordinate, *i.e.*, ΔR_p negative and ΔR_w positive, and 7 years are in the 3rd coordinate, *i.e.*, ΔR_p and ΔR_w are both negative. In the sub-period II 3 years are distributed in the 1st coordinate and 1 year is in the 4th coordinate, *i.e.*, ΔR_p positive and ΔR_w negative while 2 years are in the 2nd and 3rd coordinates, respectively. In the sub-period III 2 years out of 12 in total are in the 2nd coordinate while remaining 8 years are in the 2nd coordinate and 2 years are in the 3rd coordinate. In the sub–period IV only 1 year out of 15 in total is in the 1st coordinate while remaining 10 years are in the 3rd coordinate and only 1 year and 2 years are exceptionally in the 2nd and 4th coordinates, respectively.

Thus, we find that most sub-periods, excluding II, show dominatingly negative ΔR_p no matter how ΔR_w are valued. This trend in the sub-period I is due to the decrease (increase) in supply (import) in majorly grains. In the sub-section III similar trend comes from decreasing supply of rice, vegetables, fruits, meat and fish. Sub-period IV shows the decreasing trend of major food items excluding meat corresponds to the negative ΔR_w. Sub-period II shows that both ΔR_p and ΔR_w are close to zero.

Figure 4 graphs the time series changes in the four sub-periods, and for the 1962-2011 period as a whole, relative to a center point. We find that ΔR_p and ΔR_w follow two different paths unrelated each other. In all four sub-periods, ΔR_w tended to be stably staying around the circular line corresponding to the value zero while ΔR_p unsteadily scattered along the scale and mostly staying in the area corresponding to "negative" points. We find that the graphs reiterate ΔR_p's tendency of being mostly negative for sub-periods I and III while it is mostly close to ΔR_w's for sub-periods II and IV.

We also note that the values of the center points are all different for these figures while the scale for each sub-period is also different. This means ΔR_p and ΔR_w vary in each sub-period as we see the different shapes of

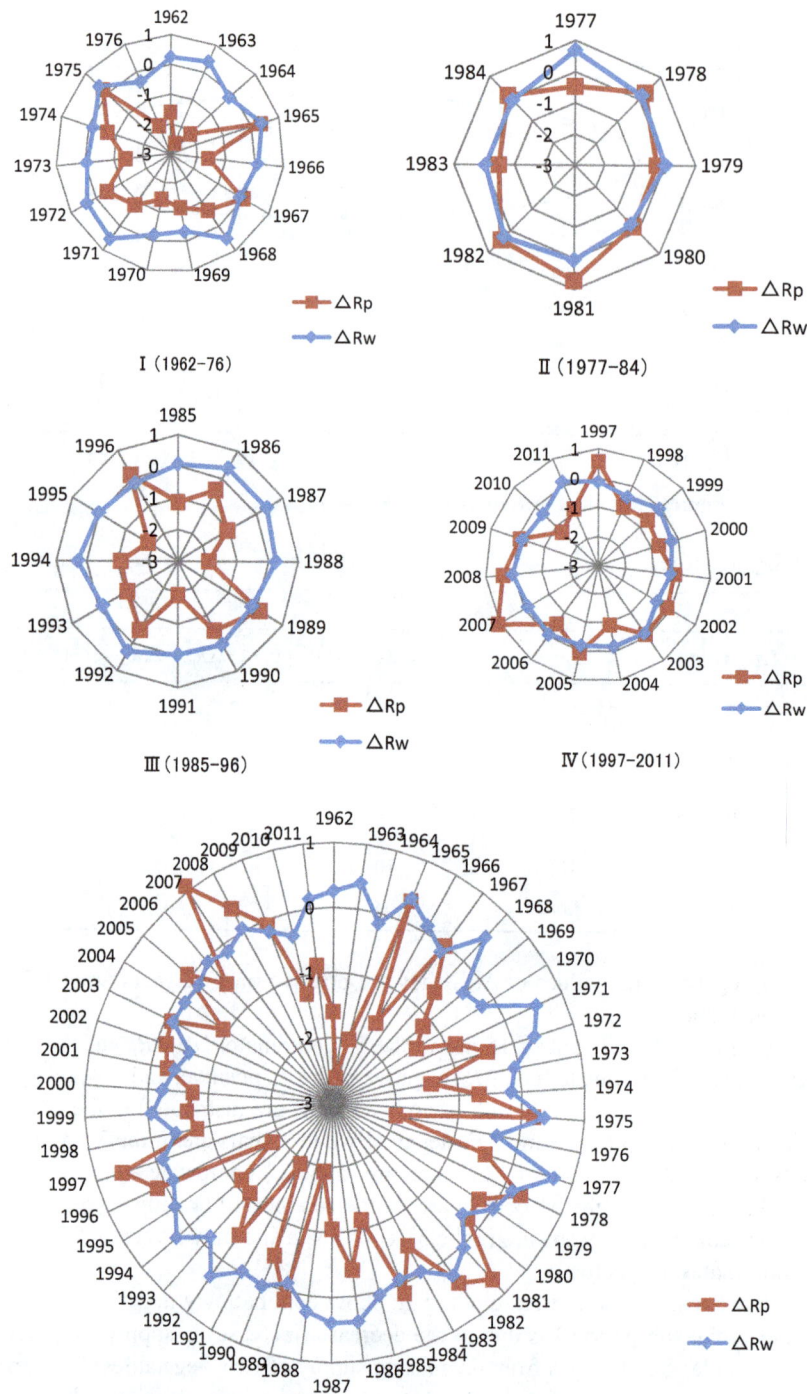

Figure 4. ΔR_p and ΔR_w in 4 sub–periods and all.

the graphs. Moreover, we learn that there were but 8 years (out of 50, or 16%) in the reference period where ΔR_p and ΔR_w took on positive values, and stayed close to each other. Those were in 1965, 1975, 1978, 1981, 1982, and 2008. It explains why the total change ΔR were mostly negative, and *WSSR* declined during the 1961-2011 period.

4.3. Impact of *MFG* on the Factorial Component Changes

Of all 20 *MFG* concerned, the impacts they have on ΔR_p and ΔR_w (and, therefore, on ΔR) are quite different. We compute minimum, median and maximum of ΔR_p and ΔR_w for all 20 *MFG* in the reference period to observe such impacts. **Table 5** presents these minimums, medians and maximum, and **Table 6** lists the top 10 *MFG* with the greatest impacts on ΔR_p and ΔR_w, respectively. The ranking is based on (1) the scale of the impact with respect to the median, and (2) the nature of *MFG*'s fluctuation pattern. These **Table 5** and **Table 6** are obtained from calculating numerical results given in **Table A** in the **Appendix**.

From **Table 5** and **Table 6** we find that regarding the median of ΔR_w, changes of supply for "Cereals", "Alcoholic Beverages", and "Fish, Seafood" give rather positive impacts while only "Starchy Roots" gives rather negative impact. Negative impact due to "Starchy Roots", even though small, may result from the fact that they are transformed into some other consuming foods rather than directly consumed in the market. Also regarding the median of ΔR_p, only "Sugar & Sweeteners" brings small positive impact while other "Cereals", "Fruits", and "Fish, Seafood" bring large negative impacts. These large negative impacts may result from the fact that we have been importing these foods such as "Cereals", "Fruits", and "Fish, Seafood" constantly and largely as major foods in Japan.

As shown in **Table 5** and **Table 6**, we see that maximum values with respect to ΔR_p are very high for "Cereals", "Starchy Roots", "Sugar & Sweeteners", "Alcoholic Beverages", and "Fish, Seafood", and their minimum values are generally very small, which implies that these foods including "Vegetables" and "Meat" have both large positive and large negative impacts on our ΔR. The "Cereals" *MFG* has the second largest and highly important impact on both ΔR_p and ΔR_w. The "Fish, Seafood" *MFG* has more impact on ΔR_p and less on ΔR_w while "Meat" and "Vegetable Oils" have almost the same importance in affecting ΔR_p and ΔR_w. The *MFG* "Cereals", "Starchy Roots", "Fish, Seafood", "Meat", "Vegetables", "Vegetable Oils", "Milk", "Alcoholic Beverages" are presented in both columns, implying their importance in Japanese diets. The missing *MFG* in the list ("Sugarcrops", "Treenuts", "Oilcrops", "Spices", "Offals", "Animal Fats", "Aquatic Products, Other") reveals that food policy-making process can target these *MFG* without much impact on *WSSR*. On the other hand, policy makers should be careful when targeting *MFG* in the Top 10 list in order to avoid an unwanted impact on ΔR_p and/or ΔR_w, and therefore on ΔR and *WSSR*.

Table 5. Impact of *MFG* on ΔR_p and ΔR_w.

MFG	ΔR_p			ΔR_w		
	Min	Median	Max	Min	Median	Max
Cereals-Excluding Beer	−1.809	−0.118	0.696	−0.233	0.040	0.312
Starchy Roots	−0.844	−0.030	0.566	−0.232	−0.019	0.136
Sugar Crops	0.000	0.000	0.000	0.000	0.000	0.000
Sugar & Sweeteners	−0.328	0.021	0.610	−0.357	0.000	0.238
Pulses	−0.140	−0.010	0.133	−0.018	0.000	0.024
Tree Nuts	−0.020	−0.003	0.012	−0.014	0.000	0.013
Oil Crops	−0.123	−0.004	0.016	−0.059	0.000	0.058
Vegetable Oils	−0.118	−0.010	0.056	−0.017	0.008	0.036
Vegetables	−0.454	−0.075	0.291	−0.266	−0.023	0.375
Fruits-Excluding Wine	−0.414	−0.115	0.323	−0.173	0.004	0.186
Stimulants	−0.036	−0.004	0.014	−0.037	−0.006	0.040
Spices	−0.012	0.000	0.030	−0.022	0.000	0.020
Alcoholic Beverages	−0.549	−0.010	0.535	−0.330	0.051	0.269
Meat	−0.438	−0.060	0.338	−0.010	0.012	0.060
Offals	−0.022	−0.002	0.059	−0.033	0.002	0.013
Animal Fats	−0.079	−0.002	0.081	−0.013	0.000	0.018
Eggs	−0.040	−0.003	0.034	−0.047	0.006	0.071
Milk-Excluding Butter	−0.793	0.009	0.561	−0.124	0.000	0.109
Fish, Seafood	−0.794	−0.183	0.613	−0.140	0.012	0.178
Aquatic Products, Other	−0.062	−0.002	0.065	−0.030	0.000	0.016

Table 6. Top 10 *MFG* affecting ΔR_p and ΔR_w.

Rank	Impact on ΔR_p	Impact on ΔR_w
1	Fish, Seafood	Alcoholic Beverages
2	Cereals-Excluding Beer	Cereals-Excluding Beer
3	Fruits-Excluding Wine	Vegetables
4	Vegetables	Starchy Roots
5	Meat	Meat
6	Starchy Roots	Milk-Excluding Butter
7	Sugar & Sweeteners	Fish, Seafood
8	Vegetable Oils	Vegetable Oils
9	Milk-Excluding Butter	Stimulants
10	Alcoholic Beverages	Eggs

5. Summary and Conclusions

It is inarguable that food self-sufficiency has become a very serious policy issue in Japan. Local scholars and bureaucrats believe the country's Food *SSR* has been as low as 40% in the recent trend from 2005 until now on a calorie basis. The results in this paper show a higher value, 53%, on a weight basis. Yet, it is still low, compared to that of 87% in 1961. It is a drop of nearly 34 percentage points in 50 years. This situation puts a pressure on the country's agriculture sector and leaves national food security and national safety vulnerable.

By breaking ΔR into two major factorial component changes, ΔR_p and ΔR_w, we were able to analyze and find that it was the change in the major food group's self-sufficiency ratio, ΔR_p, that drove the change in ΔR, not that of the change in the major food group's quantity supply, ΔR_w.

The trend analysis concludes that ΔR_p and ΔR_w, and that ΔR_p have a stronger and a greater impact on ΔR while there is no explicit relation between them. Thus we concluded that the decline of ΔR was mostly due to the decline in major food group's self-sufficiency ratio (ΔR_p). We also found that the values for ΔR_p and those for ΔR can be well explained by a linear equation, with an almost 90% goodness of fit. On the other hand, we could not find any satisfactory expression to explain the values of ΔR_w and those for ΔR.

We also divided the reference period into four sub-periods to investigate the characteristics of the changes and to explain *WSSR*'s overall declining trend. We noted the differences in the characteristics of each sub-per- iod, and found the commodities that contributed to such characteristics.

We also computed and listed the top 10 *MFG* that had the greatest impacts on both ΔR_p and ΔR_w in particular and ΔR as a whole. Among them, the "Cereals" *MFG* proved to have had the major and most highly important impact on both ΔR_p and ΔR_w. The *MFG* "Starchy Roots", "Fish, Seafood", "Meat", "Vegetables", "Vegetable Oils", "Milk", "Alcoholic Beverages" all displayed their importance in the Japanese diets. These are the *MFG* that the policy-making process needs to pay more attention to in order to avoid a negative impact on ΔR_p and/or ΔR_w, and thus on ΔR and *WSSR*.

By studying ΔR_p and ΔR_w—the endogenous factors of *WSSR*, the paper partly fills the gap in the literature on Japan's food problems. The findings in this paper lead to many suggestions and implications for both policy makers and food researchers. From both the research and policy-making points of view, one obvious question, among others, would be to what extent *WSSR* can be recovered. What would be the maximum sustainable *WSSR*? In an attempt to find the answer to such questions, a study on food network flow programming is conducted. The optimal results from such optimization model would be significant for further investigation of food security in Japan.

References

[1]   Ogura, T. (1976) Implication of Japan's Declining Food Self-Sufficiency Ratio. *The Development Economics*, **14**, 419-448.

[2]   Higuchi, T. (1991) Japanese Dietary Habits and Food Consumption. *Committee for the Japanese Agriculture Session, XXI IAAE Conference*, Agriculture and Agricultural Policy in Japan, University of Tokyo Press, Tokyo, 87-104.

[3]   Saeki, N. (1991) Development of Trade in Agricultural Products and Border Adjustment in Agriculture. *Committee for the Japanese Agriculture Session, XXI IAAE Conference*, Agriculture and Agricultural Policy in Japan, University of Tokyo Press, Tokyo, 121-142.

[4]   Kako, T. (2009) Sharp Decline in the Food Self-Sufficiency Ratio in Japan and Its Future Prospects. Contributed Paper Prepared for Presentation at the International Association of Agricultural Economists Conference, Beijing, 16-22 August 2009.
http://ageconsearch.umn.edu/bitstream/51570/2/kako%20Sharp%20decline%20in%20food%20self-sufficiency1.pdf

[5]   Tanaka, T. and Hosoe, N. (2008) Productivity Shocks and National Food Security for Japan. RIETI Discussion Paper Series 09-E-004.

[6]   Mashimo, T. (2008) To What Level Could Japan's Food Self-Sufficiency Recover? Paper Presented at the World Foodless Day in Tokyo. http://www.nishoren.org/en/?p=287

[7]   Oyama, T. and Yoshii, K. (2006) Evaluation and Systematization of Policy for the Ministry of Agriculture, Forestry and Fisheries: Current Situation and Problems. In: Oyama, T., Ed., *Theory and Practice for Public Policy Evaluation*, Gendaitosho, Sagamihara, 281-295.

[8]   Trung, N.H., Yoshii, K. and Oyama, T. (2009) Quantitative Data Analyses for the Recent Change of the Japanese Food Self-Sufficiency Ratios. *The 8th International Symposium on Operations Research and Its Application* (ISORA), Zhangjiajie, 20-22 September 2009, 372-386.

[9]   Hayami, Y. and Godo, Y. (1997) Economics and Politics of Rice Policy in Japan: A Perspective on the Uruguay round. In: Ito, T. and Kruger, A.O., Eds., *Regionalism versus Multilateral Trade Arrangements*, University of Chicago Press, Chicago, 371-399.

[10]   Hayami, Y. (2000) Food Security: Fallacy or Reality? In: Chern, S.W., Carter, C.A. and Shei, S., Eds., *Food Security in Asia*: Edward Elgar, 11-17.

[11]   Japan Ministry of Agriculture, Forestry and Fisheries (2000-2008) Various Reports and Papers.

Appendix

Table A. 20 Major food groups' self-sufficiency ratios, Japan 1961-2011.

	1961	1962	1963	1964	1965	1966	1967	1968	1969	1970
SSR Cereals–Excluding Beer +	70.2	67.5	60.3	55.6	55.0	50.6	53.5	52.1	46.1	39.0
SSR Starchy Roots +	99.4	97.1	101.7	99.2	99.4	99.0	97.1	97.2	97.3	97.0
SSR Sugarcrops +	100.0	100.0	100.0	100.0	100.0	100.0	100.0	100.0	100.0	100.0
SSR Sugar & Sweeteners +	27.7	30.6	31.2	38.8	37.0	32.8	31.9	28.5	25.7	23.5
SSR Pulses +	84.2	66.1	65.0	46.7	63.3	50.6	67.7	59.4	51.1	67.1
SSR Treenuts +	84.8	75.7	70.6	67.4	60.5	58.6	62.5	64.0	68.1	62.3
SSR Oilcrops +	28.8	24.8	—	15.0	13.9	10.6	9.3	7.9	6.4	5.1
SSR Vegetable Oils +	100.8	98.2	95.9	95.2	97.9	97.9	96.9	96.0	94.7	96.7
SSR Vegetables +	100.0	100.2	99.8	99.7	99.8	99.7	99.5	99.4	99.7	99.3
SSR Fruits–Excluding Wine +	101.1	99.1	94.4	92.7	92.9	91.7	90.9	89.7	87.0	84.5
SSR Stimulants +	73.2	65.5	57.0	56.8	56.1	47.4	48.0	45.9	43.5	40.8
SSR Spices +	83.3	150.0	85.7	71.4	83.3	42.9	28.6	25.0	28.6	6.3
SSR Alcoholic Beverages +	100.1	100.1	100.1	100.1	100.0	100.1	100.1	100.1	100.0	99.9
SSR Meat +	94.8	95.6	91.4	91.1	93.7	88.0	88.9	87.5	84.8	88.4
SSR Offals +	100.0	100.0	98.6	97.6	98.9	99.0	99.1	96.4	96.8	98.0
SSR Animal Fats +	69.1	82.4	78.0	65.2	62.4	52.4	53.1	52.5	52.6	60.1
SSR Eggs +	100.8	100.6	100.2	100.1	99.9	99.7	99.2	98.3	98.3	97.9
SSR Milk–Excluding Butter +	85.5	82.1	78.1	77.4	80.5	79.5	74.0	80.2	83.2	84.8
SSR Fish, Seafood +	108.1	106.0	101.2	96.3	97.7	99.7	99.5	94.9	98.7	100.2
SSR Aquatic Products, Other +	88.7	90.2	88.6	84.8	86.0	77.8	84.4	85.7	82.2	82.0
	1971	**1972**	**1973**	**1974**	**1975**	**1976**	**1977**	**1978**	**1979**	**1980**
SSR Cereals–Excluding Beer +	34.1	33.2	31.1	31.3	33.2	28.6	29.6	28.6	26.8	22.1
SSR Starchy Roots +	96.6	103.0	96.6	88.5	84.3	93.1	92.0	90.7	92.3	92.4
SSR Sugarcrops +	100.0	100.0	100.0	100.0	100.0	100.0	100.0	100.0	100.0	100.0
SSR Sugar & Sweeteners +	22.0	22.9	25.5	18.9	21.0	23.3	23.2	35.3	36.3	43.9
SSR Pulses +	44.7	73.3	67.0	60.7	47.7	49.5	58.5	52.2	45.6	30.6
SSR Treenuts +	56.8	57.7	53.8	60.2	53.1	42.1	47.2	46.9	50.4	43.1
SSR Oilcrops +	4.6	4.3	3.8	4.4	4.0	3.5	3.4	4.5	4.5	4.1
SSR Vegetable Oils +	100.1	93.8	87.2	84.2	85.4	82.1	84.4	84.8	84.1	85.6
SSR Vegetables +	99.0	98.8	98.3	97.9	98.0	97.3	97.4	96.6	96.3	96.0
SSR Fruits–Excluding Wine +	83.3	82.0	83.7	84.0	83.8	82.3	83.0	80.7	82.2	81.9
SSR Stimulants +	38.3	34.7	35.2	35.1	36.1	33.4	32.3	34.2	27.0	27.6
SSR Spices +	0.0	0.0	0.0	0.0	0.0	0.0	0.0	0.0	0.0	0.0
SSR Alcoholic Beverages +	99.4	99.3	99.2	98.4	99.0	96.3	96.4	96.0	95.5	94.9
SSR Meat +	86.5	83.1	78.4	87.4	83.9	80.1	82.4	82.9	82.5	85.3
SSR Offals +	97.6	96.8	94.4	97.2	90.6	84.7	84.0	81.5	83.4	83.5
SSR Animal Fats +	61.4	58.8	60.8	77.8	76.8	69.4	84.4	84.0	92.4	91.0
SSR Eggs +	98.1	98.1	97.8	98.1	97.9	97.7	97.6	98.1	98.1	98.4
SSR Milk – Excluding Butter +	86.9	86.1	84.9	81.0	82.4	74.0	72.8	71.3	72.8	74.9
SSR Fish, Seafood +	104.7	101.4	97.8	98.8	98.8	98.9	91.2	95.7	93.6	93.8
SSR Aquatic Products, Other +	80.4	77.0	73.3	77.5	81.1	66.6	65.3	71.5	65.6	82.8

	1981	1982	1983	1984	1985	1986	1987	1988	1989	1990
SSR Cereals−Excluding Beer +	23.6	23.8	23.6	25.8	25.5	24.8	22.5	22.1	22.6	22.8
SSR Starchy Roots +	91.7	92.1	94.1	89.8	75.2	84.0	89.5	80.3	75.2	77.8
SSR Sugarcrops +	100.0	100.0	100.0	100.0	100.0	100.0	100.0	100.0	100.0	100.0
SSR Sugar & Sweeteners +	53.0	49.8	48.6	52.4	51.3	51.4	50.3	51.2	51.7	51.3
SSR Pulses +	31.5	51.8	33.0	59.7	47.2	40.7	42.4	41.1	43.4	48.6
SSR Treenuts +	51.3	44.8	42.5	40.3	37.8	32.3	32.4	25.4	25.2	25.8
SSR Oilcrops +	4.8	4.5	4.5	4.4	3.9	4.2	4.7	4.3	4.3	3.5
SSR Vegetable Oils +	81.7	81.6	82.7	84.3	83.8	83.2	81.4	79.8	77.7	76.7
SSR Vegetables +	95.4	96.3	95.7	94.4	95.1	94.5	93.9	91.5	91.8	91.7
SSR Fruits−Excluding Wine +	80.2	80.7	82.4	77.8	78.8	75.5	75.8	72.6	70.5	68.8
SSR Stimulants +	26.1	24.2	24.0	20.7	20.4	18.5	17.3	15.8	15.4	15.0
SSR Spices +	0.0	0.0	0.0	0.0	0.0	0.0	0.0	0.0	0.0	25.0
SSR Alcoholic Beverages +	95.2	95.7	95.0	94.3	94.8	95.1	94.5	93.5	93.1	93.2
SSR Meat +	82.6	84.1	83.6	83.1	83.8	81.8	79.4	76.5	74.0	73.2
SSR Offals +	80.8	81.9	81.6	80.4	78.4	75.5	75.4	72.7	72.9	73.2
SSR Animal Fats +	99.3	104.7	110.9	121.1	113.6	110.3	101.5	122.5	101.7	106.0
SSR Eggs +	98.3	98.6	98.9	99.1	99.0	98.4	99.1	99.0	98.9	98.8
SSR Milk−Excluding Butter +	78.1	76.5	77.4	77.6	76.6	78.2	74.4	73.6	78.4	80.4
SSR Fish, Seafood +	94.6	97.2	92.4	95.1	92.9	88.8	87.1	83.7	85.3	78.2
SSR Aquatic Products, Other +	90.1	90.0	88.9	90.6	90.0	90.4	88.8	87.9	87.2	93.6
	1991	1992	1993	1994	1995	1996	1997	1998	1999	2000
SSR Cereals−Excluding Beer +	20.3	22.1	20.3	24.4	22.3	21.7	21.0	19.2	19.9	21.2
SSR Starchy Roots +	77.9	76.9	75.4	78.9	82.2	79.3	81.5	80.0	78.1	76.6
SSR Sugarcrops +	100.0	100.0	100.0	100.0	100.0	100.0	100.0	100.0	100.0	100.0
SSR Sugar & Sweeteners +	53.9	50.8	52.3	51.3	54.7	50.6	54.0	55.2	54.1	54.0
SSR Pulses +	43.7	32.6	22.8	36.5	50.0	37.5	38.4	37.9	36.9	38.0
SSR Treenuts +	19.9	20.6	15.8	19.6	18.2	17.2	19.0	15.8	18.0	13.5
SSR Oilcrops +	3.3	3.1	1.7	1.9	2.2	2.6	2.1	2.3	2.6	3.2
SSR Vegetable Oils +	75.5	75.0	74.3	73.7	73.1	73.8	72.8	73.7	76.0	75.7
SSR Vegetables +	90.5	90.2	88.5	86.0	85.1	85.4	85.5	83.3	81.9	81.5
SSR Fruits−Excluding Wine +	65.2	65.8	62.0	57.2	55.6	54.5	58.6	55.8	55.2	51.1
SSR Stimulants +	14.1	15.1	14.9	13.3	13.8	13.0	13.5	12.9	12.8	11.6
SSR Spices +	22.3	27.3	26.4	24.1	17.9	22.0	19.9	19.6	16.1	13.7
SSR Alcoholic Beverages +	93.4	93.6	93.3	90.9	90.5	91.6	91.1	89.8	91.5	90.1
SSR Meat +	71.0	67.9	66.3	63.3	56.5	54.3	56.8	56.2	53.8	51.6
SSR Offals +	70.9	69.0	71.9	70.8	67.1	66.5	68.7	67.8	67.5	65.4
SSR Animal Fats +	92.3	84.8	83.9	79.6	67.7	63.2	66.5	75.5	74.8	69.6
SSR Eggs +	98.3	98.9	98.8	98.7	98.5	98.4	98.4	98.7	98.6	98.4
SSR Milk−Excluding Butter +	77.2	78.8	80.4	78.9	77.0	79.8	79.7	80.4	80.0	79.3
SSR Fish, Seafood +	73.1	67.5	65.8	60.7	53.3	56.4	56.2	57.9	54.8	54.0
SSR Aquatic Products, Other +	92.6	92.6	90.9	86.5	85.9	83.9	83.9	84.7	82.4	83.0

	2001	2002	2003	2004	2005	2006	2007	2008	2009	2010
SSR Cereals−Excluding Beer +	20.7	20.5	18.7	21.0	21.7	20.5	21.1	21.9	19.9	20.0
SSR Starchy Roots +	76.4	78.0	76.9	74.5	76.2	73.7	75.0	74.5	73.3	72.8
SSR Sugarcrops +	100.0	100.0	100.0	100.0	100.0	100.0	100.0	100.0	100.0	100.0
SSR Sugar & Sweeteners +	55.7	56.2	57.1	57.5	57.6	58.8	56.5	58.9	60.1	58.5
SSR Pulses +	34.7	36.2	30.0	44.6	40.8	32.8	34.8	36.5	28.1	32.6
SSR Treenuts +	13.4	13.2	11.8	10.8	10.0	10.6	10.6	13.2	11.1	10.9
SSR Oilcrops +	3.9	3.6	3.1	2.4	3.4	3.5	3.5	4.1	3.9	3.7
SSR Vegetable Oils +	75.0	74.7	73.4	68.8	65.8	65.8	66.3	63.2	62.1	61.8
SSR Vegetables +	80.8	82.2	81.0	79.1	78.5	79.5	81.2	82.7	83.2	80.5
SSR Fruits−Excluding Wine +	52.2	47.5	46.2	41.9	42.9	40.7	41.8	43.3	44.6	40.8
SSR Stimulants +	11.6	11.5	12.4	12.7	12.5	11.3	11.8	13.3	11.7	11.3
SSR Spices +	13.2	19.6	17.8	21.8	23.6	23.5	26.4	28.4	34.0	34.6
SSR Alcoholic Beverages +	94.4	93.8	93.4	93.2	93.2	85.9	93.4	85.5	83.6	81.7
SSR Meat +	50.7	52.5	52.0	53.0	50.2	52.6	52.3	52.2	54.7	52.3
SSR Offals +	66.6	72.8	70.3	86.4	85.0	85.3	84.0	84.3	83.9	82.5
SSR Animal Fats +	65.0	71.9	71.6	72.3	70.7	72.3	74.0	70.6	79.2	76.2
SSR Eggs +	98.4	98.2	98.4	98.3	97.0	98.1	98.5	98.5	98.7	98.7
SSR Milk−Excluding Butter +	79.6	80.6	81.1	80.2	80.8	81.3	79.8	82.0	81.8	81.2
SSR Fish, Seafood +	49.1	46.2	52.1	47.9	49.1	49.9	52.7	53.9	54.6	54.6
SSR Aquatic Products, Other +	84.2	84.7	83.1	80.6	82.2	81.7	86.1	85.7	86.8	86.8

	2011
SSR Cereals−Excluding Beer +	20.9
SSR Starchy Roots +	71.2
SSR Sugarcrops +	100.0
SSR Sugar & Sweeteners +	53.7
SSR Pulses +	29.2
SSR Treenuts +	8.6
SSR Oilcrops +	3.9
SSR Vegetable Oils +	60.6
SSR Vegetables +	79.3
SSR Fruits−Excluding Wine +	40.1
SSR Stimulants +	10.6
SSR Spices +	34.2
SSR Alcoholic Beverages +	80.9
SSR Meat +	49.9
SSR Offals +	80.5
SSR Animal Fats +	75.3
SSR Eggs +	98.1
SSR Milk−Excluding Butter +	80.1
SSR Fish, Seafood +	54.6
SSR Aquatic Products, Other +	86.8

23

A Modified Interactive Stability Algorithm for Solving Multi-Objective NLP Problems with Fuzzy Parameters in Its Objective Functions

Mohamed Abd El-Hady Kassem[1], Ahmad M. K. Tarabia[2], Noha Mohamed El-Badry[2*]

[1]Mathematics Department, Faculty of Science, Tanta University, Tanta, Egypt
[2]Mathematics Department, Faculty of Science, Damietta University, Damietta, Egypt
Email: mohd60_371@hotmail.com, a_tarabia@yahoo.com, *nooha_moh@yahoo.com

Abstract

This paper presents a modified method to solve multi-objective nonlinear programming problems with fuzzy parameters in its objective functions and these fuzzy parameters are characterized by fuzzy numbers. The modified method is based on normalized trade-off weights. The obtained stability set corresponding to α-Pareto optimal solution, using our method, is investigated. Moreover, an algorithm for obtaining any subset of the parametric space which has the same corresponding α-Pareto optimal solution is presented. Finally, a numerical example to illustrate our method is also given.

Keywords

Multi-Objective Nonlinear Programming, Stability, Trade-Off Method, Fuzzy Parameters

1. Introduction

Many real-life optimization problems have several conflicting objective functions that should be minimized or maximized simultaneously. Researchers and practitioners use various approaches to solve these multi-objective problems. Many authors such as Shih and Lee [1] proposed many approaches that integrated the simulation models with stochastic multiple objective optimization algorithms, many of which used the Pareto-based ap-

---
*Corresponding author.

proaches that generated a finite set of compromise or tradeoff solutions.

In such multi-objective optimization problems, finding the best possible solution means trading off between the different objectives. Instead of a single optimal solution, we have a set of compromise solutions so-called Pareto optimal solutions where none of the objective function values can be improved without impairing at least one of the others. Indeed, in most multi-objective nonlinear programming problems, multi-objective functions usually conflict with each other, which means any improvement of one objective function can be achieved only at the expense of another.

Accordingly, our aim is to find the satisficing solution for the decision maker which is also Pareto optimal. However, formulating the multi-objective nonlinear programming problem which closely describes and represents the real decision situation reflects various factors of the real system. The description of the objective function and constraints involves many parameters whose possible values may be assigned by the experts.

Fuzzy nonlinear programming problem (FNLPP) is very useful in solving problems which are difficult to solve due to the imprecise, subjective nature of the problem formulation or have an accurate solution.

In an earlier work, Osman [2] introduced the notions of the stability set of the first kind and the second kind, and analyzed these concepts for parametric convex nonlinear programming problems. Osman and El-Banna [3] presented the qualitative analysis of the stability set of the first kind for fuzzy parametric multi-objective nonlinear programming problems. Kassem [4] dealt with the interactive stability of multi-objective nonlinear programming problems with fuzzy parameters in the constraints. Sakawa and Yano [5] introduced the concept of α-multi-objective nonlinear programming and α-Pareto optimality. Katagiri and Sakawa [6] dealt with fuzzy random programming, Loganathan and Sherali [7] presented an interactive cutting plane algorithm for determining a best-compromise solution to a multi-objective optimization problem in situations with an implicitly defined utility function. Jameel and Sadeghi [8] solved nonlinear programming problem in fuzzy enlivenment. Recently, Elshafei [9] and Parag [10] gave an interactive stability compromise programming method for solving fuzzy multi-objective integer nonlinear programming problems.

Our motivation depends on developing the method given by Kassem [11] by adding normalized trade-off weights. The modified technique uses an interactive compromise programming algorithm to obtain the optimal stable solution of multi-objective nonlinear programming problems with fuzzy parameters in the objective function. In this algorithm a normalized trade-off weight for each objective function is considered. Moreover, our strategy is, if the decision maker (DM) is unsatisfied with the corresponding α-Pareto optimal solution with the degree α, the DM updates the degree α of α-level set by considering the stability set which has the same corresponding α-Pareto optimal solution.

In this paper, preliminary results are given in Section 2. Section 3 shows the stability set of the first kind. The problem formulation is given in Section 4. Section 5 deals an interactive stability compromise programming algorithm for solving multi-objective nonlinear programming problems with fuzzy parameters in the objective functions. Finally, an illustrative example is given to clarify the obtained results.

2. Preliminary Results

In this section several necessary basic concepts are recalled.

In general, the fuzzy multi-objective nonlinear programming problem (FMONLP) is represented as the following problem:

$$\textbf{(FMONLP)} \quad \max\left(f_1\left(x,\tilde{a}_1\right), f_2\left(x,\tilde{a}_2\right),\cdots, f_m\left(x,\tilde{a}_m\right)\right)$$
$$\text{subject to} \quad X=\left\{x\in R^n \mid g_j\left(x\right)\le 0, j=1,\cdots,k\right\}, x\in X,$$

where $f_i\left(x,\tilde{a}_i\right)$ and $g_j\left(x\right)$ are continuously differentiable and concave functions for all $i=1,2,\cdots,m$ and $j=1,2,\cdots,k$. A nonempty set X is a convex and compact set, and $\tilde{a}=\left(\tilde{a}_1,\tilde{a}_2,\cdots,\tilde{a}_m\right)$ represents a vector of fuzzy parameters in the objective function $f_i\left(x,\tilde{a}_i\right), i=1,2,\cdots,m$.

These fuzzy parameters are assumed to be characterized as the fuzzy numbers as given by Dubois and Prade [12]. It is appropriate to recall here that a real fuzzy number $\tilde{a}$ is a convex continuous fuzzy subset of the real line whose membership function $\mu_{\tilde{a}}\left(a\right), a\in R$ and satisfies the following properties:

1. A continuous mapping from R to the closed interval [0, 1].
2. $\mu_{\tilde{a}}\left(a\right)=0$ for all $a\in\left(-\infty,a_1\right]$.
3. Strict increase on $\left[a_1,a_2\right]$.

4. $\mu_{\tilde{a}}(a) = 1$ for all $a \in [a_2, a_3]$.
5. Strict decrease on $[a_3, a_4]$.
6. $\mu_{\tilde{a}}(a) = 0$ for all $a \in [a_4, +\infty)$.

A possible shape of fuzzy number $\tilde{p}$ is illustrated in **Figure 1**.

Definition 1. (Dubois and Prade [12]).

The α-level set of the fuzzy numbers $\tilde{a}_i$ is defined as the ordinary set $L_\alpha(\tilde{a})$ for which degree of their membership function exceeds, the level α:

$$L_\alpha(\tilde{a}) = \{a \mid \mu_{\tilde{a}}(a_i) \geq \alpha, i = 1, \cdots, m\}$$

For a certain degree of α, the problem FMONLP can be rewritten by using Sakawa and Yano's method [5]. In the following nonfuzzy α-multi-objective nonlinear programming problem form:

$$(\alpha\text{-MONLP}) \quad \max\left(f_1(x, a_1), f_2(x, a_2), \cdots, f_m(x, a_m)\right)$$
$$\text{subject to } x \in X, \ a \in L_\alpha(\tilde{a}).$$

Problem (α-MONLP) can be rewritten as the following form:

$$(\alpha\text{-MONLP}) \quad \max\left(f_1(x, a_1), f_2(x, a_2), \cdots, f_m(x, a_m)\right)$$
$$\text{subject to } x \in X, \ A_i \leq a_i \leq B_i, \ i = 1, \cdots, m,$$

where A_i, B_i are lower and upper bounds on a_i for $i = 1, 2, \cdots, m$.

Indeed, Sakawa, and Yano [5] consider that the membership function $\mu_{\tilde{a}}(a)$ is differentiable on $[a_1, a_4]$ and the problem FMONLP is stable, hence our new problem α-MONLP is also stable.

Definition 2. (α-Pareto optimal solution).

$x^* \in X$ is said to be an α-Pareto optimal solution to the (α-MONLP), if and only if there does not exists another $x \in X$, $a \in L_\alpha(\tilde{a})$. Such that $f_i(x, a_i) \geq f_i(x^*, a_i^*), i = 1, 2, \cdots, m$, with strictly inequality holding for at least one i, where the corresponding values of parameters a_i^* are called α-level optimal parameters.

For some (unknown) implicit utility function we have the following problem:

$$(\alpha M) \quad \max U\left[\left(f_1(x, a_1), f_2(x, a_2), \cdots, f_m(x, a_m)\right)\right]$$
$$\text{subject to } x \in X, \ A_i \leq a_i \leq B_i, \ i = 1, \cdots, m.$$

where $U(.)$ is a concave and continuously differentiable function. It is clear that (x^*, a^*) is an α-Pareto optimal solution of (α-MONLP) if and only if (x^*, a^*) is optimal solution of (αM).

3. Stability Set of the First Kind [11]

In this section we discuss the stability set of the first kind of the nonlinear programming problem with fuzzy parameters in the objective function.

Definition 3. (Stability set of the first kind).

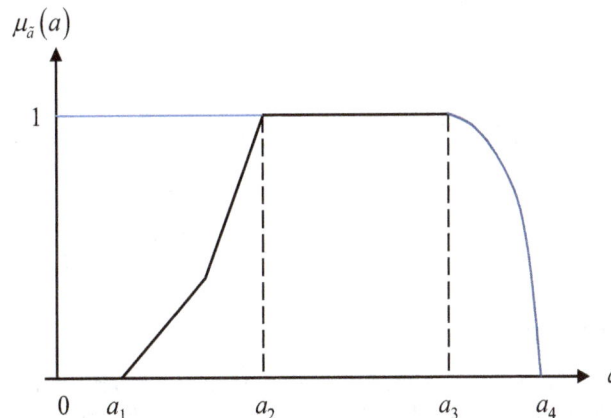

Figure 1. Membership function of fuzzy number $\tilde{a}$.

Suppose that $(\bar{A}, \bar{B}) \in R^{2m}$ with a corresponding α-Pareto optimal solution $(\bar{x}, \bar{a})$, then the stability set of the first kind of (α-MONLP) corresponding to $(\bar{x}, \bar{a})$, denote by $S(\bar{x}, \bar{a})$, is defined by

$$S(\bar{x}, \bar{a}) = \{(A, B) \in R^{2m} \mid (\bar{x}, \bar{a}) \text{ is an } \alpha\text{-Pareto optimal solution of } (\alpha\text{-MONLP})\}.$$

Let a certain $(\bar{A}, \bar{B})$ with a corresponding α-Pareto optimal solution $(\bar{x}, \bar{a})$ be given, then from the stability of problem (α-MONLP) there exist.

$(A, B) \in R^{2m}$, $v \in R^k$, λ, δ and $\mu \in R^m$ such that the following Kuhn-Tucker conditions are satisfied:

$$\sum_{i=1}^{m} \lambda_i \frac{\partial f_i}{\partial x_r}(\bar{x}, \bar{a}) + \sum_{j=1}^{k} v_j \frac{\partial g_j}{\partial x_r}(\bar{x}) = 0, \quad r = 1, \cdots, n, \tag{1}$$

$$\lambda_i \frac{\partial f_i}{\partial \tilde{a}_i}(\bar{x}, \bar{a}) - \delta_i + \mu_i = 0, \quad i = 1, \cdots, m, \tag{2}$$

$$\sum_{i=1}^{m} \lambda_i = 1, \tag{3}$$

$$g_j(\bar{x}) \le 0, \quad j = 1, \cdots, k, \tag{4}$$

$$\bar{a}_i - B_i \le 0, \quad i = 1, 2, \cdots, m, \tag{5}$$

$$A_i - \bar{a}_i \le 0, \quad i = 1, 2, \cdots, m, \tag{6}$$

$$v_j g_j(\bar{x}) = 0, \quad j = 1, 2, \cdots, k \tag{7}$$

$$\delta_i(\bar{a}_i - B_i) = 0, \quad i = 1, \cdots, m, \tag{8}$$

$$\mu_i(A_i - \bar{a}_i) = 0, \quad i = 1, 2, \cdots, m, \tag{9}$$

$$\lambda_i, \mu_i, \delta_i, v_j \ge 0, \quad i = 1, 2, \cdots, m; \ j = 1, 2, \cdots, k \tag{10}$$

The determination of the stability set of the first kind $S(\bar{x}, \bar{a})$ depends only whether any of the variables $\delta_i, i = 1, \cdots, m$, and any of the variables $\mu_i, i = 1, 2, \cdots, m$, given as Equation (2) and Equation (10), are positives or zero.

Given $\mu_i = 0, i \in I_1 \subseteq \{1, 2, \cdots, m\}; \mu_i > 0, i \notin I_1$, as Equation (2) and Equation (10), then in order that the other Kuhn-Tucker conditions (6) and (9) yield

$$A_i \le \bar{a}_i, \ i \in I_1; \ A_i = \bar{a}_i, \ i \notin I_1.$$

Given $\delta_i = 0, i \in I_2 \subseteq \{1, 2, \cdots, m\}; \lambda_i > 0, i \notin I_2$, as Equation (2) and Equation (10), then in order that the other Kuhn-Tucker conditions (5) and (8) yield

$$B_i \ge \bar{a}_i, \ i \in I_2; \ B_i = \bar{a}_i, \ i \notin I_2.$$

Let

$$S_{I_1, I_2}(\bar{x}, \bar{a}) = \{(A, B) \in R^{2m} \mid A_i \le \bar{a}_i, i \in I_1; A_i = \bar{a}_i, i \notin I_1$$

$$B_i \ge \bar{a}_i, i \in I_2; B_i = \bar{a}_i, i \notin I_2\}$$

$$S(\bar{x}, \bar{a}) = \bigcup_{I_1, I_2} S_{I_1, I_2}(\bar{x}, \bar{a})$$

4. Problem Formulation

A modified interactive stability method is the interactive nonlinear programming with fuzzy parameters in the objective functions to obtain the solution of (FMONLP) problem. In this method the DM asks to specify the degree α of the α-level set. For the DM's degree α, the corresponding optimal solution is given by solving the following problem:

$$\max_{x \in X} \min_{1 \le i \le m} \sum_{i=1}^{m} \tilde{w}_{ij}^{l} \left(f_i(x, a_i) - \overline{f}_i \right),$$

subject to $x \in X$, $a \in L_{\alpha}(\tilde{a})$, $i = 1, \cdots, m$,

or equivalently

$$\left(\alpha W^{l} \right) \quad \max \eta$$

subject to $\eta \le \sum_{i=1}^{m} \tilde{w}_{ij}^{l} \left(z_i - z_i^{h} \right)$, $h = 0, 1, 2, \cdots, l$,

$$\left(x, a \right) \in X \times L_{\alpha}(\tilde{a}),$$

where $\eta \approx \min_{1 \le i \le m} \sum_{i=1}^{m} \tilde{w}_{ij}^{l} \left(f_i(x, a_i) - \overline{f}_i \right)$, and $\tilde{w}_{ij}^{l}$ normalized trade-off weights for $i \le j$

$$\tilde{w}_{ij}^{l} = \frac{\left(\dfrac{\partial U(z)}{\partial z_j} \right) \Big/ \left(\dfrac{\partial U(z)}{\partial z_i} \right)}{\sum_{j=1}^{k} \left(\dfrac{\partial U(z)}{\partial z_j} \right) \Big/ \left(\dfrac{\partial U(z)}{\partial z_i} \right)}, \quad i \le j,$$

where $z = (z_1, z_2, \cdots, z_m)$, $z_i \equiv f_i(x, a_i)$ for $i = 1, 2, \cdots, m$, and $z^{l} \equiv \left(f_1(x^{l}, a_1^{l}), \cdots, f_m(x^{l}, a_m^{l}) \right)$.

The following lemma establishes an important characteristic of the problem αP^{l}.

Lemma 1.

Let $(\overline{x}, \overline{a}, \overline{\eta})$ be a solution for the problem αW^{l}, and let $\overline{z} = \left(f_1(\overline{x}, \overline{a}_1), \cdots, f_m(\overline{x}, \overline{a}_m) \right)$. Then $(\overline{x}, \overline{a})$ is an α-Pareto optimal solution to the α-MONLP problem.

Proof:

Suppose that $(\overline{x}, \overline{a})$ is not α-Pareto optimal solution, then there exists $(x^{*}, a^{*}) \in X \times L_{\alpha}(\tilde{a})$ such that $z^{*} > \overline{z}$ and $z^{*} \ne \overline{z}$, where $z^{*} \equiv \left(f_1(x^{*}, a_1^{*}), \cdots, f_m(x^{*}, a_m^{*}) \right)$, then

$$\overline{\eta} = \min_{h=0,1,\cdots,l} \sum_{i=1}^{m} \tilde{w}_{ij}^{l} \left(\overline{z} - z^{h} \right) < \min_{h=0,1,\cdots,l} \sum_{i=1}^{m} \tilde{w}_{ij}^{l} \left(z^{*} - z^{h} \right),$$

which contradicts with the optimality of $(\overline{x}, \overline{a}, \overline{\eta})$. This completes the proof.

5. Interactive Algorithm

Following the above discussion, the interactive algorithm to derive the satisficing solution for the DM from among the α-Pareto optimal solution set is constructed. The steps are marked with numbers involving interaction with the DM.

Step 1: Ask the DM to select the initial values of $\alpha, (0 \le \alpha \le 1)$.

Step 2: Determine the α-level set of the fuzzy numbers.

Step 3: Convert the FMONLP in the form of α-MONLP and select an initial feasible point (x^{l}, a^{l}), set $l = 0$ and let $z^{l} = \left(f_1(x^{l}, a_1^{l}), \cdots, f_m(x^{l}, a_m^{l}) \right)$.

Step 4: The DM specify values of $\tilde{w}_{ij}^{l}$ at (x^{l}, a^{l}) for $i = 1, 2, \cdots, m$.

Step 5: Solve problem αW^{l}. Suppose $(x^{l+1}, a^{l+1}, \eta^{l+1})$ be an optimal solution, and let $z^{l+1} = \left(f_1(x^{l+1}, a_1^{l+1}), \cdots, f_m(x^{l+1}, a_m^{l+1}) \right)$. Set $l = l + 1$.

Step 6: If $\eta^{l} = 0$, go to Step 7. Otherwise go to Step 4.

Step 7: Determine the stability set of the first kind $S(\overline{x}, \overline{a})$.

Step 8: If the DM is satisfied with the current values of the objective functions and α of α-Pareto solution, go to Step 9. Otherwise, ask the DM to update the degree α and return to Step 2.

Step 9: Terminate with (x^{l}, a^{l}) as the final solution.

6. Numerical Illustration Example

In this section, we give an example to illustrate the theoretical part have mention above. Our computation is carried out by utilizing MAPLE 13 (see **Appendix**).

Consider the following fuzzy multi-objective nonlinear programming problem

$$\text{(FMONLP)} \quad \max\left(x_1 + \tilde{a}_1, x_2 + \tilde{a}_2\right)$$

$$\text{subject to} \quad x \in X = \left\{(x_1, x_2) \mid -x_1 + x_2 \le 3, x_1^2 + x_2^2 \le 25, x_1, x_2 \ge 0\right\},$$

with

$$\mu_{\tilde{a}_i}(a_i) = \begin{cases} 0 & -\infty < a_i \le c_1 \\ \dfrac{a_i - c_1}{c_2 - c_1} & c_1 < a_i < c_2 \\ 1 & c_2 \le a_i \le c_3 \\ \dfrac{a_i - c_4}{c_3 - c_4} & c_3 < a_i < c_4 \\ 0 & c_4 \le a_i < \infty \end{cases}$$

where $I = 1, 2$ and the values of $c_j, (j = 1, 2, 3, 4)$ are given as in the following table:

$\tilde{a}_i$	c_1	c_2	c_3	c_4
$\tilde{a}_1$	3.8	4	4.8	5
$\tilde{a}_2$	1	2	3	4

Let $z_1 = f_1(x, a_1) = x_1 + a_1$ and $z_2 = f_2(x, a_2) = x_2 + a_2$. The DM's utility function U is assumed to be as the following:

$$U = -(z_1 - 20)^2 - 2(z_2 - 10)^2$$

Step 1: Suppose that the DM select $\alpha = 0.9$.

Step 2: $L_{0.9}(\tilde{a}) = \left\{(a_1, a_2) \mid 3.98 \le a_1 \le 4.82 \text{ and } 1.9 \le a_2 \le 3.1\right\}$

Step 3: The α-MONLP problem becomes:

$$\max\left(x_1 + a_1, x_2 + a_2\right)$$

$$\text{subject to} \quad x \in X, \, a \in L_{0.9}(\tilde{a})$$

Let us select the feasible point $(x^0, a^0) = (3.18, 2.9, 4.82, 3.1)$ as the starting solution. Hence, the corresponding point in the objective space is: $z^0 \equiv (8, 6)$.

Step 4: The normalized trade off weights w^0 at (x^0, a^0) is: $w^0 = (0.6, 0.4)$.

Step 5: Solve the following problem:

$$\left(\alpha W^0\right) \quad \max \eta$$

$$\text{subject to} \quad \eta \le 0.6 x_1 + 0.4 x_2 + 0.6 a_1 + 0.4 a_2 - 7.2$$

$$x \in X, \, a \in L_{0.9}(\tilde{a})$$

The outputs of our calculation are show in **Table 1**.

Step 6: Since $\eta^l \ne 0$, then return to Step 4.

Repeat Steps 4 - 6 till $\eta^l \approx 0$, otherwise, go to Step 7.

Step 7: The corresponding stability set of the first kind is:

$$S\left(x^9, a^9\right) = \left\{(\overline{A}, \overline{B}) \mid \overline{A}_1 \le 4.82, \overline{B}_1 = 4.82, \overline{A}_2 \le 3.1, \overline{B}_2 = 3.1\right\}.$$

Step 8: Suppose that the DM is satisfied with the z^9 and $\alpha = 0.9$. Go to Step 9.

Table 1. The solution of the problem $\left(\alpha W^{l}\right), l = 1, 2, \cdots, 9$.

Iteration l	$\left(x^{l}, a^{l}\right)$	η^{l}	z^{l}
1	$(4.16025, 2.7735, 4.82, 3.1)$	0.53755	$(8.98025, 5.8735)$
2	$(4.003358, 2.9955173, 4.82, 3.1)$	0.00528124	$(8.823358, 6.0955173)$
3	$(4.1004068, 2.8612347, 4.82, 3.1)$	0.0019716	$(8.9204068, 5.9612347)$
4	$(4.038235, 2.948331, 4.82, 3.1)$	0.0008195	$(8.858235, 6.048331)$
5	$(4.078066, 2.892987, 4.82, 3.1)$	0.00033366	$(8.898066, 5.992987)$
6	$(4.0497106, 2.93254906, 4.82, 3.1)$	0.0001699	$(8.8697106, 6.03254906)$
7	$(4.0724333, 2.9009114, 4.82, 3.1)$	0.000108795	$(8.8924333, 6.0009114)$
8	$(4.0554198, 2.9246487, 4.82, 3.1)$	0.0000611	$(8.8754198, 6.0246487)$
9	$(4.0665451, 2.9091599, 4.82, 3.1)$	≈ 0	$(8.8865451, 6.0091599)$

Step 9: The final solution is $\left(x^{9}, a^{9}\right) = (4.0665451, 2.9091599, 4.82, 3.1)$.

7. Conclusions

In this paper we have integrated different methods like normalized trade-off in [9] and compromised method by Kassem [11] to find a modified method which can be used to solve and obtain an optimal solution to the assumed problem.

The modified technique is based on the reformulation of the given problem and in such way that enables us to solve it easily. A new form of the assumed problem is obtained; hence using a computer program the stability of its solution is studied. Moreover, depending on stability calculation, the validity of the optimal solution has been ensured.

References

[1]   Shih, C.J. and Lee, H.W. (2004) Level-Cut Approaches of First and Second Kind for Unique Solution Design in Fuzzy Engineering Optimization Problems. *Tamkang Journal of Science and Engineering*, **7**, 189-198.

[2]   Osman, M. (1977) Qualitative Analysis of Basic Notions in Parametric Convex Programming I (Parameters in the Constraints). *Aplikace Matematiky*, **22**, 318-332.

[3]   Osman, M. and El-Banna, A. (1993) Stability of Multiobjective Nonlinear Programming Problems with Fuzzy Parameters. *Mathematics and Computers in Simulation*, **35**, 321-326. http://dx.doi.org/10.1016/0378-4754(93)90062-Y

[4]   Kassem, M. (1995) Interactive Stability of Multiobjective Nonlinear Programming Problems with Fuzzy Parameters in the Constraints. *Fuzzy Sets and Systems*, **73**, 235-243. http://dx.doi.org/10.1016/0165-0114(94)00317-Z

[5]   Sakawa, M. and Yano, H. (1989) Interactive Decision Making for Multiobjective Nonlinear Programming Problems with Fuzzy Parameters. *Fuzzy Sets and Systems*, **29**, 315-326. http://dx.doi.org/10.1016/0165-0114(89)90043-2

[6]   Katagiri, H. and Sakawa, M. (2011) Interactive Multiobjective Fuzzy Random Programming through the Level Set-Based Probability Model. *Information Sciences*, **181**, 1641-1650. http://dx.doi.org/10.1016/j.ins.2011.01.003

[7]   Loganathan, G.V. and Sherali, H.D. (1987) A Convergent Interactive Cutting-Plane Algorithm for Multiobjective Optimization. *Operations Research*, **35**, 365-377.

[8]   Jameel, A. and Sadeghi, A. (2012) Solving Nonlinear Programming Problem in Fuzzy Enlivement. *International Journal of Contemporary Mathematical Sciences*, **7**, 159-170.

[9]   Elshafei, M.M. (2006) Interactive Stability of Multiobjective Integer Nonlinear Programming Problems. *Applied Mathematics and Computation*, **176**, 230-236.

[10] Pendharkar, P.C. (2013) Scatter Search Based Interactive Multi-Criteria Optimization of Fuzzy Objectives for Coal Production Planning. *Engineering Applications of Artificial Intelligence*, **26**, 1503-1511. http://dx.doi.org/10.1016/j.engappai.2013.01.001

[11] Kassem, M. (2007) Interactive Stability Cutting-Plane Algorithm for Multiobjective Nonlinear Programming Problems. *Applied Mathematics and Computation*, **192**, 446-456. http://dx.doi.org/10.1016/j.amc.2007.03.037

[12] Duboism, D. and Prade, H. (1980) Fuzzy Sets and Systems: Theory and Application. Academic Press, New York.

Appendix

A maple program for solving multi-objective nonlinear programming (MONLP) problems and stability of this solution.
>restart:

With (Optimization):

>N1: = solve ({(a[1] − 3.8)/(4 − 3.8)> = 0.9}, [a[1]]);

>N2: = solve ({(a[1] − 5)/(4.8 − 5)> = 0.9}, [a[1]]);

N3: = solve ({(a[2] − 1)/(2 − 1)> = 0.9}, [a[2]]);

>N4: = solve ({(a[2] − 4)/(3 − 4)> = 0.9}, [a[2]]);

>U: = - (z[1] − 20) ^ 2 − 2 * (z[2] − 10) ^ 2;

r[1]: = diff (U, z[1])/(diff (U, z[1]) + diff (U, z[2]));

r[2]: = diff (U, z[2])/(diff (U, z[1]) + diff (U, z[2]));

r[1]: = sub s ({z[1] = 3.18 + 4.82, z[2] = 2.9 + 3.1}, r[1]);

r[2]: = subs ({z[1] = 3.18 + 4.82, z[2] = 2.9 + 3.1}, r[2]);

Q1: = **Maximize** (e, {e < = 0.6 * a[1] + 0.6 * x[1] + 0.4 * a[2] + 0.4* x[2] − 7.2, −x[1] + x[2] < = 3, x[1] ^ 2 + x[2] ^ 2 < = 25, 3.98 < = a[1], a[1] < = 4.82, 1.9 < = a[2], a[2] < = 3.1}, assume = nonnegative);

r[1]: = diff (U, z[1])/(diff (U, z[1]) + diff (U, z[2]));

r[2]: = diff (U, z[2])/(diff (U, z[1]) + diff (U, z[2]));

r[1]: = subs ({z[1] = 4.16025 + 4.82, z[2] = 2.7735 + 3.1}, r[1]);

r[2]: = subs ({z[1] = 4.16025 + 4.82, z[2] = 2.7735 + 3.1}, r[2]);

Q2: = **Maximize**(e, {e < = 0.572 * x[1] + 0.428 * x[2] + 0.572 * a[1] + 0.428 * a[2] − ((0.572 * 8.98025) + (0.428 * 5.8735)), −x[1] + x[2] < = 3, x[1] ^ 2 + x[2] ^ 2 < = 25, 3.98 < = a[1], a[1] < = 4.82, 1.9 < = a[2], a[2] < = 3.1}, assume = nonnegative);

r[1]: = diff (U, z[1])/(diff(U, z[1]) + diff (U, z[2]));

r[2]: = diff (U, z[2])/(diff(U, z[1]) + diff (U, z[2]));

r[1]: = subs ({z[1] = 4.003358 + 4.82, z[2] = 2.9955173 + 3.1}, r[1]);

r[2]: = subs ({z[1] = 4.003358 + 4.82, z[2] = 2.9955173 + 3.1}, r[2]);

Q3: = Maximize (e, {e < = 0.589 * x[1] + 0.411 * x[2] + 0.589 * a[1] + 0.411 * a[2] − ((0.589 * 8.823358) + (0.411 * 6.0955173)), −x[1] + x[2] < = 3, x[1] ^ 2 + x[2] ^ 2 < = 25, 3.98 < = a[1], a[1] < = 4.82, 1.9 < = a[2], a[2] < = 3.1}, assume = nonnegative);

r[1]: = diff (U, z[1])/(diff (U, z[1]) + diff (U, z[2]));

r[2]: = diff (U, z[2])/(diff (U, z[1]) + diff (U, z[2]));

r[1]: = subs ({z[1] = 4.1004068 + 4.82, z[2] = 2.8612347 + 3.1}, r[1]);

r[2]: = subs ({z[1] = 4.1004068 + 4.82, z[2] = 2.8612347 + 3.1}, r[2]);

Q4: = **Maximize** (e, {e < = 0.578 * x[1] + 0.422 * x[2] + 0.578 * a[1] + 0.422* a[2] − ((0.578 * 8.9204068) + (0.422 * 5.9612347)), −x[1] + x[2] < = 3, x[1] ^ 2 + x[2] ^ 2 < = 25, 3.98 < = a[1], a[1] < = 4.82, 1.9 <= a[2], a[2] < = 3.1}, assume = nonnegative);

r[1]: = diff (U, z[1])/(diff(U, z[1]) + diff (U, z[2]));

r[2]: = diff (U, z[2])/(diff(U, z[1]) + diff (U, z[2]));

r[1]: = subs ({z[1] = 4.038235 + 4.82, z[2] = 2.948331 + 3.1}, r[1]);

r[2]: = subs ({z[1] = 4.038235 + 4.82, z[2] = 2.948331 + 3.1}, r[2]);

Q5: = Maximize (e, {e< = 0.585 * x[1] + 0.415 * x[2] + 0.585 * a[1] + 0.415 * a[2] − ((0.585 * 8.858235) + (0.415 * 6.048331)), −x[1] + x[2] < = 3, x[1] ^ 2 + x[2] ^ 2 < = 25, 3.98 < = a[1], a[1] < = 4.82, 1.9 < = a[2], a[2] < = 3.1}, assume = nonnegative);

>r[1]: = diff (U, z[1])/(diff (U, z[1]) + diff (U, z[2]));

r[2]: = diff (U, z[2])/(diff (U, z[1]) + diff (U, z[2]));

r[1]: = subs ({z[1] = 4.078066 + 4.82, z[2] = 2.892987 + 3.1}, r[1]);

r[2]: = subs ({z[1] = 4.078066 + 4.82, z[2] = 2.892987 + 3.1}, r[2]);

Q6: = Maximize (e, {e < = 0.58 * x[1] + 0.42 * x[2] + 0.58 * a[1] + 0.42 * a[2] − ((0.58 * 8.898066) + (0.42 * 5.992987)), −x[1] + x[2] < = 3, x[1] ^ 2 + x[2] ^ 2 < = 25, 3.98 < = a[1], a[1] < = 4.82, 1.9 < = a[2], a[2] < = 3.1}, assume = nonnegative);

>r[1]: = diff (U, z[1])/(diff (U, z[1]) + diff (U, z[2]));

r[2]: = diff (U, z[2])/(diff (U, z[1]) + diff (U, z[2]));

r[1]: = **subs ({z[1] = 4.0497106 + 4.82, z[2] = 2.93254906 + 3.1}, r[1]);**

r[2]: = **subs ({z[1] = 4.0497106 + 4.82, z[2] = 2.93254906 + 3.1}, r[2]);**

Q7: = Maximize (e, {e < = 0.584 * x[1] + 0.416 * x[2] + 0.584 * a[1] + 0.416 * a[2] − ((0.584 * 8.8697106) + (0.416 * 6.03254906)), −x[1] + x[2] < = 3, x[1] ^ 2 + x[2] ^ 2 < = 25, 3.98 < = a[1], a[1] < = 4.82, 1.9 < = a[2], a[2] < = 3.1}, assume = nonnegative);

>**r[1]: = diff (U, z[1])/(diff (U, z[1]) + diff (U, z[2]));**

r[2]: = diff (U, z[2])/(diff (U, z[1]) + diff (U, z[2]));

r[1]: = subs ({z[1] = 4.0724333 + 4.82, z[2] = 2.9009114 + 3.1}, r[1]);

r[2]: = subs ({z[1] = 4.0724333 + 4.82, z[2] = 2.9009114 + 3.1}, r[2]);

Q8: = Maximize (e, {e < = 0.581 * x[1] + 0.419 * x[2] + 0.581 * a[1] + 0.419 * a[2] − ((0.581 * 8.8924333) + (0.419 * 6.0009114)), −x[1] + x[2] < = 3, x[1] ^ 2 + x[2] ^ 2 < = 25, 3.98 < = a[1], a[1] < = 4.82, 1.9 < = a[2], a[2] < = 3.1}, assume = nonnegative);

>**r[1]: = diff(U, z[1])/(diff (U, z[1]) + diff (U, z[2]));**

r[2]: = diff (U, z[2])/(diff (U, z[1]) + diff (U, z[2]));

r[1]: = subs ({z[1] = 4.0554198 + 4.82, z[2] = 2.9246487 + 3.1}, r[1]);

r[2]: = subs ({z[1] = 4.0554198 + 4.82, z[2] = 2.9246487 + 3.1}, r[2]);

Q9: = Maximize (e, {e < = 0.583 * x[1] + 0.417 * x[2] + 0.583 * a[1] + 0.417 * a[2] − ((0.583 * 8.8754198) + (0.417 * 6.0246487)), −x[1] + x[2] < =3, x[1] ^ 2 + x[2] ^ 2 < = 25, 3.98 < = a[1], a[1] < = 4.82, 1.9 < = a[2], a[2] < = 3.1}, assume = nonnegative);

24

Solving the Binary Linear Programming Model in Polynomial Time

Elias Munapo

School of Accounting, Economics and Decision Sciences, North West University, Mafeking, South Africa
Email: emunapo@gmail.com

Abstract

The paper presents a technique for solving the binary linear programming model in polynomial time. The general binary linear programming problem is transformed into a convex quadratic programming problem. The convex quadratic programming problem is then solved by interior point algorithms. This settles one of the open problems of whether P = NP or not. The worst case complexity of interior point algorithms for the convex quadratic problem is polynomial. It can also be shown that every liner integer problem can be converted into binary linear problem.

Keywords

NP-Complete, Binary Linear Programming, Convex Function, Convex Quadratic Programming Problem, Interior Point Algorithm and Polynomial Time

1. Introduction

The binary linear programming (BLP) model is *NP*-complete and up to now we have not been aware of any polynomial algorithm for this model. See for example Fortnow [1] [2] for more on complexity. In this paper we present a technique for transforming the BLP model into a convex quadratic programming (QP) problem. The optimal solution of the resultant convex QP is also the optimal solution of the original problem BLP. This solves one of the famous open problems of whether P = NP or not.

2. The BLP Model

Let any BLP model be represented by

Maximize CX^{T},

$$AX^{\mathrm{T}} \leq B^{\mathrm{T}}, \ X^{\mathrm{T}} \leq I^{\mathrm{T}}, \ X^{\mathrm{T}} \geq 0, \ \text{where } I = \begin{pmatrix} 1 & 1 & \cdots & 1 \end{pmatrix}, \ A = \begin{pmatrix} a_{11} & \cdots & a_{1n} \\ \vdots & \ddots & \vdots \\ a_{m1} & \cdots & a_{mn} \end{pmatrix}, \tag{1}$$

$$B = \begin{pmatrix} b_1 & b_2 & \cdots & b_m \end{pmatrix}, \ C = \begin{pmatrix} c_1, c_2, \cdots, c_n \end{pmatrix}, \ X = \begin{pmatrix} x_1 & x_2 & \cdots & x_n \end{pmatrix}.$$

Any minimization BLP can be converted into maximization form and vice versa. There are several strategies for solving mixed 0 - 1 integer problems that are presented in Adams and Sherali [3].

3. Convex Quadratic Programming Model

Let a quadratic programming problem be represented by (2).

$$\text{Maximize } f(X) = CX^{\mathrm{T}} + \frac{1}{2} XQX^{\mathrm{T}},$$

$$\text{Subject to } AX^{\mathrm{T}} \leq B^{\mathrm{T}},$$

$$X^{\mathrm{T}} \leq I^{\mathrm{T}}, \tag{2}$$

$$X^{\mathrm{T}} \geq 0.$$

$$\text{where } Q = \begin{pmatrix} q_{11} & \cdots & q_{1n} \\ \vdots & \ddots & \vdots \\ q_{n1} & \cdots & q_{nn} \end{pmatrix}.$$

We assume that:
1) matrix Q is symmetric and positive definite,
2) function $f(X)$ is strictly convex,
3) since constraints are linear then the solution space is convex,
4) any maximization quadratic problem can be changed into a minimization and vice versa.

When the function $f(X)$ is strictly convex for all points in the convex region then the quadratic problem has a unique local minimum which is also the global minimum [4]-[6].

4. Transforming BLP into a Convex/Concave Quadratic Programming Problem

Our problem is to transform problem (1) into (2) and once that is done then (2) can be solved in polynomial time implying P = NP. Interior point algorithms can solve the convex/concave QP problem in polynomial time.

4.1. Rules with Binary Variables

Binary variables have certain special features that we can capitalize on when solving.

4.1.1. Rule 1
Given any binary variable x_j then slack variable s_j is also binary in the optimal solution.
Proof

$$x_j + s_j = 1. \tag{3}$$

Case 1: When $x_j = 1$ then $s_j = 0$.
Case 2: When $x_j = 0$ then $s_j = 1$.

4.1.2. Rule 2
For any binary variable x_j and slack variable s_j the following must hold at optimality for BLPs.

$$x_j^2 + s_j^2 = 1. \tag{4}$$

The proof is the same as the one given in 4.1.1. Note that it is only binary variables that can satisfy (4). Even though none binary values such as $x_j = 0.9$ then $s_j = 0.1$ can satisfy (3) the same values cannot satisfy (4), *i.e.,* $0.9^2 + 0.1^2 = 0.82 \neq 1$. The binary variable slack relationship given in (4) is the backbone of this paper.

4.2. Forcing Variables to Assume Binary Variables

The main weakness of the objective function given in (1) is that it does not force variables to assume binary values. In this paper we alleviate this challenge by adding a nonlinear extension to the objective function as given in (5).

$$\text{Maximize } C\overline{X}^{\mathrm{T}} + \ell\overline{XX}^{\mathrm{T}} \tag{5}$$

where $\overline{X} = \begin{pmatrix} x_1 & x_2 & \cdots & x_n & s_1 & s_2 & \cdots & s_n \end{pmatrix}$. and ℓ is a very large constant. The constant ℓ is very large in terms of its size compared to any of the coefficients in the objective function. This large value can be approximated as:

$$\ell = 1000\left(|c_1| + |c_2| + \cdots + |c_n|\right) \tag{6}$$

Proof

$$\ell\overline{XX}^{\mathrm{T}} = \ell\left(x_1^2 + x_2^2 + \cdots + x_n^2 + s_1^2 + s_2^2 + \cdots + s_n^2\right), \tag{7}$$

$$\ell\overline{XX}^{\mathrm{T}} = \ell\left(\left(x_1^2 + s_1^2\right) + \left(x_2^2 + s_2^2\right) + \cdots + \left(x_n^2 + s_n^2\right)\right). \tag{8}$$

Since from Rule 2, $x_j^2 + s_j^2 = 1$, then $C\overline{X}^{\mathrm{T}} + \ell\overline{XX}^{\mathrm{T}}$ is minimized when

$$\left(x_1^2 + s_1^2 = 1\right),\left(x_2^2 + s_2^2 = 1\right),\cdots,\left(x_n^2 + s_n^2 = 1\right). \tag{9}$$

In other words $C\overline{X}^{\mathrm{T}} + \ell\overline{XX}^{\mathrm{T}}$ is maximized when variable x_j and slack variable s_j are integers. In this paper the nonlinear extension $\ell\overline{XX}^{\mathrm{T}}$ is called an *enforcer*. An enforcer is a function, a set of constraint(s) or combination of both added to a problem to force an optimal solution with desired features such integrality.

4.3. Convexity of $C\overline{X}^{\mathrm{T}} + \ell\overline{XX}^{\mathrm{T}}$

A function $f\left(\overline{X}\right) = f\left(x_1, x_2, \cdots, x_n, s_1, s_2, \cdots, s_n\right)$ is convex if and only if it has second-order partial derivatives for each point $\overline{X} = \left(x_1, x_2, \cdots, x_n, s_1, s_2, \cdots, s_n\right) \in S$ and for each $\overline{X}' \in S$ all principal minors of the Hessian matrix are none negative.
Proof

In this case

$$\begin{aligned} f\left(\overline{X}\right) &= f\left(x_1, x_2, \cdots, x_n, s_1, s_2, \cdots, s_n\right) \\ &= c_1 x_1 + c_2 x_2 + \cdots + c_n x_n + \ell\left(x_1^2 + x_2^2 + \cdots + x_n^2 + s_1^2 + s_2^2 + \cdots + s_n^3\right). \end{aligned} \tag{10}$$

This has continuous second order partial derivatives and the $2n$ by $2n$ Hessian matrix is given by

$$H\left(x_1, x_2, \cdots, x_n, s_1, s_2, \cdots, s_n\right) = \begin{bmatrix} 2\ell & 0 & \cdots & 0 \\ 0 & 2\ell & \cdots & 0 \\ \vdots & \vdots & & \vdots \\ 0 & 0 & \cdots & 2\ell \end{bmatrix}. \tag{11}$$

Since all principal minors of $H\left(x_1, x_2, \cdots, x_n, s_1, s_2, \cdots, s_n\right)$ are nonnegative then $f\left(x_1, x_2, \cdots, x_n, s_1, s_2, \cdots, s_n\right)$ is convex. See Winston [7] for more on convex functions.

4.4. Convex Quadratic Programming Form

The function $C\overline{X}^{\mathrm{T}} + \ell\overline{XX}^{\mathrm{T}}$ can be expressed in the convex quadratic programming form

$$C\overline{X}^{\mathrm{T}} + \frac{1}{2}\left(2\ell\overline{XX}^{\mathrm{T}}\right) = C\overline{X} + \frac{1}{2}\overline{X}\tilde{Q}\overline{X}^{\mathrm{T}} \tag{12}$$

where matrix $\tilde{Q}$ is of dimension $2n$ by $2n$, symmetric and positive definite as given in (13).

$$\tilde{Q} = \begin{bmatrix} 2\ell & 0 & \cdots & 0 \\ 0 & 2\ell & \cdots & 0 \\ \vdots & \vdots & & \vdots \\ 0 & 0 & \cdots & 2\ell \end{bmatrix} \tag{13}$$

Thus matrix $\tilde{Q}$ is symmetric and positive definite. Note that $\overline{X}\tilde{Q}\overline{X}^{\mathrm{T}} \geq 0, \forall \overline{X}^{\mathrm{T}} \geq 0$.

4.5. Complexity of Convex Quadratic Programming

The main reason for converting a BLP into a convex quadratic programming model is to take advantage of the availability of interior point algorithms which can solve convex QPs in polynomial time [8]. If any BLP can be converted into a convex quadratic problem, then any BLP can be solved in polynomial time.

4.6. Proof of Optimality

The proof is easily shown by reducing the convex quadratic objective function to the original linear form given in (1). The proposed objective function of the convex QP is reduced as follows:

$$\text{Maximize } c_1 x_1 + c_2 x_2 + \cdots + c_n x_n + \ell\left(x_1^2 + x_2^2 + \cdots + x_n^2 + s_1^2 + s_2^2 + \cdots + s_n^3\right)$$

$$= \text{Maximize } c_1 x_1 + c_2 x_2 + \cdots + c_n x_n + \ell\left(\left(x_1^2 + s_1^2\right) + \left(x_2^2 + s_2^2\right) + \cdots + \left(x_n^2 + s_n^2\right)\right).$$

Since $x_j^2 + s_j^2 = 1, \forall j$ then,

$$\text{Maximize } c_1 x_1 + c_2 x_2 + \cdots + c_n x_n + \ell\left((1) + (1) + \cdots + (1)\right) \tag{14}$$

$$= \text{Maximize } c_1 x_1 + c_2 x_2 + \cdots + c_n x_n + n\ell. \tag{15}$$

In other words $\ell\left(x_1^2 + x_2^2 + \cdots + x_n^2 + s_1^2 + s_2^2 + \cdots + s_n^3\right)$ is a constant and the objective function is the same as:

$$\text{Maximize } c_1 x_1 + c_2 x_2 + \cdots + c_n x_n.$$

where x_j is binary for $j = 1, 2, \cdots, n$ this is the original form given in (1).

4.7. Infeasible Binary Integer Solution Space

In this case the solution of the convex OP will not be integer. The objective,

$$\text{Maximize } c_1 x_1 + c_2 x_2 + \cdots + c_n x_n + \ell\left(x_1^2 + x_2^2 + \cdots + x_n^2 + s_1^2 + s_2^2 + \cdots + s_n^3\right).$$

forces variables to binary or integral values if an integer point exists in the solution space. If an integer point does not exists in the solution space the large constant ℓ in the objective forces variables to assume values whose sum of squares are near one and not necessarily one. In other words the variables will assume values x_j' and s_j' such that

$$\left(x_j'\right)^2 + \left(s_j'\right)^2 < 1, \tag{16}$$

$$\left(x_j'\right)^2 + \left(s_j'\right)^2 \approx 1. \tag{17}$$

4.8. Mixed BLP Models

In some BLP problems that occur in real life, a fraction of some of the variables may not be restricted to integer values. In this case the enforcer $\ell\overline{X}\overline{X}^{\mathrm{T}}$ is composed of only those variables that are supposed be binary and integer.

4.9. Interior Point Algorithm for Convex QP

Any maximization BLP problem can be converted into a minimization BLP and vice versa. This can be done by the substitution given in (18).

$$x_j = 1 - \bar{x}_j. \tag{18}$$

where $\bar{x}_j$ is also a binary variable.

Suppose the primal-dual pair of the convex QP is given by (19) and (20).

Primal:

$$
\begin{aligned}
&\text{Minimize } CX^T + \frac{1}{2}XQX^T, \\
&\text{Subject to } AX^T = B^T, \\
&\qquad X^T \geq 0.
\end{aligned} \tag{19}
$$

Dual:

$$
\begin{aligned}
&\text{Minimize } B^T Y - \frac{1}{2}XQX^T, \\
&\text{Subject to } A^T Y + \mu - QX^T = C^T.
\end{aligned} \tag{20}
$$

where Y is free, $\mu \geq 0$ and μ is a diagonal matrix.

The first order optimality conditions for (19) and (20) are given by (21)

$$
\begin{aligned}
AX^T &= B^T, \\
A^T Y + \mu - QX^T &= C^T, \\
X^T \mu e &= 0, \\
X^T &\geq 0, \\
\mu &\geq 0.
\end{aligned} \tag{21}
$$

where e is a vector of ones. The primal-dual central path method can be used to solve the convex QP. Detailed information on this interior point algorithm and other variants can be obtained in Gondzio [8].

5. BLP and Convex QP Relationship

$$
\left.
\begin{aligned}
&\text{Maximize } c_1 x_1 + c_2 x_2 + \cdots + c_n x_n \\
&a_{11}x_1 + a_{12}x_2 + \cdots + a_{1n}x_n \leq b_1 \\
&a_{21}x_1 + a_{22}x_2 + \cdots + a_{2}x_n \leq b_2 \\
&\qquad\qquad \vdots \\
&a_{m1}x_1 + a_{m2}x_2 + \cdots + a_{mn}x_n \leq b_m \\
&x_j \leq 1, j = 1, 2, \cdots, n \\
&x_j = \text{integer}
\end{aligned}
\right\} \text{NP-Complete form}
$$

$$
\left.
\begin{aligned}
&\text{Maximize } c_1 x_1 + c_2 x_2 + \cdots + c_n x_n + \ell\left(x_1^2 + x_2^2 + \cdots + x_n^2 + s_1^2 + s_2^2 + \cdots + s_n^3\right), \\
&a_{11}x_1 + a_{12}x_2 + \cdots + a_{1n}x_n \leq b_1 \\
&a_{21}x_1 + a_{22}x_2 + \cdots + a_{2}x_n \leq b_2 \\
&\qquad\qquad \vdots \\
&a_{m1}x_1 + a_{m2}x_2 + \cdots + a_{mn}x_n \leq b_m \\
&x_j + s_j = 1, j = 1, 2, \cdots, n \\
&\ell = 1000\left(|c_1| + |c_2| + \cdots + |c_n|\right) \text{ and } x_j \geq 0.
\end{aligned}
\right\} \text{P form}
$$

From the two versions of the same problem

$$\text{NP} = \text{P} \tag{22}$$

6. Numerical Illustration

The following numerical illustration shows how a BLP problem is transformed into convex quadratic programming model and then solved.

6.1. Pure Binary Linear Programming

$$\text{Maximize } 3x_1 + 14x_2 + 3x_3 + 8x_4 + 4x_5,$$
$$\text{Such that } 10x_1 + 12x_2 + 4x_3 + 6x_4 + 13x_5 \leq 20,$$
$$17x_1 - 22x_2 + 35x_3 + 8x_4 + 18x_5 \leq 25, \tag{23}$$
$$-10x_1 + 8x_2 + 23x_3 + 11x_4 - 6x_5 \geq 18.$$

where $x_1, x_2, x_3, x_4, x_5 \geq 0$ are binary variables.

Transforming into a convex quadratic programming problem becomes (24)

$$\text{Maximize } 3x_1 + 14x_2 + 3x_3 + 8x_4 + 4x_5 + 32000\left(x_1^2 + x_2^2 + x_3^2 + x_4^2 + x_5^1 + s_1^2 + s_2^2 + s_3^2 + s_4^2 + s_5^2\right);$$
$$\text{Such that } 10x_1 + 12x_2 + 4x_3 + 6x_4 + 13x_5 \leq 20,$$
$$17x_1 - 22x_2 + 35x_3 + 8x_4 + 18x_5 \leq 25, \tag{24}$$
$$-10x_1 + 8x_2 + 23x_3 + 11x_4 - 6x_5 \geq 18,$$
$$x_1 + s_1 = 1, \, x_2 + s_2 = 1, \, x_3 + s_3 = 1, \, x_4 + s_4 = 1, \, x_5 + s_5 = 1.$$

where $s_1, s_2, s_3, s_4, s_5 \geq 0$ are also binary variables.

The solution to the convex quadratic problem is given in (25).

$$x_2 = x_4 = s_1 = s_3 = s_5 = 1 \quad \text{and} \quad x_1 = x_3 = x_5 = s_2 = s_4 = 0. \tag{25}$$

6.2. Mixed Binary Linear Programming Problem

In the case of a mixed binary linear programming problem, only the binary integer variables occupy the enforcer. In other words, if only the r binary variables $x_1, x_2, \cdots, x_r$ are integer then use

$$\text{Maximize } c_1 x_1 + c_2 x_2 + \cdots + c_n x_n + \ell\left(\left(x_1^2 + s_1^2\right) + \left(x_2^2 + s_2^2\right) + \cdots + \left(x_r^2 + s_r^2\right)\right) \tag{26}$$

Suppose in 5.1, the variables x_1 and x_2 are not restricted to integer but both variables are less than 1. The transformation becomes as shown in (27).

$$\text{Maximize } 3x_1 + 14x_2 + 3x_3 + 8x_4 + 4x_5 + 32000(x_3^2 + x_4^2 + x_5^2 + s_3^2 + s_4^2 + s_5^2),$$
$$\text{Such that } 10x_1 + 12x_2 + 4x_3 + 6x_4 + 13x_5 \leq 20,$$
$$17x_1 - 22x_2 + 35x_3 + 8x_4 + 18x_5 \leq 25, \tag{27}$$
$$-10x_1 + 8x_2 + 23x_3 + 11x_4 - 6x_5 \geq 18,$$
$$x_3 + s_3 = 1, \, x_4 + s_4 = 1, \, x_5 + s_5 = 1, \, x_1 \leq 1, \, x_2 \leq 1.$$

The solution to the convex quadratic problem is:

$$x_2 = 0.833, x_3 = x_4 = s_5 = 1 \quad \text{and} \quad x_1 = x_5 = s_3 = s_4 = 0. \tag{28}$$

7. From Mixed Integer Problem to BLP

The problems that occur in real life do not have binary variables only. These practical problems occur as general mixed integer problem (MIP) where variables assume integer values greater than 1. There are methods that can be used to solve these problems but we are not aware of any method that can solve these mixed integer problems in polynomial time up now. The obvious strategy is to expand the general mixed integer variable into binary ones.

7.1. Converting MIP into BLP

Any MIP variable $\left(x_j^g\right)$ can be expanded into binary variables as given in (29).

$$x_j^g = x_0^j + 2^1 x_1^j + 2^2 x_2^j + \cdots + 2^k x_k^j. \tag{29}$$

where x_i^j is a binary variable for $i = 0, 1, 2, \cdots, k$. This procedure is explained in Owen and Mehrotra [9].

7.2. Numerical Illustration

Convert the following MIP into a BLP.

$$\begin{aligned}
&\text{Maximize } 6x_1 + 10x_2 + 14x_3 + 5x_4, \\
&\text{Such that } \quad 8x_1 + 12x_2 + 7x_3 + 15x_4 \le 52.
\end{aligned} \tag{30}$$

where $x_1, x_2, x_3, x_4 \ge 0$ are integers.
   The following substitutions change the problem into a BLP.

$$\begin{aligned}
x_1 &= x_0^1 + 2x_1^1 + 4x_2^1, \\
x_2 &= x_0^2 + 2x_1^2 + 4x_2^2, \\
x_3 &= x_0^3 + 2x_1^3 + 4x_2^3, \\
x_4 &= x_0^4 + 2x_1^4.
\end{aligned} \tag{31}$$

where x_i^j is a binary variable for $i = 0, 1, 2$ and $j = 1, 2, 3.4$.

8. Conclusion

The general BLP problem has been given so much attention by researchers all over the world for over half a century without a breakthrough. A difficult category of BLP models includes the traveling salesman, generalized assignment, quadratic assignment and set covering problems. The paper presented a technique to solve BLP problems by first transforming them into convex QPs and then applying interior point algorithms to solve them in polynomial time. We also showed that the proposed technique worked for both pure and mixed BLPs and also for the general linear integer model where variables were expanded into BLPs. We hope the proposed approach will give more clues to researchers in the hunt for efficient solutions to the general difficult integer programming problem.

Acknowledgements

The author is thankful to the referees for their helpful and constructive comments.

References

[1] Fortnow, F. (2009) The Status of the P versus NP Problem. *Communications of the ACM*, **52**, 78-86. http://dx.doi.org/10.1145/1562164.1562186

[2] Fortnow, F. (2013) The Golden Ticket: P, NP, and the Search for the Impossible. Princeton University Press, Princeton. http://dx.doi.org/10.1515/9781400846610

[3] Adams, W.P. and Sherali, H.D. (1990) Linearization Strategies for a Class of Zero-One Mixed Integer Programming Problems. *Operations Research*, **38**, 217-226. http://dx.doi.org/10.1287/opre.38.2.217

[4] Freund, R.M. (2002) Solution Methods for Quadratic Optimization: Lecture Notes. Massachusetts Institute of Technology, Cambridge, MA.

[5] Jensen, P.A and Bard, J.F. (2012) Operations Research Models and Methods. John Wiley &Sons, Inc., Hoboken, NJ.

[6] Taha, H.A. (2004) Operations Research: An Introduction. 7th Edition, Pearson Educators, New Delhi.

[7] Winston, W.L. (2004) Operations Research Applications and Algorithms. 4th Edition, Duxbury Press, Ontario.

[8] Gondzio, J. (2012) Interior Point Methods 25 Years Later. *European Journal of Operational Research*, **218**, 587-601. http://dx.doi.org/10.1016/j.ejor.2011.09.017

[9] Owen, J.H. and Mehrotra, S. (2002) On the Value of Binary Expansions for General Mixed-Integer Linear Programs. *Operations Research*, **50**, 810-819. http://dx.doi.org/10.1287/opre.50.5.810.370

Permissions

All chapters in this book were first published in AJOR, by Scientific Research Publishing; hereby published with permission under the Creative Commons Attribution License or equivalent. Every chapter published in this book has been scrutinized by our experts. Their significance has been extensively debated. The topics covered herein carry significant findings which will fuel the growth of the discipline. They may even be implemented as practical applications or may be referred to as a beginning point for another development.

The contributors of this book come from diverse backgrounds, making this book a truly international effort. This book will bring forth new frontiers with its revolutionizing research information and detailed analysis of the nascent developments around the world.

We would like to thank all the contributing authors for lending their expertise to make the book truly unique. They have played a crucial role in the development of this book. Without their invaluable contributions this book wouldn't have been possible. They have made vital efforts to compile up to date information on the varied aspects of this subject to make this book a valuable addition to the collection of many professionals and students.

This book was conceptualized with the vision of imparting up-to-date information and advanced data in this field. To ensure the same, a matchless editorial board was set up. Every individual on the board went through rigorous rounds of assessment to prove their worth. After which they invested a large part of their time researching and compiling the most relevant data for our readers.

The editorial board has been involved in producing this book since its inception. They have spent rigorous hours researching and exploring the diverse topics which have resulted in the successful publishing of this book. They have passed on their knowledge of decades through this book. To expedite this challenging task, the publisher supported the team at every step. A small team of assistant editors was also appointed to further simplify the editing procedure and attain best results for the readers.

Apart from the editorial board, the designing team has also invested a significant amount of their time in understanding the subject and creating the most relevant covers. They scrutinized every image to scout for the most suitable representation of the subject and create an appropriate cover for the book.

The publishing team has been an ardent support to the editorial, designing and production team. Their endless efforts to recruit the best for this project, has resulted in the accomplishment of this book. They are a veteran in the field of academics and their pool of knowledge is as vast as their experience in printing. Their expertise and guidance has proved useful at every step. Their uncompromising quality standards have made this book an exceptional effort. Their encouragement from time to time has been an inspiration for everyone.

The publisher and the editorial board hope that this book will prove to be a valuable piece of knowledge for researchers, students, practitioners and scholars across the globe.

List of Contributors

Peiqi Ma
School of Management, Jinan University, Guangzhou, China

Zu'bi M. F. Al-Zu'bi, Ekhleif Tarawneh, Ayman Bahjat Abdallah and Mahmoud A. Fidawi
School of Business, The University of Jordan, Amman, Jordan

Nguyen Khac Minh
Economics and Management Faculty, Thang Long University, Hanoi, Vietnam

Pham Van Khanh
Institute of Economics and Corporate Group, Hanoi, Vietnam

Nguyen Viet Hung
Faculty of Economics, National Economics University, Hanoi, Vietnam

Mazen Arafeh
The Department of Industrial Engineering, Faculty of Engineering & Technology, The University of Jordan, Amman, Jordan

Ventepaka Yadaiah
Department of Mathematics, Osmania University, Hyderabad, India

V. V. Haragopal
Department of Statistics, Osmania University, Hyderabad, India

Zhenping Li and Zhiguo Wu
School of Information, Beijing Wuzi University, Beijing, China

Bo Li
Ashland University, Ashland, OH, USA

Pillaiboothamgudi Sundararaghavan and Udayan Nandkeolyar
University of Toledo, Toledo, OH, USA

Mohammad Sadegh Pakkar
Faculty of Management, Laurentian University, Sudbury, Canada

Bharti Sharma
Department of Mathematics, University of Delhi, New Delhi, India

Promila Kumar
Department of Mathematics, Gargi College, University of Delhi, New Delhi, India

Nejmaddin A. Sulaiman and Rebaz B. Mustafa
Department of Mathematics, College of Education, University of Salahaddin, Erbil, Iraq

Pablo Aarón Anistro Jiménez and Carlos E. Escobar Toledo
Department of Chemical Engineering, National University of Mexico (UNAM), Mexico City, Mexico

Faisal Alkaabneh, Mahmoud Barghash and Yousef Abdullat
University of Jordan, Amman, Jordan

Ruzayn Quaddoura
Department of Computer Science, Faculty of Information Technology, Zarqa University, Zarqa, Jordan

Maged George Iskander
Faculty of Business Administration, Economics and Political Science, The British University in Egypt, El-Sherouk City, Egypt

Ventepaka Yadaiah
Department of Mathematics, Osmania University, Hyderabad, India

V. V. Haragopal
Department of Statistics, Osmania University, Hyderabad, India

William P. Fox
Department of Defense Analysis, Naval Postgraduate School, Monterey, USA

Wissam M. Alobaidi, Hussain M. Al-Rizzo and Eric Sandgren
Systems Engineering Department, Donaghey College of Engineering & Information Technology, University of Arkansas at Little Rock, Little Rock, Arkansas, USA

Entidhar A. Alkuam
Department of Physics and Astronomy, College of Arts, Letters, and Sciences, University of Arkansas at Little Rock, Little Rock, Arkansas, USA

Lamees M. Al-Durgham and Mahmoud A. Barghash
Industrial Engineering Department, Faculty of Engineering and Technology, The University of Jordan, Amman, Jordan

İbrahim Şahbaz
Department of Opticianry, Üsküdar University, Istanbul, Turkey

Mehmet Tolga Taner
Department of Business Administration, Doğuş University, Istanbul, Turkey

Gamze Kağan
Department of Occupational Health and Safety, Üsküdar University, Istanbul, Turkey

Engin Erbaş
Institute of Health Sciences, Üsküdar University, Istanbul, Turkey

Irinel Dragan
University of Texas, Mathematics, Arlington, TX, USA

Shirajul Islam Ukil and Md. Sharif Uddin
Jahangirnagar University, Savar, Bangladesh

Kunihisa Yoshii
Ministry of Agriculture, Forestry and Fisheries (MAFF), Tokyo, Japan

Tatsuo Oyama
National Graduate Institute for Policy Studies (GRIPS), Tokyo, Japan

Mohamed Abd El-Hady Kassem
Mathematics Department, Faculty of Science, Tanta University, Tanta, Egypt

Ahmad M. K. Tarabia and Noha Mohamed El-Badry
Mathematics Department, Faculty of Science, Damietta University, Damietta, Egypt

Elias Munapo
School of Accounting, Economics and Decision Sciences, North West University, Mafeking, South Africa